Precision Conservation: Geospatial Techniques for Agricultural and Natural Resources Conservation

J.A. Delgado, G.F. Sassenrath, and T. Mueller, Editors

Book and Multimedia Publishing Committee

Shuyu Liu, Chair
Daniel Sweeney, ASA Editor-in-Chief
C. Wayne Smith, CSSA Editor-in-Chief
David D. Myrold, SSSA Editor-in-Chief

Managing Editor: Danielle Lynch

Agronomy Monograph 59

American Society of Agronomy
Crop Science Society of America
Soil Science Society of America
5585 Guilford Rd., Madison, WI 53711-5801 USA

agronomy.org • crops.org • soils.org
dl.sciencesocieties.org
SocietyStore.org

Agronomy Monograph Series
ISSN: 2156-3276 (online)
ISSN: 0065-4663 (print)

ISBN: 978-0-89118-356-3 (electronic)
ISBN: 978-0-89118-355-6 (print)
doi: 10.2134/agronmonogr59

Cover design: Patricia Scullion

Printed in the United States of America.

Contents

James C. Ascough II
Dennis C. Flanagan
John Tatarko
Mark A. Nearing
Holm Kipka

Yongping Yuan
Ronald Bingner
Henrique Momm

John Fulton
Matthew Darr

Mark Tomer
D.B. Jaynes
Sarah Porter
David E. James
T.M. Isenhart

Foreword

Agronomy Monograph 59, titled *"Precision Conservation: Geospatial Techniques for Agricultural and Natural Resources Conservation"*, comes at a time of rapid transition towards more spatial and temporal precision in agricultural management. It makes a strong case for conservation planners and practitioners to avail themselves of modern GIS, GPS and decision support tools available for other agronomic practices such as GPS tractor and implement guidance, precision planting, and fertilization and pesticide applications. These geospatial technologies are fast becoming mainstream tools in conservation planning, assessment and implementation. At the same time, these technologies are enabling collaboration across silos of divergent expertise in conservation and precision agronomic practices including irrigation and drainage. The 17 chapters of this book detail existing technologies and management methods and look forward to those on the horizon. The Editors, ASA/CSSA members Jorge Delgado, Gretchen Sassenrath, and Tom Mueller, masterfully bring together key experts to cover use of GIS and GPS tools to plan, design and place conservation structures and to explore economic consequences and opportunities for payback. These include water quality incentives, greenhouse gas offset markets, and established incentive programs offered by federal and state governments.

Much of the new work and methods covered by the book were engendered by the NRCS Conservation Effects Assessment Project (CEAP) established in 2002, which quantifies the impacts of conservation practices across U.S. croplands. From the beginning, CEAP integrated geospatial databases, partner monitoring data, and practice implementation data with analytical models. In so doing, it highlighted the potential for precision conservation now detailed in this book. Indeed, many of the authors are USDA ARS and university partners in CEAP. With its focus on environmental impacts at the watershed level, CEAP projects must take a geospatial approach and consider multiple conservation practices within a watershed, which in turn require a geospatial modeling approach much evident in this book. The work is also informed by outcomes from the Long Term Agro-ecosystems Research (LTAR) network. The LTAR network is a USDA-ARS-University-NGO joint effort at 18 sites across the United States, each of which contrasts "business-as-usual" practices with advanced "aspirational" practices, such as the precision conservation methods described in the book's several chapters.

The editors argue effectively that we must go beyond 4 R nutrient management principles and practices (Right place, Right time, Right amount and Right kind) and include precise planning and siting of conservation practices if we are to achieve sustainable reductions in nitrogen and phosphorus in streams, rivers, and water bodies from lakes to seas. They advocate for a 7 R approach, which expands the 4 Rs to include Precision Conservation, including additional aspects of variable rate irrigation management and drainage management. The additional components are timely and essential since there has been a steady increase in irrigated area in watersheds most prone to nutrient outflows and since surface and subsurface drainage systems are ubiquitous to these vulnerable areas. This book will be a valuable resource as the American Society of Agronomy rolls out the new Certified Crop Adviser specialization in Precision Agriculture. It will hold value for those with specializations in Sustainability and 4 R Nutrient Management as well. As a science-based guide to the present and future of precision conservation, the book has relevance for practitioner, researcher, and policy maker alike.

Steven R. Evett
President, American Society of Agronomy

Mark E. Westgate
President, Crop Science Society of America

Contributors

R. Cibin	319 Forest Resource Lab., Penn State University, University Park, PA 16802 (craj@psu.edu)
Q. Feng	Dep. of Agricultural and Biological Engineering, Purdue Univ., West Lafayette, IN 47907 (feng37@purdue.edu)
I. Chaubey	Dep. of Agricultural and Biological Engineering, Purdue Univ., West Lafayette, IN 47907 (ichaubey@purdue.edu)
L.W. Burger Jr.	Mississippi Agricultural and Forestry Experiment Station, 211F Bost Center, 190 Bost Extension Dr., Mississippi State, MS 39762-9740 (lwb6@msstate.edu)
M.D. McConnell	Warnell School of Forestry & Natural Resources, Univ. of Georgia, 180 E. Green St., Athens, GA 30602-2152 (mdm@uga.edu)
A. Saleh	Texas Institute for Applied Environmental Research, 1333 W. Washington St., Tarleton State Univ., Stephenville, TX (saleh@tiaer.tarleton.edu)
E. Osei	Texas Institute for Applied Environmental Research, 1333 W. Washington St., Tarleton State Univ., Stephenville, TX (osei@tiaer.tarleton.edu)
D.J. Mulla	Dep. of Soil, Water & Climate, 1991 Upper Buford Circle, University of Minnesota, St. Paul, MN 55108 (mulla003@umn.edu)
S. Belmont	Belmont, Dep. of Environment & Society, 5215 Old Main Hill, Utah State Univ., Logan, UT 84322-5215 (swb.in.ut@gmail.com)
P. Fiener	Institut für Geographie, Universität Augsburg, Alter Postweg 118, 86159 Augsburg, Germany (fiener@geo.uni-augsburg.de)
K. Auerswald	Lehrstuhl für Grünlandlehre, Technische Universität München, Alte Akademie 12, 85354 Freising, Germany (auerswald@wzw.tum.de)
D.L. Bjornesberg	USDA-ARS, 3793 N 3600 E, Kimberly, ID 83341-5076
R.G. Evans	32606 W. Knox Road, Benton City, WA 99320 (robertevans910@gmail.com).
E. John Sadler	Rm. 269 Agric. Eng. Bldg., University of Missouri, Columbia, MO 65211 (john.sadler@ars.usda.gov)
M.D. Tomer	USDA-ARS, National Laboratory for Agriculture and the Environment, Ames, IA

D.B. Jaynes   USDA-ARS, National Laboratory for Agriculture and the Environment, Ames, IA

S.A. Porter   USDA-ARS, National Laboratory for Agriculture and the Environment, Ames, IA

D.E. James   USDA-ARS, National Laboratory for Agriculture and the Environment, Ames, IA

T.M. Isenhart   Department of Natural Resource Ecology and Management, Iowa State University, Ames, IA.

M.G. Dosskey   USDA Forest Service, National Agroforestry Center, 1945 North 38th Street, Lincoln, NE, USA 68583

S. Neelakantan   Washington State Department of Natural Resources, MS 47020, Olympia, WA 98504

T. Mueller   Decision Science and Modeling Team, Deere & Company, Mercury Building, 4140 NW 114th Street, Urbandale, IA 50263

Z. Qiu   Department of Chemistry and Environmental Science, New Jersey Institute of Technology, University Heights, Newark, NJ 07102. (mdosskey@fs.fed.us)

J. Fulton   Agricultural Engineering Department, Ohio State University, Columbus, OH 43210 (fulton.20@osu.edu)

M. Darr   Agriculture and Biosystems Engineering Department, Iowa State University, Ames, IA 50011.

J.C. Ascough II   USDA-ARS-PA, Water Management and Systems Research Unit, Fort Collins, CO 80528

D.C. Flanagan   USDA-ARS-MWA, National Soil Erosion Research Laboratory, West Lafayette, IN 47907

J. Tatarko   USDA-ARS-PA, Rangeland Resources and Systems Research Unit, Fort Collins, CO 80528

M.A. Nearing   USDA-ARS-PWA, Southwest Watershed Research Unit, Tucson, AZ 85719

H. Kipka   H. Kipka, Dep. of Civil and Environmental Engineering, Colorado State University, Fort Collins, CO 80528

Y. Yuan   Environmental Protections Agency, Office of Research and Development, P.O. Box 93478, Las Vegas, NV 89119

R. Bingner   USDA Agricultural Research Service, 598 McElroy Drive, Oxford, MS 38655

H. Momm   Middle Tennessee State University, Department of Geosciences, MTSU P.O. Box 9, Kirksey Old Main 325B, Murfreesboro, TN 37132

D.E. Clay — South Dakota State University, Brookings, SD 57007

J. Chang — South Dakota State University, Brookings, SD 57007

G. Reicks — South Dakota State University, Brookings, SD 57007

S.A. Clay — South Dakota State University, Brookings, SD 57007

C. Reese — South Dakota State University, Brookings, SD 57007

V. Shedekar — Ohio State University, Food, Agricultural and Biological Engineering Dept., Columbus, OH 43210

L. Brown — Ohio State University, Food, Agricultural and Biological Engineering Dept., Columbus, OH 43210

P.J.A. Kleinman — USDA-ARS, Pasture Systems and Watershed Management Research Unit, University Park, PA

A. Buda — USDA-ARS, Pasture Systems and Watershed Management Research Unit, University Park, PA

A.N. Sharpley — Dep. Crop, Soil and Environmental Sciences, Division of Agriculture, Univ. Arkansas, Fayetteville, AR

R. Khosla — Dep. Soil and Crop Sciences, Colorado State University, Fort Collins, CO

Acknowledgments

Thanks and acknowledgments are extended to Nicole Sandler (American Society of Agronomy, Crop Science Society of America, and Soil Science Society of America), and to Donna Neer (USDA Agricultural Research Service) for their contributions in helping edit chapters of the book. The editors would especially like to thank Danielle Lynch (American Society of Agronomy, Crop Science Society of America, and Soil Science Society of America), who helped with the final editing of the book and layout of the chapters. The editors would also like to express their appreciation to all of the reviewers that contributed to the review of the book chapters, with special thanks to David Clay, who led the review of the summary chapter.

Dedication

With the challenges that humanity faces in the 21st century, including a changing climate and an expected increase in the occurrence of extreme events, conservation of soil and water resources will be critical for food security and environmental quality. This book is dedicated to farmers and ranchers across the world who are contributing to food security and conservation of land and water resources; as well as to conservation professionals, nutrient managers, and other technical personnel helping implement conservation practices on the ground to conserve the quality of air, water, and especially soil resources, for future generations.

Precision conservation can be applied not only to agricultural systems, but also to other soil and water conservation systems such as urban areas (e.g. parks), forests, and areas under construction, among other examples, in a given watershed. In the great majority of cases, they are part of watersheds that include agricultural systems. As described in this book, riparian buffers, sedimentation ponds, constructed wetlands and other resources and/or practices can be used with precision conservation to increase the effectiveness of conservation efforts across a watershed. This book is therefore also dedicated to foresters, biologists, and conservation professionals that are working to conserve these resources.

In short, this book is dedicated to everyone working in these conservation efforts.

Soil Erosion Modeling and Conservation Planning

James C. Ascough II, Dennis C. Flanagan,* John Tatarko, Mark A. Nearing, and Holm Kipka

Abstract

Accelerated soil erosion induced by human activities is a primary cause of soil degradation worldwide. The main driver behind the problem is agriculture, and at stake is the long-term sustainability of global agricultural production capacity. Barring major, unforeseen scientific advances in the future, and if soil erosion and population growth remain unconstrained from their current rates, humankind may eventually lose the ability to feed itself. A significant problem associated with soil erosion is off-site sediment pollution. Costs related to the delivery of sediment to streams and other water bodies worldwide are enormous. Simulation modeling is one of the most powerful scientific tools available to address questions, assess alternatives, and support decision making for natural resource management. Over the past three decades, a number of erosion models with varying levels of complexity have been developed to address a wide range of environmental issues and describe/understand the behavior of land and water systems under prevailing and projected land use and climate conditions. They are increasingly used to evaluate potential water and air quality impacts of pollution control strategies, including agricultural practices. For example, process-based models are commonly used to assess field- to watershed-scale management practice effects (including conservation practices) on both on-site and off-site erosion concerns. Selection, development, and application of erosion models should be conducted according to the intended modeling purpose, i.e., prediction, exploratory analysis, or knowledge building. Whatever the modeling purpose, the key issue is establishing reasonable practical credibility in the model-based analysis. This book chapter focuses on models of soil erosion as they are used for purposes of soil conservation planning. In particular, we concentrate here primarily on soil erosion by water but will also briefly touch on models of other agricultural erosion processes, such as wind erosion, that apply similar principles for modeling and conservation planning.

Abbreviations: AnnAGNPS, Annualized Agricultural Non-Point Source; APEX, Agricultural Policy/Environmental eXtender; GIS, geographic information systems; RHEM, Rangeland Hydrology and Erosion Model; RS, remote sensing; RUSLE, Revised Universal Soil Loss Equation; SWAT, Soil and Water Assessment Tool; USLE, Universal Soil Loss Equation; WEPP, Water Erosion Prediction Project; WEPS, Wind Erosion Prediction System.

J.C. Ascough II, USDA-ARS-PA, Water Management and Systems Research Unit, Fort Collins, CO 80528; D.C. Flanagan, USDA-ARS-MWA, National Soil Erosion Research Laboratory, West Lafayette, IN 47907; J. Tatarko, USDA-ARS-PA, Rangeland Resources and Systems Research Unit, Fort Collins, CO 80528; M.A. Nearing, USDA-ARS-PWA, Southwest Watershed Research Unit, Tucson, AZ 85719; H. Kipka, Dep. of Civil and Environmental Engineering, Colorado State University, Fort Collins, CO 80528. *Corresponding author (dennis.flanagan@ars.usda.gov).

doi:10.2134/agronmonogr59.2013.0011

Accelerated soil erosion induced by human activities is a primary cause of soil degradation worldwide. The main driver behind the problem is agriculture, and at stake is the long-term sustainability of global agricultural production capacity. Barring major, unforeseen scientific advances in the future, and if soil erosion and population growth remain unconstrained from their current rates, humankind may eventually lose the ability to feed itself. A significant problem associated with soil erosion is off-site sediment pollution. Costs related to the delivery of sediment to streams and other water bodies worldwide are enormous (Pimentel et al., 1995). Simulation modeling is one of the most powerful scientific tools available to address questions, assess alternatives, and support decision making for natural resource management. Erosion models are used to describe and understand the behavior of land and water systems under prevailing and projected land use and climate conditions. They are increasingly used to evaluate potential water and air quality impacts of pollution control strategies, including agricultural practices. This chapter focuses on models of soil erosion as they are used for purposes of soil conservation planning. We will primarily examine soil erosion by water but will also briefly touch on models of other agricultural erosion processes, such as wind erosion, that apply similar principles for modeling and conservation planning.

Assessment of individual and cumulative conservation practice effects on sediment transport using monitoring (measured) data is ideal, but can be costly, and in some cases, highly impractical. Well-ordered experiments at field and watershed scales have frequently been used to characterize how long-term changes in land use or implementation of management actions affect hydrologic and sediment quality fluxes (Bishop et al., 2005). The wider use of these approaches, however, faces challenging issues including: i) identifying baseline or control landscapes; ii) quantifying the role of type, location, and timing of individual practices when multiple practices are implemented over time; iii) overcoming delay or lag time in erosion response to changes in management due to large pools of legacy sediments in soils and streams; and iv) addressing the uncertainty of methods for transferability of results to other spatial and temporal scales. Simulation modeling in combination with strategic monitoring can help address these issues (Easton et al., 2008; Tomer and Locke, 2011).

Erosion models used in application for conservation planning fall into two basic categories: empirical and process-based. These models facilitate dynamic hydrologic and erosion assessments under scenarios that cannot be investigated practically using actual field experiments. They can simulate "what-if" scenarios at varying spatial (field to landscape) and temporal (subdaily to yearly) scales (Fig. 1). Erosion models typically embody well-tested scientific hypotheses. As such, they are suitable for application within formal statistical hypothesis testing frameworks to answer questions relevant to conservation assessment and planning. For example, modeling scenarios can provide essential information to examine the hypothesis that a targeted approach to implementation of conservation practices is significantly more effective in reducing sediment losses than a random approach. Similarly, the likelihood that the assimilative capability of a particular water body will significantly decrease under predicted climate change scenarios can be tested with simulation models.

Over the past three decades, a number of erosion models with varying levels of complexity have been developed to address a wide range of environmental

Fig. 1. Processes (continuous blue line), forms (dashed red line) and features (dotted brown line) placed within a space–time continuum (showing extent vs. duration) that can be simulated by erosion models (Karydas et al., 2012).

issues. Significant advances in modeling technologies have been made to study generation, fate, and transport of sediment. Empirical models, such as the Universal Soil Loss Equation (USLE) (Wischmeier and Smith, 1978), the Revised Universal Soil Loss Equation (RUSLE) (Renard et al., 1997), and the Chinese Soil Loss Equation (CSLE) (Liu et al., 2002) are ordinarily used for on-site field scale assessment of soil loss. Process-based models, such as the Soil and Water Assessment Tool (SWAT) (Arnold et al., 1998), Limburg Soil Erosion Model (LISEM) (De Roo et al., 1994), European Soil Erosion Model (EUROSEM) (Morgan et al., 1998), Annualized AGricultural Non-Point Source (AnnAGNPS) (Bingner and Theurer, 2003), Agricultural Policy/Environmental eXtender (APEX) (Williams and Iza-urralde, 2006), Water Erosion Prediction Project (WEPP) (Flanagan et al., 2007), Wind Erosion Prediction System (WEPS) (Wagner, 2013), the Rangeland Hydrology and Erosion Model (RHEM) (Nearing et al., 2011), and the Kinematic Runoff and Erosion Model (KINEROS2) (Goodrich et al., 2012) are frequently used for field- to watershed scale assessment of management practice effects (including conservation practices) on both on-site and off-site concerns. Selection, development, and application of erosion models should be conducted according to the intended modeling purpose, (i.e., prediction, exploratory analysis, or knowledge building) (Brugnach and Pahl-Wostl, 2007). Whatever the purpose, the key issue is establishing reasonable practical credibility in the model-based analysis.

Conservation Planning for Erosion Control

Conservation planning for erosion control is the assessment of alternative systems according to a defined criteria and choosing the one that best fits the situation. An acceptable conservation system should fit multiple requirements. In general, the system must protect both on-site resources (which include the landscape and the soils that make up the landscape) and off-site resources (which include air and water quality, water conveyance structures, reservoirs, and other locations where sedimentation and deposition occurs). The conservation system must accommodate a desired land use, which might be crop production, recreation, or waste disposal. The proposed system should be cost effective, unless supported by funding from outside sources, and fit within the financial, managerial, and other resources available to implement new conservation technology. For example, without access to financial capital, implementation of technology that requires purchase of new farm implements or installation of conservation systems (e.g., terraces) cannot occur. In addition, the conservation system must fit the scale of the land user. A system requiring large equipment does not fit small fields on steep landscapes. The system also must be culturally and socially acceptable; "What will neighbors think and say?" is a valid concern. Far more important than is normally recognized are the personal preferences of the land user and/or land owner. Is the system expedient to farm and to manage? Will the land user or owner be comfortable with the proposed system? If the land user or owner is not satisfied with a conservation system, the likelihood of long-term use and mainte-nance of the system is dramatically reduced.

In general, conservation planning for erosion control is choosing a system so that the adverse impacts of erosion are within acceptable limits, which are often defined by indicators that address the well-being of resources impacted by ero-sion. Several indicators are typically used because no single one addresses all erosional concerns associated with a resource. For example, maintenance of the ecological well-being of rangelands is a high priority and soil resources are criti-cally important for supporting plant life. Protection of the soil against excessive erosion is necessary to maintain productivity and a diverse plant community. However, because rangeland ecology is related to many other factors besides soil, consideration must be given to other processes besides erosion. That is, erosion control may well not be the foremost interest in holistic conservation planning for any land use. Therefore, an indicator for well-being of a resource must be: i) understandable (and acceptable) to those interested in the resource; ii) mea-surable (with both repeatable and consistent measurements); and iii) based on sound scientific knowledge, including how the indicator represents the quality of the resource and how environmental processes and land use affect the indicator. The conservation planning process therefore often involves the use of an indica-tor related to maintenance of the soil and a mathematical model to estimate soil erosion rates. A set of acceptable conservation systems are proposed related to site-specific conditions and preferences of the land user. The mathematical soil erosion model is used to estimate an erosion rate for each alternative and this rate is compared to values for indicators used to represent the well-being of on-site and off-site resources. The land user selects from those practices having estimated soil erosion rates that are lower than those considered to be unacceptable. Thus, two technologies are important. One is the erosion prediction technology and the second is the technology for the indicators used to describe resource well-being.

In addition, conservation planning can occur on many spatial scales. If the concern is off-site (such as for water quality in a reservoir or air quality in a com-munity), the area of concern is the upstream or up-wind areas that produce the sediment or deposition and the delivery system. In this case, planning is on a broad area (i.e., watershed or basin) basis and considers the variation of weather, soil, topography, and land use over the contributing region and how individual land units interact to affect sediment production and delivery. However, at some point in the planning process, attention may become focused on individual land units because land users make management decisions at that level. If the concern is on-site, the attention is immediately focused on individual land units. Thus, soil conservation planning is conducted at both the land unit and catchment or watershed levels and a range of soil erosion prediction technology exists that is applicable at these levels.

Soil Erosion Prediction Technology

Soil erosion prediction technologies are mathematical procedures that estimate rates of erosion and sediment delivery and sediment characteristics for specific sites as a function of weather, soil, topography, and land use. These technologies vary according to underlying concepts on which the procedures are based, the processes and effects represented by the procedures, the governing equations, the mathematical structure that connects the equations, and the variables for which computations are made. Estimated values for the same variable can fluctuate significantly among soil erosion models. As previously stated, soil erosion prediction technologies can be loosely classified as being either empirical or fundamentally process-based. In the empirical models, the underlying mathematical structure typically represents main effects with a set of indicators and parameter values that have been empirically derived. The intent is not to describe erosion processes but to elucidate the main effects of the variables that affect erosion processes. In process-based models, individual erosion processes are explicitly described and effects of variables on soil erosion are quantified by how these variables affect the fundamental processes represented in the model. It is important to note that all erosion models, regardless of their underlying structure, require empirical data. Even the most scientifically advanced models require empirical data to determine values for parameters such as soil erodibility (if these models are to be used in conservation planning). The requirements for erosion models vary according to the intended purpose of the model. The requirements for a model used in a scientific study of erosion differ greatly from those for a model used in routine conservation planning by field personnel. Below we describe erosion models used for predictive, exploratory, and science advancement purposes.

Predictive Modeling

The purpose of predictive erosion modeling is to closely approximate the biogeophysical responses of a system under current or future climate and land-use conditions. Predictive tools are also used to assess outcomes of certain actions, including the option to take no action. Current outcomes in development and application of predictive erosion models for conservation management include the following:

- Predicting fluxes of water and sediment in surface water or airborne particulate concentrations, including predicting consequences of future climate change;
- Identifying critical sediment or emission source areas within watersheds;
- Quantifying sediment and particulate loading effects of management alternatives, (e.g. agricultural practices); and
- Tracking and/or predicting trends and identifying implications of long-term changes in land use on water quantity and air or water quality.

Numerous studies have applied models to determine erosion impacts of agricultural conservation practices. Prominent studies include the national CEAP cropland studies in the Upper Mississippi River Basin (USDA NRCS, 2010), the Chesapeake Bay region (USDA NRCS, 2011a), and the Great Lakes system (USDA NRCS, 2011b). These studies used the SWAT, APEX, and/or Environment Policy

Integrated Climate (EPIC) models for two main reasons. First, the models were applied to quantify the effects of conservation practices currently present on the landscape in the regions. Second, potential benefits that could be gained by implementation of additional conservation treatment in under-treated agricultural areas were projected. The outputs of these studies were expressed deterministically in terms of numeric reduction estimates. For example, the Upper Mississippi River Basin study for a stream location at Grafton, IL, projected that implemented conservation practices to date have resulted in a 37% reduction of sediment loads (USDA NRCS, 2010). Despite best efforts to calibrate the modeling component of the study, the projected reductions of sediment were typically within the range of modeling errors. Therefore, the reliability of these predictions remains to be investigated. A noteworthy goal for additional studies would be establishing higher confidence in modeling outcomes by reducing uncertainties (to ensure that predicted reduction of sediment can be directly attributed to implementation of practices and not an artifact of modeling uncertainties).

A complementary and somewhat more realistic approach to the assessment of conservation practices at the watershed scale has been ongoing for a considerable period of time in 14 USDA-Agricultural Research Service (ARS) benchmark watersheds. The primary goals of these studies are to advance the science of process-based models used for simulating agricultural pollutants, conduct strategic monitoring to collect and analyze data for conservation effect assessment, and assist with validation of models used for the national assessment component of CEAP. The selected models for these studies are SWAT and AnnAGNPS (Harmel et al., 2008; Heathman et al., 2008; Richardson et al., 2008; Yuan et al., 2008; Cho et al., 2010). Simultaneous application of the two models in the benchmark watersheds presents the opportunity to evaluate the role of model structure uncertainty on the simulated effects of practices. When fully available, data collected from USDA-ARS benchmark watersheds could be used to develop methods for enhancing the degree of confidence in modeling environmental benefits of conservation efforts. Other studies have predicted erosion impacts of agricultural practices in different regions of the world using RUSLE2, SWAT (Behera and Panda, 2006; Bracmort et al., 2006; Gassman et al., 2007; Lam et al., 2011), AnnAGNPS (Mostaghimi et al., 1997; Yuan et al., 2002; Licciardello et al., 2007; Polyakov et al., 2007; Kuhnle et al., 2008; Zhou et al., 2009), WEPP (Renschler and Lee, 2005; Flanagan et al., 2007; Pandey et al., 2009), WEPS (Wagner, 2013; Sharratt et al., 2015; Blanco-Canqui et al., 2016) and APEX (Tuppad et al., 2010). One important and often neglected aspect of assessing conservation benefits arises from modeling uncertainties due to the stochastic nature of multi-scale physical processes and our incomplete knowledge of the system. When models are used in the predictive mode, errors must be minimized, uncertainties need to be reduced, and confidence in model-based analysis should be explicitly documented. While not directly related to erosion, several studies have demonstrated that the estimated benefits of pollution control strategies could markedly differ as a function of uncertainty in model parameters, incomplete representation of critical processes within the model structure, and more importantly, the numerical procedure for representation of management actions (e.g., Harmel et al., 2006; Harmel and Smith, 2007; Arabi et al., 2007b, 2008).

Exploratory Modeling

Exploratory modeling is used to provide insights about potential outcomes, opportunities, and risks associated with alternative management strategies. It is useful for designing effective policy instruments and exploring the attainability of various policy objectives. Examples of the use of erosion models for exploratory purposes include modeling the following:

- Critical natural processes and key management actions that influence fate and transport of sediment at multiple scales;
- Geospatial and temporal factors that control effectiveness of conservation practices;
- The likelihood that desired water and air quality targets can be achieved;
- Tradeoffs and targets associated with alternative sediment control strategies; and
- Possible outcomes of new regulations on land use and cropping patterns within fields or watersheds.

In the context of field-to-watershed management and conservation planning, it is often desired to highlight tradeoffs and targets associated with various management options. Integrated modeling optimization systems can facilitate incorporation of environmental, socioeconomic, and institutional factors in the planning process. Over the past two decades, spatially explicit multiobjective optimization methods have been developed and demonstrated to explore optimal cropping systems or cost-effective agricultural practices within watersheds to meet targets for sediment and other water quality indicators (Srivastava et al., 2002; Veith et al., 2003; Whittaker et al., 2003; Bekele and Nicklow, 2005; Rabotyagov et al., 2010; Rodriguez et al., 2011). For example, Arabi et al. (2006a) applied SWAT within a genetic algorithm-based optimization framework in two small (< 10 km^2 [2.92 mi^2]) agricultural watersheds in Indiana and showed that optimal allocation of practices would yield three-fold greater environmental benefits in terms of sediment and nutrient reductions at the same cost compared with an existing suite of practices. The use of models for identifying optimal conservation options is an example of model application to investigate the possibility of certain outcomes without assigning a given probability to them. In this context, modeling uncertainties could serve as a source of innovation and could unveil possibilities that are consistent with internal system behavior. Therefore, modeling uncertainties for exploratory purposes does not necessarily need to be eliminated, but uncertainty must be explicated and clearly communicated.

Science Advancement through Modeling

Erosion models may be applied to enrich scientific knowledge of critical natural processes and key management actions that control movement of sediment within a natural system. Examples of model application for science advancement purposes include the following:

- Evaluating impacts of spatial and temporal scales on modeling erosion processes;
- Evaluating appropriate spatial and temporal scales for assessment of conservation practices;
- Evaluating suitability of different conceptual approaches and algorithms for representation of conservation practices; and
- Evaluating efficacy of erosion models for representation of agricultural practices.

In general, models have been used extensively to develop knowledge about important issues that affect assessment, planning, and cost-effective implementation of conservation practices. For example, models have been used to determine critical areas within a watershed that are major sources of nonpoint-source pollution. Targeting critical areas for implementation of conservation practices is considered an efficient approach to nonpoint-source pollution control at the watershed scale. Determining landscape position of these areas for a pollutant such as sediment depends on the magnitude of sources and transport pathways from the source to a stream location or a water body of concern. Alexander et al. (2000) used the SPARROW (SPAtially Referenced Regressions on Watershed Attributes) model to estimate the importance of in-stream transport processes on removal of sediment (and chemicals) within the Mississippi River Basin. They concluded that major sources of sediment that drain into large tributaries (i.e., higher order streams) contribute at significantly higher magnitudes to total sediment delivered to the Gulf of Mexico than other major sources on small tributaries.

Erosion Prediction Technology Requirements and Challenges for Conservation Assessment

Requirements

Important requirements for erosion prediction technology related to conservation assessment and planning include:

1. Applicability to the situation. The model must apply to the erosion processes being considered in the planning process. For example, a model designed to compute sheet and rill erosion generally cannot be used to estimate gully erosion.

2. Consistent and repeatable results. The model must use parameters where consistent input values are chosen by the same or multiple users for the same or similar situations. While land users/owners may not be able to judge absolute values of soil loss estimates, they can easily judge the consistency of estimated values. Inconsistent results dramatically diminish the creditability of a model.

3. Coverage of a full range of applications. Users strongly prefer a single model that applies to all of the situations where erosion is a concern in their conservation planning activities. Having to use separate models can cause problems in situations where the models overlap because results from models typically differ.

4. Easily used with available resources. Models require the use of resources including computing infrastructure, expertise and time, and availability of input data. If the resources required to use a model exceed the perceived value of model results, the conservation planner will resist using that model.

5. The model must be valid. Attributes of a valid model can be described by the following from Beck et al. (1997): i) soundness of mathematical representation of processes; ii) sufficient correspondence between model outputs and past observations; and iii) fulfillment of the designated task. Validity of erosion prediction technology for conservation planning is oftentimes misunderstood. The proper definition of validity is that the model serves its intended purpose. Too frequently and incorrectly, validity is judged solely on how well estimates from a model fit measured research data. While measures of how well estimates from a model fit measured data are important, these measures alone are incomplete and inadequate for determining how well a particular model fits the needs of conservation planning.

Challenges

Application of erosion models for conservation planning and assessment, sediment source identification and assessment of management options is often hampered by lack of credibility, insufficient data, challenges in determination of model parameters, and absence of widely accepted modeling algorithms for explicit representation of conservation practices.

Establishing Model Application Credibility

Processes that control the generation, fate, and transport of water and sediment, as well as airborne particulates, within fields and watersheds are complex and tend to interact in a highly nonlinear fashion. Development of erosion models requires making pragmatic assumptions and simplification of the real system. When models are used to make decisions that affect stakeholders, factors that could conspire against reliable model outcomes should be considered. Reliability or credibility of a model can be defined as "a sufficient degree of belief in the validity of the model" (Rykiel, 1996). Establishing credibility for the development, selection, and application of models is an important component of modeling studies.

Advances in computer technology have enhanced our capacity to measure and model erosion processes at finer spatial and temporal resolutions. However, any modeling process entails uncertainties from both model abstractions and input data. To this uncertainty, we can add the uncertainty that exists with the decision-making process. Consequently, it may be tempting to postpone decisions until more data of higher quality become available or our knowledge of watershed or airshed processes is more complete. But failing to act comes at a cost to the health of the ecosystem and can lead to unacceptable environmental degradation. Therefore, some levels of uncertainty in our scientific knowledge, including insights grasped from model applications, must be tolerated but should be recognized. Erosion modeling is increasingly embedded in the conservation planning decision-making process, but establishing confidence in a model prior to its application is vital for the acceptance of the model and its outcomes in support of environmental decisions.

Peer review is a well-accepted mechanism for appraising the soundness of model algorithms. Model calibration (i.e., parameter estimation) is typically conducted to establish the performance validity of a model. When model simulations fail to match measured observations according to a set of predefined criteria, model outcomes for future conditions or management options bear little credibility. Process-based erosion models can include numerous parameters that represent, for example, various soil-water-plant processes. Many of these parameters do not have a direct physical interpretation and thus cannot be directly measured in the field. Instead, they are estimated using manual or autocalibration procedures whereby model parameters are changed until an acceptable level of agreement between model predictions and observed behavior of the system is achieved. The set of model parameters that produces the best performance is called the calibrated or optimal set and is used for future predictions. The model calibration process can also expose shortcomings in conceptual elements of a model. Then the analyst must consider enhancing or revising the structure of the model which may necessitate collection of additional observational data for calibration and testing of new algorithms. This cycle of data collection and model refinement is typically terminated at a reasonable point, i.e., when "essential characteristics" of the system under study are captured. The term "essential characteristics" encompasses both critical natural processes and key anthropogenic activities and is highly dependent on professional judgment and interpretation.

Data Availability

Sufficient data must be available to properly develop an erosion model for a field or watershed of interest, estimate model parameters, and corroborate model performance. Soil, land cover, elevation, and hydrography datasets have become easily accessible due to the advancement of geographic information systems (GIS) and remote sensing (RS) technology. Long-term measurements of climate, including precipitation, temperature, wind speed and solar radiation are generally available for most areas in the United States. However, other important data are difficult (if not impossible) to collect and include locations of existing conservation practices within watersheds and their characteristics, farm-level information about management operations and timings, and information on legacy sediments in the channel network.

In particular, access to spatially explicit information on conservation practices is essential for evaluation of their impacts but is rarely available. Performance of practices could vary substantially according to their landscape positions, soil properties, and other geospatial factors. Practices must be represented in the model at the precise locations where they are implemented; otherwise modeling their impacts on sediment transport processes is fraught with uncertainty. In general, research data used to develop erosion models are often incomplete, and frequently the data do not cover a sufficiently broad range of conditions. Furthermore, erosion research data are highly variable, and far too few replications are available to achieve narrow confidence intervals for statistical measures of goodness of fit. However, the data are entirely adequate for defining the trends produced by major effects when judged as a whole and when analyzed using proven scientific principles. Finally, accurately describing trends based on accepted scientific principles is often more important than how well the model fits the research data.

Determination of Model Parameter Values

The agreement between model outputs and observed data is typically evaluated using some predefined statistical criteria. Many different goodness of fit measures have been proposed to provide guidelines for proper model calibration (e.g., Gupta et al., 1998; Legates and McCabe, 1999; Moriasi et al., 2015). Many researchers (e.g., Konikow and Bredehoeft, 1992) believe that environmental models cannot be proven or validated, only tested and invalidated. As such, it cannot be ascertained with complete certainty that optimal parameter values are unique or can be unambiguously identified. Literature is replete with studies (e.g., Beven and Binley, 1992; Beven, 1993; Spear et al., 1994) attesting to the feasibility of obtaining multiple sets of model parameters that produce equally good fits (i.e., the concept of equifinality whereby a given end state can be reached by many potential means) between model simulations and observations. This consideration is particularly important for application of complex erosion models with a large number of parameters. In the context of erosion modeling, the non-uniqueness of optimal values for many parameters (e.g., soil erodibility parameters) poses a challenge. Model parameters are surrogates for the processes that they represent.

Therefore, different parameter sets indicate the importance of different processes, identify different critical runoff and sediment source areas, and highlight different key pathways for movement of water and sediments. A subsequent question then arises: which set of parameter values should be used to identify sediment sources and to select cost-effective control strategies? The choice of cost-effective sediment control strategies could substantially vary depending on the choice of calibrated parameter values. Therefore, identification of critical source areas, assessment of conservation practice impacts, and targeting of effective abatement strategies remains elusive.

Erosion Model Selection for Conservation Assessment and Planning

Selection of an existing model of appropriate complexity or development of a new model structure should be a function of the modeling purpose, characteristics of the natural system under study, and data availability (Wagener et al., 2001). Due to inherent trade-offs between data uncertainty and model structure uncertainty, an optimal level of complexity exists for erosion models. With the temptation to incorporate more processes in erosion models comes an increase in their structural complexities. To the contrary, data availability and identifiability among other pragmatic considerations tend to favor adopting simple model structures. The analyst must select a proper level of complexity that can be supported by available data from the area of interest. Model complexity is characterized by the level of detail in the processes that constitute the model. Sediment processes and their nonlinear relationships are captured by state variables and parameters. Sufficient data must be available to estimate model parameters and adjust initial conditions for each state variable. Therefore, data requirements increase as the number of model parameters and state variables increase.

Erosion models suitable for conservation assessment and planning should have certain characteristics. First, the model should be able to represent essential characteristics of the system and capture the impacts of practices on hydrologic and sediment transport processes. Second, model parameters should be distributed in space to represent practices in a spatially explicit fashion. Third, continuous simulations are desired for assessing long-term benefits of management alternatives. Daily or subdaily time steps could facilitate simulation of the fate and transport of sediment that moves across fields and watersheds in response to discrete climate events. Finally, capabilities for routing water and sediment overland and within streams are essential when the goal is to characterize interactions among management practices. Several considerations are involved in choosing a model for use in conservation planning (the requirements mentioned above provide a partial basis for the choice). It is important to select a model that serves the intended purpose because many models don't provide specific outputs (i.e., characteristics of sediment leaving a land unit) or apply to a particular process (i.e., erosion caused by surface irrigation). Another important consideration is whether a potential model describes the main effects of interest. Skill of the model developer and model user are typically more important than the model itself and the science on which it is based. The purpose of a model for conservation planning is not to describe erosion processes but to describe the main effects of variables that affect erosion. All models are based on a particular set of assumptions. A model based on the assumption of a uniform slope can have elaborate equations for erosion processes; however, it will perform poorly when applied to nonuniform slopes.

A process-based model can generally be thought of as one that uses mass and sometimes momentum balance (or continuity) equations. For the purposes of this book chapter, we designate a process-based erosion model as one that uses conservation of mass to route sediment along the slope or runoff pathway (with detachment or deposition acting as the source or sink term in the mass balance equation). Even process-based models are partly empirical in that experimental data are required to calibrate parameter values, with the result that differences among models are significantly reduced when fitted to and evaluated with the same experimental data. Also, erosion models are typically extrapolated beyond the research data used to derive them. When extrapolated, a well-developed empirical model may outperform a highly theoretical model if the theoretical model does not use robust relationships. Although the problem of extrapolating empirical models beyond their research data is well recognized, similar difficulties exist with fundamental process-based models but are often not recognized. Construction of the governing equations and the model structure, which reflect the experience and skill of the model developer rather than the model type, are key factors. The chosen model should be consistent with the available resources required to use the model in terms of available parameter values, input data, technical ability to run the model, training opportunities, computer equipment, and time required to make a set of computations for developing a conservation plan.

In summary, considerable art and judgement is involved in using models for conservation planning. Model users learn from experience how to choose parameter values and how to apply the model in relation to field conditions. The skill of the model user often overcomes model deficiencies. Just as land users prefer "comfortable" conservation practices, conservationists prefer "comfortable" erosion prediction technology. "Comfort" sometimes relates to being able to understand how the model works, but it simply may be no more than a personal preference for or previous experience with a particular model. For example, in making a final decision in the choice of an erosion model, if two models result in the same planning decision the choice is often based on other considerations (i.e., model preference or ease of use).

Erosion Model Development Skill and Knowledge Requirements

Considerable skill and knowledge can be required to properly develop erosion models. In general, this expertise can be grouped into three categories: (i) comprehensive knowledge of hydrologic and sediment transport processes and essential characteristics of the physical system under study; (ii) computer programming and GIS skills; and (iii) adequate knowledge of statistical concepts for exploratory data analysis. Therefore, erosion modeling activities typically require close collaboration between researchers and scientists from traditionally disparate disciplines. Application of comprehensive erosion models by action agencies or scientific organizations requires availability of similar expertise. Agencies that choose to use complex erosion models must ensure that their personnel are trained for the development, calibration, and credible application of the models. Otherwise, significant investments may be made in ill-advised adoption of modeling tools that either do not serve their intended goals or that cannot be used due to lack of adequate expertise. The consequences of making decisions based on uncorroborated models may also include investments wasted in management options that fail to perform to predicted outcomes.

Existing expertise within project teams, availability of technical support from model development teams, proper documentation, and model interface ease of use have been identified as primary reasons for the selection of particular erosion models. The increased temptation to link models or to incorporate new algorithms for decision-making purposes drives rising model complexity. A more complex model that includes more processes and parameters may perform better. However, in doing so, the propensity for parameter interactions and nonuniqueness tends to increase. A serious issue then arises: are available flow and sediment data sufficient for parameter identifiability and reduction of uncertainties associated with the more complex model?

Scientific Foundation of Existing Erosion Models for Conservation Planning

In general, the complexity and nonlinear nature of field- to watershed scale processes can overwhelm the capacity of existing erosion modeling tools to reveal the water or air quality impacts of conservation practices. Overland routing, that is, moving water and pollutants downgradient from field to field, is a key modeling issue. Overland flow processes play a significant role in the delivery of sediment from upland or upwind areas to the stream network or adjacent land areas. Incorporation of overland processes or lack thereof could be particularly consequential for capturing interactions between conservation practices. Many existing erosion models trade off representation of overland processes for improving computational efficiency. Accurate appraisal of the effects of buffer strips and riparian zones or windbreaks and filter strips calls for inclusion of better overland routing process representation in erosion models.

Other major modeling needs for credible assessment of conservation practices at the watershed scale include the following water driven processes: gully erosion, legacy sediments in the channel network, and sound numerical algorithms for representation of conservation practices. Terrain effects, canopy density and aerodynamic roughness, friction velocity changes across surface boundaries, and robust numerical representation of conservation practices are needed as well for wind driven erosion. Consider the Cheney Lake Watershed in Kansas as an example. According to field observations and measurements, gully erosion from agricultural fields has been identified as a primary mechanism for transport of sediment to the lake. The SWAT model was used to simulate hydrologic and sediment fluxes in the system (Parajuli et al., 2009), but it does not have the capacity to simulate ephemeral gully erosion, one of the essential characteristics of the system. Similarly, nearly 15% of sediment loads originate from stream banks in the watershed. The SWAT model simply cannot adequately capture movement of legacy sediments in the channel network, that is, long-term evolution of the channel network due to sediment deposition and bed or bank erosion cannot be captured in SWAT. Therefore, calibration of the model for sediment yield at the watershed outlet will most likely result in an overestimation of sediment loadings from upland areas (e.g., sheet and rill erosion). Overestimating sheet erosion, in turn, could affect the estimated benefits of conservation practices such as conservation tillage and terraces.

Overall, the current suite of available erosion models cannot adequately capture the interactions between upland (hillslope and field-scale) sediment loadings and within-channel sediment processes. Reduction of sheet erosion from upland areas may expedite channel degradation and erosion. Therefore, at least at the watershed scale, analysis of long-term benefits of conservation practices through modeling exercises necessitate improving the channel routing component of the available widely used erosion models or development of new models capable of capturing such interactions.

Numerical Representation of Conservation Practices

Incorporation of explicit procedures for representation of land treatment options should be considered in the development of the next generation of erosion models. The inconsistency in conservation practice representation in models can result in markedly different estimated environmental benefits. Availability of explicit functionalities for simulation of management options within erosion models could minimize the subjectivity of the procedure. For example, many studies have used the USLE support practice (P) factor for representation of grassed waterways. This factor was originally developed to account for the influence of terraces, strip-cropping, and residue management on sheet erosion; suggested values for grassed waterways or filter strips, among other practices, are generally not available. Thus, the appropriateness of representing grassed waterways by changing the USLE P factor is not widely accepted and could be a source of considerable error. Additionally, the numerical procedures used to represent conservation practices can influence their estimated benefits. Few, if any, comprehensive erosion models include explicit options for representation of conservation practices, particularly structural practices. Therefore, relevant model parameters (such as the USLE P factor) are altered to mimic the impacts of practices on hydrologic and sediment transport processes. This approach is somewhat subjective and results in inconsistent evaluation of management options, even when within the same geographic region. Various studies (e.g., Gassman et al., 2007; Arabi et al., 2008) have focused on development of standard protocols for representation of practices with models such as SWAT. However, until these methods are well established and commonly accepted, numerical representation of conservation practices will be a source of considerable uncertainty. Finally, implementation of conservation practices is not necessarily followed by proper operation and maintenance in many cases. Lack of proper maintenance can degrade the benefits of practices (Bracmort et al., 2006); therefore, greater emphasis should be placed on incorporating the degradation of practices over time in the modeling process.

Modeling Outcomes vs. Monitoring Trends

Often, the environmental benefits of land treatment options from many modeling studies are severely overestimated in comparison with monitored trends in reduction of sediments. Several factors may be responsible including modeling uncertainties, lack of sensitivity and/or statistical power in the monitoring program, lag time between implementation of conservation practices and reduction of sediment at the watershed outlet, degradation of practices due to operation and maintenance issues, and unaccounted disturbances in the systems. Enhancements in the scientific basis of modeling tools along with adoption of proper application protocols can address many of these issues. Several approaches have been suggested to reduce parametric and structural modeling uncertainties, while making best use of available data. Ongoing model development efforts have focused on incorporating more accurate representation of overland and channel processes in commonly used erosion models. Ideally, these improvements will address the issues of lag time and legacy sediment.

How Robust is Erosion Prediction Technology?

Erosion prediction technology is somewhat controversial, that is, the technology has been criticized in general and certain models have been criticized in particular. While some of the condemnation is valid, it also points in the wrong direction. Erosion prediction technology should be used as a guide for conservation planning, that is, the planner does the planning and erosion prediction technology provides information to assist in the planning that would not be available otherwise. Also, almost all erosion prediction technologies do a good job of describing main effects. Problems arise when erosion prediction technology becomes too rigidly used in regulatory-type applications. Sometimes models are misused in conservation planning such as trying to estimate ephemeral gully erosion with a model that only estimates sheet and rill erosion. Criticisms of this misuse are certainly valid. Erosion models have also been criticized for not fitting measured research data very well (e.g., small storm estimation) without consideration of how well the model estimates main effects that are well accepted by the scientific community or without consideration for the quality of the observed data. Such criticism is misdirected. Additionally, erosion prediction technology has been censured when the concern seems more related to a particular public policy than to the adequacy of the technology. Conservation planning is to support and encourage conservation, identify areas where erosion is excessive, and to develop conservation plans that are readily accepted. In reality, erosion prediction technology has proven to be a valuable tool in guiding these types of conservation activities and it should be judged in that context.

Enhancing Erosion Modeling Capacity for Conservation Assessment and Planning

Enhancing erosion modeling capacity for conservation assessment and planning can only be achieved via a concerted effort among model development teams, users of models, and agencies that require models to support management actions or decisions. Enhancing the scientific basis of erosion models is necessary for conservation effect assessment but is not sufficient to ensure their proper applications. Users must adopt good modeling practices to successfully achieve study objectives. Funding organizations can also play a critical role by designing proper program synopses and questions that can be addressed using erosion models.

Model Development

The primary role of model development teams and the broader research community is establishing credibility for model algorithms. In this context, development, incorporation, and communication of sound algorithms are the key tasks. Pressing algorithm development needs include representation of gully erosion processes, fate and transport of legacy sediments in the stream network, overland routing processes, and land treatment options. Developers and researchers must insist on and facilitate proper use of models. The model user community may not have the capacity or resources to link commonly used erosion models with calibration (manual or automatic), sensitivity analysis, and uncertainty analysis procedures (among other relevant analytical tools). Such linkages will ensure users have access to tools needed for proper application of models. A significant number of researchers continue to improve existing erosion models, in particular the WEPP and WEPS models. These improvements are usually seamlessly reconciled with the previous model code base after a proper verification process and application of quality control and/or quality assurance protocols. Finally, the importance of modeling support for selection and application of erosion models should be noted. Model development teams should provide technical support via help-desk services, user manual and technical documentation guides, and training for model application. An inventory of successful case studies could also serve as a mechanism for learning good modeling practices.

Graphical User Interface Development

In addition to core science model components that simulate fundamental erosional processes, a number of interface programs and databases have also been created to allow erosion models to be more easily applied by practitioners such as field engineers, foresters, and soil conservation personnel. For example, GIS-based interfaces allow much easier model setup for larger and more complex watershed simulations since digital elevation data and other land use and/or land cover and soil GIS data layers can be automatically processed to create the watershed structure and slope inputs. Web-based geospatial interfaces are powerful tools that can allow even a novice model user to quickly and easily create and complete a simulation for land management impacts on runoff, soil erosion, and sediment loss. Examples of web-based erosion model interfaces include the WEPP web interface (Fig. 2, http://milford.nserl.purdue.edu/wepp/weppV1.html), the RHEM web interface (Fig. 3, http://dss.tucson.ars.ag.gov/rhem/) and the Iowa Daily Erosion Project (DEP) (Fig. 4, http://dailyerosion.org) which is based on WEPP hillslope model technology. Although the customized interfaces for various erosion model simulations are easy to set up and run for different types of users (e.g., researchers, action agency personnel, ranchers and farmers, etc.), further work is needed to be able to customize the applications for users requiring more specific scenario building capabilities.

Model Application

Model users should establish credibility for model application in a system of interest according to the purpose of the analysis. This entails selection of a proper model and demonstration of sufficient correspondence between model outputs and the observed behavior of the system under study. To this end, adoption of a standard procedure for model application is highly recommended. Following a modeling protocol serves multiple purposes including reducing potential modeler bias, providing a roadmap to be followed, allowing others to repeat the study, and fostering acceptance of model results. The model application process is iterative and should be conducted according to the overall modeling purpose. The Guidance for Quality Assurance Project Plans for Modeling reports (USEPA, 2002) describes the level of quality assurance required for different modeling purposes. Similar to monitoring, all modeling activities must be coordinated in a comprehensive quality assurance and quality control program that assures the credibility of modeling steps. Erosion model application for conservation assessment and planning at the watershed scale could include the following steps:

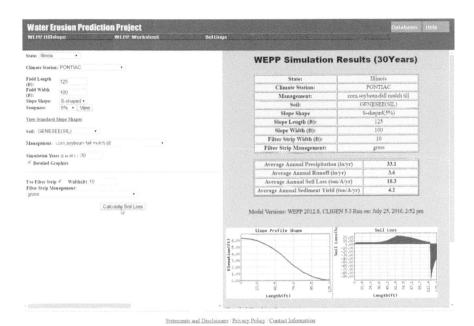

Fig. 2. Water Erosion Prediction Project (WEPP) model hillslope profile web interface with a grass filter strip. Simulation results from a 30-yr simulation shown for a slope profile near Pontiac, IL, with a corn–soybean crop rotation upslope and a grass filter strip downslope.

Fig. 3. Rangeland Hydrology and Erosion Model (RHEM) web tool, showing results from a simulation in Arizona.

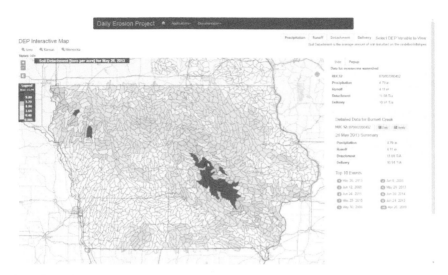

Fig. 4. Daily Erosion Project (DEP) website at Iowa State University, providing near real-time estimates of soil loss using the Water Erosion Prediction Project (WEPP) model and NexRad storm precipitation estimates. This screen capture shows estimated soil loss on a HUC-12 basis for 26 May 2013.

1. Define the modeling problem and application mode, and clearly specify modeling goals, objectives, and hypotheses.

2. Collect data at the appropriate scale of interest and identify needs for future data collection (data requirements may change according to the modeling goals and objectives).

3. Select a model appropriate for the study that includes capabilities to represent essential characteristics of the system and management or conservation options at desired spatial and temporal scales.

4. Create the model input files. An important representation issue for conservation assessment is availability of information on field boundaries or common land units. Ideally, model configuration should permit mapping of outputs to land units where management decisions are made.

5. Calibrate and evaluate the model. A diverse set of performance measures and error statistics in the calibration process should be evaluated to examine model performance under different hydrologic and climatic conditions (e.g., Nash and Sutcliffe, 1970; Gupta et al., 1998; Legates and McCabe, 1999; Moriasi et al., 2015) and assumed (or calibrated) input parameter values should be reasonable for the system.

6. Interpret modeling outcomes, test hypotheses, and summarize results. Modeling outputs should be generated according to the target audience.

7. Document the modeling process, including decisions made throughout the effort. The required level of documentation should vary with the model application.

Precision Conservation and Erosion Prediction Models

In the forthcoming decades, society will encounter significant challenges to maintain sustainability of agricultural systems currently stressed by increasing food/

biofuels demands and global warming. Precision conservation will be needed to support concomitant increases in soil and water conservation practices that will contribute to sustainability of these very intensively-managed systems while ensuring a corresponding increase in conservation of natural areas. Delgado and Berry (2008) provide a definition of precision conservation that is technologically based and advocates the integration of one or more spatial technologies [e.g., global positioning systems (GPS), GIS, and RS] and the ability to analyze spatial relationships within and among mapped data according to three broad categories: simulation modeling, spatial data mining, and map analysis. These tools and geospatial techniques have contributed to new evaluation and application of precision conservation management practices and afforded new insight into site-specific conservation applications.

Historically, erosion models have demonstrated the potential to assess agricultural and environmental tradeoffs and determine environmental benefits from strategically targeted conservation practices (e.g., use of conservation buffers or conversion of sensitive agricultural lands to natural vegetation). For example, Thomas et al. (2011) used the GLEAMS-NAPRA model to conduct precision conservation studies in Indiana that evaluated the effects of corn residue removal at different rates (i.e., 38%, 52.5%, and 70%) on different soils and tillage management on erosion rates. They found that even with no-till practices, removing crop residue increased erosion rates compared to no residue removal and concluded that future studies were necessary to adequately assess the effects of biofuel production systems on water quality (e.g., erosion, surface transport, nitrate leaching). Meki et al. (2011) used the APEX model to assess the environmental impacts of corn residue removal across 3703 farm fields in the Upper Mississippi River Basin. Similar to findings by Thomas et al. (2011), they found that the effects of corn stover removal were site-specific and varied with soils and management practices. The research by Meki et al. (2011) suggests that site-specific model evaluations could be used to help guide precision conservation practice implementation by identifying areas where the harvesting of crop residue will have minimal impacts on erosion and soil organic matter content. Bonner et al. (2016) utilized precision conservation and multicriteria decision analysis (MCDA) techniques to model the integration of switchgrass (*Panicum virgatum* L.), a perennial bioenergy crop, into a corn (*Zea mays* L.)-producing field in Iowa, United States. The impacts of energy crop integration were quantified in terms of productivity, economics, and environmental performance. Management areas identified using a multi-objective optimization method were modeled using the Landscape Environmental Assessment Framework (LEAF) to calculate biomass availability and impacts to soil health, while the Water Quality Index for Agricultural Lands (WQIag) was used to assess the risk to surface water quality. The modeling results showed that subfield management zones, optimized to maximize environmental performance, were capable of improving the annual rate of soil organic carbon (SOC) gain by 69%, reducing annual soil erosion by 63%, and increasing sustainable biomass availability by 35%.

In summary, erosion modeling research conducted during the last 5 yr is helping improve understanding of many of the processes that contribute to spatial and temporal variability of erosion and soil productivity and how to use field and off-site practices for precision conservation of agricultural and natural areas. As previously discussed, erosion modeling tools are: (i) becoming more user friendly and often utilize GIS platforms to help analysts make better management decisions that target management practices at the most sensitive areas of a watershed and/or at a given site-specific field; (ii) helping to connect and integrate management of agricultural and natural areas (e.g., fields, buffers, native grasses, trees, wetlands); and (iii) facilitating a more precise identification of the landscape locations where grass waterways, buffers, filter strips, and water collection furrows, for example, should be applied. The significance is that we are now deploying erosion models that integrate more complex layers of information (i.e., variable source area hydrology, variable soil properties, and more accurate topographies) to help better assess hydrologic and sediment fluxes, thus increasing the efficiency of conservation practices by identifying better locations to apply them.

Erosion Prediction Technology Lessons and Implications

As stated above, many applications for erosion prediction technology involving GIS, GPS, and data acquisition systems already exist or can be imagined (Goodrich et al., 2012). The complexity of these applications is rapidly increasing, but stand alone, simple erosion prediction technology will still continue to be widely used, especially in certain global sectors. Development of the underlying science that supports erosion prediction technology will continue to improve as erosion science seems to be lagging in many areas. Erosion research technology needs development so that erosion data can be collected that has far less variability between replications than is found with current data. Effects of how soil drainage status (seepage, saturated, drainage) and landscape position impact erodibility, detachment, sediment transport, and deposition need to be incorporated into process-based water erosion models. While not cutting-edge science, information is needed on several "small" items for both water and wind such as decay of roughness and ridge height as a function of weather, soil, and management variables, how soil erodibility changes over time after a mechanical disturbance of the soil, and how management affects soil erodibility. While sufficient scientific information may be available to indicate that an effect exists, far too little data are usually present to develop mathematical relationships needed in erosion models that are intended for application over large, highly variable geographic regions. Improved methods of quantifying non-uniformity and scale are needed, especially in dealing with infiltration, runoff, hydraulics, and wind mechanics in relation to variations in surface conditions and topography. New creative thinking is needed for model structure (i.e., component-based modeling), which could result in major improvements toward developing models customized for site-specific conditions and processes. Perhaps more important than research related to erosion prediction is development of a modern procedure to replace soil loss tolerance. Research in the whole area of indices needed to evaluate the impact of erosion has also seriously lagged.

Complexity and uncertainty are key issues relative to the development, understanding, and use of erosion models for conservation purposes. They are inevitable considerations because of the many complex interactions inherent in the erosional system as well as the enormous inherent variability in measured erosion data. These issues do not, however, prevent us from using erosion models effectively for conservation planning. In fact, the scientific evidence indicates that choice of models, which implies choice of model complexity, is more a matter of the type of information desired and the quality and amount of data available for the specific application. If our goal is to know to a high level of accuracy the erosion rate on a particular area of ungauged land, we cannot rely on models. Natural variability is too great, and uncertainty in predictions is too high (Nearing, 2004). For appropriate and common uses, such as those discussed above, erosion models can be effective conservation tools.

Conclusions

Used as a guide, erosion prediction technology along with soil loss tolerance and other indices of impact is a powerful conservation-planning tool. Conservation planning is far superior when erosion prediction modeling is used. The quality of results from erosion prediction in conservation planning depends on having well-trained personnel, well-crafted guidelines, and readily available input data. Numerous erosion prediction models exist and no single model is best in every aspect. Finally, it is important to choose a model that best fits the situation and is preferred by users. A variety of erosion models will continue to exist and be further enhanced as a variety of users express a diversity of preferences.

Dedication

This chapter is dedicated to the first author, Dr. Jim Ascough, who passed away unexpectedly in December 2016, prior to the book's publication. Jim's drive, enthusiasm, and friendship will be greatly missed.

References

Alexander, R.B., R.A. Smith, and G.E. Schwarz. 2000. Effect of stream channel size on the delivery of nitrogen to the Gulf of Mexico. Nature 403:758–761. doi:10.1038/35001562

Arabi, M., J.R. Frankenberger, B.A. Engel, and J.G. Arnold. 2008. Representation of agricultural management practices with SWAT. Hydrol. Processes 22:3042–3055. doi:10.1002/hyp.6890

Arabi, M., R.S. Govindaraju, and M.M. Hantush. 2006a. Cost-effective allocation of watershed management practices using a genetic algorithm. Water Resour. Res. 42:W10429.

Arabi, M., R.S. Govindaraju, and M.M. Hantush. 2007b. A probabilistic approach for analysis of uncertainty in evaluation of watershed management practices. J. Hydrol. 333:459–471. doi:10.1016/j.jhydrol.2006.09.012

Arnold, J.G., R. Srinivasan, R.S. Muttiah, and J.R. Williams. 1998. Large area hydrologic modelling and assessment part I: Model development. J. Am. Water Resour. Assoc. 34:73–89. doi:10.1111/j.1752-1688.1998.tb05961.x

Beck, M.B., J.R. Ravetz, L.A. Mulkey, and T.O. Barnwell. 1997. On the problem of model validation for predictive exposure assessments. Stoch. Hydrol. Hydraul. 11:229–254. doi:10.1007/BF02427917

Behera, S., and R.K. Panda. 2006. Evaluation of management alternatives for an agricultural watershed in a sub-humid subtropical region using a physical process model. Agric. Ecosyst. Environ. 113(1-4):62–72. doi:10.1016/j.agee.2005.08.032

Bekele, E.G., and J.W. Nicklow. 2005. Multiobjective management of ecosystem services by integrative watershed modeling and evolutionary algorithms. Water Resour. Res. 41:W10406. doi:10.1029/2005WR004090

Beven, K.J. 1993. Prophecy, reality and uncertainty in distributed hydrological modelling. Adv. Water Resour. 16:41–51. doi:10.1016/0309-1708(93)90028-E

Beven, K.J., and A.M. Binley. 1992. The future of distributed models: Model calibration and uncertainty prediction. Hydrol. Processes 6:279–298. doi:10.1002/hyp.3360060305

Bingner, R.L., and F.D. Theurer. 2003. AnnAGNPS technical processes documentation, Version 3.2. Oxford, MS: USDA ARS National Sedimentation Laboratory.

Bishop, P.L., W.D. Hively, J.R. Stedinger, M.R. Rafferty, J.L. Lojpersberger, and J.A. Bloomfield. 2005. Multivariate analysis of paired watershed data to evaluate agricultural best management practice effects on stream water phosphorus. J. Environ. Qual. 34:1087–1101. doi:10.2134/jeq2004.0194

Blanco-Canqui, H., J. Tatarko, A.L. Stalker, T.M. Shaver, and S.J. van Donk. 2016. Impacts of corn residue grazing and baling on wind erosion potential in a semiarid environment. Soil Sci. Soc. Am. J. 80:1027–1037. doi:10.2136/sssaj2016.03.0073

Bonner, I., G. McNunn, D. Muth, Jr., W. Tyner, J. Leirer, and M. Dakins. 2016. Development of integrated bioenergy production systems using precision conservation and multicriteria decision analysis techniques. J. Soil Water Conserv. 71:182–193. doi:10.2489/jswc.71.3.182

Bracmort, K.S., M. Arabi, J.R. Frankenberger, B.A. Engel, and J.G. Arnold. 2006. Modeling long-term water quality impact of structural BMPs. Trans. ASABE 49:367–374.

Brugnach, M., and C. Pahl-Wostl. 2007. A broadened view on the role for models in natural resource management: Implications for model development. In: C. Pahl-Wostl, P. Kabat, and J. Moltgen, editors, Adaptive and integrated water management: Coping with complexity and uncertainty. Springer Berlin Heidelberg, New York. p. 184–203.

Cho, J., G. Vellidis, D.D. Bosch, R. Lowrance, and T. Strickland. 2010. Water quality effects of simulated conservation practice scenarios in the Little River Experimental Watershed. J. Soil Water Conserv. 65:463–473. doi:10.2489/jswc.65.6.463

De Roo, A.P.J., C.J. Wesseling, N.H.D.T. Cremers, R.J.E. Offermans, C.J. Ritsema, and K. van Oostindie. 1994. LISEM: A new physically-based hydrological and soil erosion model in a GIS-environment: Theory and implementation. In: L.J. Olive, R.J. Loughran, and J.A. Kesby, editors, Variability in stream erosion and sediment transport. Proc. Canberra Symp., December, 1994. IAHS Publ. no. 22. p. 439-448.

Delgado, J.A., and J.K. Berry. 2008. Advances in precision conservation. Adv. Agron. 98:1–44. doi:10.1016/S0065-2113(08)00201-0

Easton, Z.M., M.T. Walter, and T.S. Steenhuis. 2008. Combined monitoring and modeling indicate the most effective agricultural best management practices. J. Environ. Qual. 37:1798–1809. doi:10.2134/jeq2007.0522

Flanagan, D.C., J.E. Gilley, and T.G. Franti. 2007. Water Erosion Prediction Project (WEPP): Development, history, model capabilities, and future enhancements. Trans. ASABE 50:1603–1612. doi:10.13031/2013.23968

Gassman, P.W., M.R. Reyes, C.H. Green, and J. Arnold. 2007. The Soil and Water Assessment Tool: Historical development, applications, and future research directions. Trans. ASABE 50:1211–1250. doi:10.13031/2013.23637

Goodrich, D.C., I.S. Burns, C.L. Unkrich, D. Semmens, D.P. Guertin, M. Hernandez, S. Yatheendradas, J.R. Kennedy, and L. Levick. 2012. KINEROS2/AGWA: Model use, calibration, and validation. Trans. ASABE 55:1561–1574. doi:10.13031/2013.42264

Gupta, H., S. Sorooshian, and P. Yapo. 1998. Toward improved calibration of hydrologic models: Multiple and noncommensurable measures of information. Water Resour. Res. 34:751–763. doi:10.1029/97WR03495

Harmel, R.D., R.J. Cooper, R.M. Slade, R.L. Haney, and J.G. Arnold. 2006. Cumulative uncertainty in measured streamflow and water quality data for small watersheds. Trans. ASABE 49:689–701. doi:10.13031/2013.20488

Harmel, R.D., C.G. Rossi, T. Dybala, J. Arnold, K. Potter, J. Wolfe, and D. Hoffman. 2008. Conservation Effects Assessment Project research in the Leon River and Riesel watersheds. J. Soil Water Conserv. 63:453–460. doi:10.2489/jswc.63.6.453

Harmel, R.D., and P.K. Smith. 2007. Consideration of measurement uncertainty in the evaluation of goodness-of-fit in hydrologic and water quality modeling. J. Hydrol. 337:326–336. doi:10.1016/j.jhydrol.2007.01.043

Heathman, G.C., D.C. Flanagan, M. Larose, and B.W. Zuercher. 2008. Application of SWAT and AnnAGNPS in the St. Joseph River Watershed. J. Soil Water Conserv. 63:552–568. doi:10.2489/jswc.63.6.552

Karydas, C.G., P. Panagos, and I.Z. Gitas. 2012. A classification of water erosion models according to their geospatial characteristics. Int. J. Digit. Earth. doi: 10.1080/17538947.2012.671380

Konikow, L.F., and J.D. Bredehoeft. 1992. Ground-water models cannot be validated. Adv. Water Resour. 15:75–83. doi:10.1016/0309-1708(92)90033-X

Kuhnle, R.A., R.L. Bingner, C.V. Alonso, C.G. Wilson, and A. Simon. 2008. Conservation practice effects on sediment load in the Goodwin Creek Experimental Watershed. J. Soil Water Conserv. 63:496–503. doi:10.2489/jswc.63.6.496

Lam, Q.D., B. Schmatz, and N. Fohrer. 2011. The impact of agricultural best management practices on water quality in a north German lowland catchment. Environ. Monit. Assess. 183:351–379. doi:10.1007/s10661-011-1926-9

Legates, D.R., and G.J. McCabe. 1999. Evaluating the use of "goodness-of-fit" measures in hydrologic and hydroclimatic model validation. Water Resour. Res. 35:233–241. doi:10.1029/1998WR900018

Licciardello, F., D.A. Zema, S.M. Zimbone, and R.L. Bingner. 2007. Runoff and soil erosion evaluation by the AnnAGNPS model in a small Mediterranean watershed. Trans. ASABE 50:1585–1593. doi:10.13031/2013.23972

Liu, B.Y., K.L. Zhang, and Y. Xie. 2002. An empirical soil loss equation. In: Proceedings of Process of Soil Erosion and its Environmental Effect Vol. II. 12th International Soil Conservation Organization Conference, Beijing, China, 26-31 May 2002. p. 21-25.

Meki, M.N., J.P. Marcos, J.D. Atwood, L.M. Norfleet, E.M. Steglich, J.R. Williams, and T.J. Gerik. 2011. Effects of site-specific factors on corn stover removal thresholds and subsequent environmental impacts in the Upper Mississippi River Basin. J. Soil Water Conserv. 66:386–399. doi:10.2489/jswc.66.6.386

Morgan, R.P.C., J.N. Quinton, R.E. Smith, G. Govers, J.W.A. Poesen, K. Auerswald, G. Chisci, D. Torri, and M.E. Styczen. 1998. The European Soil Erosion Model (EUROSEM): A dynamic approach for predicting sediment transport from ðelds and small catchments. Earth Surf. Process. Landf. 23:527–544. doi:10.1002/(SICI)1096-9837(199806)23:6<527::AID-ESP868>3.0.CO;2-5

Moriasi, D.N., M.W. Gitau, N. Pai, and P. Daggupati. 2015. Hydrologic and water quality models: Performance measures and evaluation criteria. Trans. ASABE 58:1763–1785. doi:10.13031/trans.58.10715

Mostaghimi, S., S.W. Park, R.A. Cooke, and Y. Wang. 1997. Assessment of management alternatives on a small agricultural watershed. Water Res. 31:1867–1878. doi:10.1016/S0043-1354(97)00018-3

Nearing, M.A. 2004. Soil erosion and conservation. In: J. Wainwright and M. Mulligan, editors, Environmental modelling: Finding simplicity in complexity. John Wiley & Sons, New York. p. 281–294.

Nearing, M.A., H. Wei, J.J. Stone, F.B. Pierson, K.E. Spaeth, M.A. Weltz, D.C. Flanagan, and M. Hernandez. 2011. A Rangeland Hydrology and Erosion Model. Trans. ASABE 54:901–908. doi:10.13031/2013.37115

Nash, J.E., and J.V. Sutcliffe. 1970. River flow forecasting through conceptual models: Part 1. A discussion of principles. J. Hydrol. 10:282–290. doi:10.1016/0022-1694(70)90255-6

Pandey, A., V.M. Chowdary, B.C. Mal, and M. Billib. 2009. Application of the WEPP model for prioritization and evaluation of best management practices in an Indian watershed. Hydrol. Processes 23:2997–3005. doi:10.1002/hyp.7411

Parajuli, P.B., N.O. Nelson, L.D. Frees, and K.R. Mankin. 2009. Comparison of AnnAGNPS and SWAT model simulation results in USDA-CEAP agricultural watersheds in south-central Kansas. Hydrol. Processes 23(5):748–763. doi:10.1002/hyp.7174

Pimentel, D., C. Harvey, P. Resosudarmo, K. Sinclair, D. Kurz, M. McNair, S. Crist, L. Sphpritz, L. Fitton, R. Saffouri, and R. Blair. 1995. Environmental and economic costs of soil erosion and conservation benefits. Science 267:1117–1123. doi:10.1126/science.267.5201.1117

Polyakov, A., A. Fares, D. Kubo, J. Jacob, and C. Smith. 2007. Evaluation of a non-point source pollution model, AnnAGNPS, in a tropical watershed. Environ. Model. Softw. 22:1617–1627. doi:10.1016/j.envsoft.2006.12.001

Rabotyagov, S.S., M. Jha, and T.D. Campbell. 2010. Nonpoint-source pollution reduction for an Iowa watershed: An application of evolutionary algorithms. Can. J. Agric. Econ. 58:411–431.

Renard, K.G., G.R. Foster, G.A. Weesies, D.K. McCool, and D.C. Yoder. 1997. Predicting soil erosion by water–A guide to conservation planning with the revised universal soil loss equation (RUSLE). Agricultural Handbook No. 703, U.S. Gov. Print. Office, Washington, DC.

Renschler, C.S., and T. Lee. 2005. Spatially distributed assessment of short- and long-term impacts of multiple best management practices in agricultural watersheds. J. Soil Water Conserv. 60:446–456.

Richardson, C.W., D.A. Bucks, and E.J. Sadler. 2008. The Conservation Effects Assessment Project benchmark watersheds: Synthesis of preliminary findings. J. Soil Water Conserv. 63:590–604. doi:10.2489/jswc.63.6.590

Rodriguez, H.G., J. Popp, C. Maringanti, and I. Chaubey. 2011. Selection and placement of best management practices used to reduce water quality degradation in Lincoln Lake watershed. Water Resour. Res. 47:W01507. doi:10.1029/2009WR008549

Rykiel, E.J. 1996. Testing ecological models: The meaning of validation. Ecol. Modell. 90:229–244. doi:10.1016/0304-3800(95)00152-2

Sharratt, B.S., J. Tatarko, J.T. Abatzoglou, F.A. Fox, and D. Huggins. 2015. Implications of climate change on wind erosion of agricultural lands in the Columbia Plateau. Weather and Climate Extremes. 10:20–31. doi:10.1016/j.wace.2015.06.001

Spear, R.C., T.M. Grieb, and N. Shang. 1994. Parameter uncertainty and interaction in complex environmental models. Water Resour. Res. 30:3159–3169. doi:10.1029/94WR01732

Srivastava, P., J.M. Hamlett, P.D. Robillard, and R.L. Day. 2002. Watershed optimization of best management practices using AnnAGNPS and a genetic algorithm. Water Resour. Res. 38:3-1–3-14. doi:10.1029/2001WR000365

Thomas, M.A., B.A. Engel, and I. Chaubey. 2011. Multiple corn stover removal rates for cellulosic biofuels and long-term water quality impacts. J. Soil Water Conserv. 66:431–444. doi:10.2489/jswc.66.6.431

Tomer, M.D., and M.A. Locke. 2011. The challenge of documenting water quality benefits of conservation practices: A review of USDA-ARS's conservation effects assessment project watershed studies. Water Sci. Technol. 64:300–310. doi:10.2166/wst.2011.555

Tuppad, P., C. Santhi, X. Wang, J.R. Williams, R. Srinivasan, and P.H. Gowda. 2010. Simulation of conservation practices using the APEX model. Appl. Eng. Agric. 26:779–794. doi:10.13031/2013.34947

USDA NRCS. 2010. Assessment of the effects of conservation practices on cultivated cropland in the Upper Mississippi River Basin. Final Report. http://www.nrcs.usda.gov/wps/portal/nrcs/detail/national/technical/nra/ceap/na/?&cid = nrcs143_014161.

USDA NRCS. 2011a. Assessment of the effects of conservation practices on cultivated cropland in the Chesapeake Bay Region. Final Report. http://www.nrcs.usda.gov/wps/portal/nrcs/detail/national/technical/nra/ceap/na/?&cid = stelprdb1041684.

USDA NRCS. 2011b. Assessment of the effects of conservation practices on cultivated cropland in the Great Lakes Region. Final Report. http://www.nrcs.usda.gov/wps/portal/nrcs/detail/national/technical/nra/ceap/na/?&cid = stelprdb1045403.

USEPA (United States Environmental Protection Agency). 2002. Guidance for quality assurance project plans for modeling. EPA QA/G-5M. Washington, DC: United States Environmental Protection Agency, Office of Environmental Information.

Veith, T.L., M.L. Wolfe, and C.D. Heatwole. 2003. Optimization procedure for cost effective BMP placement at a watershed scale. J. Am. Water Resour. Assoc. 39:1331–1343. doi:10.1111/j.1752-1688.2003.tb04421.x

Wagener, T., D.P. Boyle, M.J. Lees, H.S. Wheater, H.V. Gupta, and S. Sorooshian. 2001. A framework for development and application of hydrological models. Hydrol. Earth Syst. Sci. 5:13–26. doi:10.5194/hess-5-13-2001

Wagner, L.E. 2013. A history of wind erosion prediction models in the United States Department of Agriculture: The Wind Erosion Prediction System (WEPS). Aeolian Res. 10:9–24. doi:10.1016/j.aeolia.2012.10.001

Whittaker, G., R. Färe, R. Srinivasan, and D.W. Scott. 2003. Spatial evaluation of alternative nonpoint nutrient regulatory instruments. Water Resour. Res. 39:1079. doi:10.1029/2001WR001119

Williams, J.R., and R.C. Izaurralde. 2006. The APEX model. In: V.P. Singh and D.K. Frevert, editors, Erosion models. CRC Press, Taylor and Francis Group, Boca Raton, FL. p. 437–482.

Wischmeier, W.H., and D.D. Smith. 1978. Predicting rainfall erosion losses– a guide to conservation planning. Agricultural Handbook No. 537, U.S. Gov. Print. Office, Washington, DC.

Yuan, Y., S.M. Dabney, and R.L. Bingner. 2002. Cost effectiveness of agricultural BMPs for sediment reduction in the Mississippi Delta. J. Soil Water Conserv. 57:259–267.

Yuan, Y., M.A. Locke, and R.L. Bingner. 2008. Annualized agricultural non-point source pollution model application for Mississippi Delta Beasley Lake Watershed conservation practices assessment. J. Soil Water Conserv. 63:542–551. doi:10.2489/jswc.63.6.542

Zhou, X., M.J. Helmers, M. Al-Kaisi, and H.M. Hanna. 2009. Cost-effectiveness and cost-benefit analysis of conservation management practices for sediment reduction in an Iowa agricultural watershed. J. Soil Water Conserv. 64:314–323. doi:10.2489/jswc.64.5.314

Nitrogen Component in Nonpoint-Source Pollution Models

Yongping Yuan,* Ronald Bingner, and Henrique Momm

Abstract

Pollutants entering a water body can be very destructive to the health of that system. Best Management Practices (BMPs) and/or conservation practices are used to reduce these pollutants, but understanding the most effective practices is very difficult. Watershed models are an effective tool to aid in the decision making process of selecting the BMPs that are most effective in reducing the pollutant loading and are also the most cost effective. The Erosion-Productivity Index Calculator (EPIC), now the Agricultural Policy/Environmental eXtender (APEX) for application at field scale, the Annualized Agricultural Nonpoint Source pollution (AnnAGNPS) and the Soil and Water Assessment Tool (SWAT) for application at watershed scale have been developed as technological tools for evaluation of the impact of agricultural management practices on water quality and long-term soil productivity. Evaluation of model performance at various scales and locations demonstrated

---

Abbreviations: AON, active organic N in the soil layer; A_{cell}, AnnAGNPS cell area; Conv, intensive unit to extensive unit conversion factor; BD, soil bulk density; BDP, current soil bulk density as affected by tillage; CMN, rate constant for N mineralization from humus; CNR, ratio of carbon to N for crop at harvest; CPR, ratio of carbon to phosphorus for crop at harvest; D, thickness of the composite soil layer; DN, denitrification rate; DCR, residue decay rate constant; decomp_coeff, crop surface residue decomposition coefficient; ER, enrichment ratio; FC, field capacity; FON, organic N addition from decomposition of residue in AnnAGNPS; FOP, fresh organic P in the soil layer; HMN, mineralization rate of humus organic N; NO_3, amount of nitrate in the soil layer; NO_{3lat}, amount of nitrate removed in lateral flow; NO_{3surf}, amount of nitrate removed in surface runoff; ON, total organic N in the soil; ON_{fresh}, fresh organic N in the soil layer; orgC, organic carbon content; orgN, concentration of organic N including fresh, stable and active N in soil layer; b_{NO3}, nitrate percolation coefficient; r_b, bulk density of composite soil layer; P_{sol}, amount of P in the solution in the soil layer; PO, porosity; Q_{lat}, water discharged from the layer by lateral flow on a given day; Q_{surf}, surface runoff generated on a given day; Res, residue in the soil layer; Res_{surf}, surface residue which is computed from RUSLE module; Res_{sub}, subsurface residue which is computed from RUSLE module; Res_decomp, crop residue mass decomposition for current day; RMN, mineralization rate of fresh organic N; RDC, decomposition rate of fresh organic N; sol_N, total soluble N lost to surface runoff; SED, the sediment yield produced on current day; sedN, mass of N attached to sediment and transported to the channel; SW, the water content of composite soil layer on a given day; SWF, soil water correction factor; T_{soil}, the average soil temperature in a day; TF, temperature correction factor; temp_f, RUSLE temperature correction factor; U, water use rate; uptN, N uptake by the plant.

Y. Yuan, Environmental Protections Agency, Office of Research and Development, P.O. Box 93478, Las Vegas, NV 89119; R. Bingner, USDA Agricultural Research Service, 598 McElroy Drive, Oxford, MS 38655; H. Momm, Middle Tennessee State University, Department of Geosciences, MTSU P.O. Box 9, Kirksey Old Main 325B, Murfreesboro, TN 37132. *Corresponding author (Yuan.Yongping@epa.gov)

doi:10.2134/agronmonogr59.2013.0012

Precision Conservation: Geospatial Techniques for Agricultural and Natural Resources Conservation
J.A. Delgado, G.F. Sassenrath, and T. Mueller, editors. Agronomy Monograph 59.

mixed results due to the complexity of nitrogen processes. In this book chapter, nitrogen processes simulated in each model were reviewed and compared. Furthermore, current research on nitrogen losses from agricultural fields were also reviewed. Finally, applications with those models were reviewed and selected successful and unsuccessful stories were described. Although components of the N cycle included in each model varied, there are a lot of similarities in simulating N processes. The estimation of nitrogen loss at the watershed scale is a complex problem; nonpoint-source pollution models need to be further improved, and novel technology is still needed to integrate nutrient models with hydrological and erosion models.

Since the 1972 amendments of the Clean Water Act have placed emphasis on nonpoint-source pollution, planning required by this legislation needed methods to assess the pollution under various management practices, supporting the selection and placement of best management practices (BMPs) designed to reduce nonpoint-source pollution. National program staff scientists of the Agricultural Research Service (ARS), motivated by the high priority needs of action agencies, began the development of mathematical models for evaluation (Knisel, 1980; Knisel and Douglas-Mankin, 2012). In October 1977, a national project to develop mathematical models for evaluating nonpoint-source pollution was initiated. A field-scale model for Chemicals, Runoff, and Erosion from Agricultural Management Systems (CREAMS) (Knisel, 1980; Knisel and Douglas-Mankin, 2012) had been developed by ARS to address field-scale water quality impacts from agricultural practices. More than three decades have passed since then and a multitude of other models have been developed, specifically to evaluate the nonpoint-source pollution at field and watershed scale. Some of the most widely adopted models by scientists and practitioners are the Erosion-Productivity Index Calculator (EPIC), originally developed to simulate the effects of soil erosion on soil productivity on a field scale, and the Agricultural Policy/Environmental eXtender (APEX), a direct extension of EPIC, which can be applied to fields as well as watersheds. Agricultural Policy/Environmental eXtender was developed for field to watershed scale evaluation of the impact of agricultural management practices and soil erosion on water quality and long-term soil productivity (Wang et al., 2012; Gassman et al., 2010; Williams, 1990; Williams, 1995); Soil and Water Assessment Tool (SWAT) (Arnold et al., 2012; Gassman et al., 2007; Arnold and Forher, 2005; Neitsch et al., 2002) and Annualized Agricultural Nonpoint Source pollution (AnnAGNPS) model (Bingner et al., 2003; Yuan et al., 2001, 2003, 2005, and 2006) are for application at a watershed scale, developed to evaluate the impact of agricultural management practices on water quality. The APEX model was developed to extend EPIC's capabilities of simulating land management impacts for small watersheds and heterogeneous farms. These models integrate different components responsible for describing different physical and chemical processes and are designed to simulate pollutant movement as it relates to water movement. A common flow diagram of nonpoint pollution models is presented in Fig. 1.

Numerous studies have been conducted to evaluate a model's performance, and many studies have evaluated nonpoint-source pollution models' capability in simulating nitrogen (N). Results from various studies indicated mixed success of N simulations (Saleh et al., 2000; Santhi et al., 2001; Yuan et al., 2003; Saleh and Du, 2004; Chu et al., 2004; White and Chaubey, 2005; Arabi et al., 2006; Grunwald and Qi, 2006; Plus et al., 2006; Hu et al., 2007; Jha et al., 2007; Niraula et al., 2012; Yuan and Chiang, 2015; Sullivan and Gao, 2016). Quite a few studies showed poor performance of N simulations such as Yuan et al. (2003), Chu et al. (2004), Grunwald and Qi (2006), Hu et al. (2007) and White and Chaubey (2005). For example, Yuan et al. (2003) examined the AnnAGNPS N loading component through comparing AnnAGNPS model simulated N loadings with field measured N loadings from a small watershed in Mississippi. Regression of the monthly predicted N loadings with the observed produced an R^2 of 0.28. A similar study using SWAT for nonpoint-source pollution assessment, demonstrated that regression of the SWAT simulated monthly N loadings with measured data produced an R^2 value of 0.27 (Chu et al., 2004).

Simulation of N processes at a watershed scale has been very challenging because of complexities and uncertainties related to the processes. A complete understanding of various N pools, including their chemical, physical, and biological interactions in the soil profile, is essential for a full description of N cycle in soils and plants. A model based on mathematical descriptions of fundamental chemical, physical, and biological mechanisms of soil N behavior would be ideal for N modeling. However, fundamental chemical, physical, and biological mechanisms of soil N behavior and their interactions are not completely understood yet. In addition, because N movement depends on runoff and sediment movement, N simulation models have to be integrated with hydrological and erosion and/or sediment models. How those processes are integrated would also affect the simulation results. David et al. (2009) examined the predicted N results with several models (DAYCENT, DRAINMOD-N II, DNDC82a, DNDC82h, EPIC and SWAT) for a typical Illinois, tile-drained, corn and soybean watershed using the same environmental dataset; particularly they examined the differences in denitrification predictions among those models. They concluded that crop based models (SWAT, DRAINMOD-N II, and EPIC) were generally different from the biogeochemistry focused models (DNDC and DAYCENT). The results from their study suggest our ability to accurately predict N in the Midwestern Illinois region is

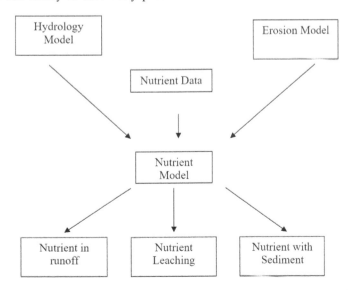

Fig. 1. Flow diagram of input and output for the nutrient model (Knisel, 1980).

quite uncertain at this time. However, no detailed information was introduced in their study about how N processes simulated in those models. This book chapter contains: (i) a review of current N research; (ii) a description of the N simulations within APEX/EPIC, AnnAGNPS and SWAT; (iii) comparisons of simulation of N components within those models; and (iv) examples of model applications to illustrate how these models can be used for precision conservation.

Current Research on Nitrogen Transformation and Transport

Nitrogen Cycle

Nitrogen has the most complex nutrient cycle of all the mineral nutrients because N can exist as a gas in ammonia (NH_3) or N_2 (Jones and Jacobsen, 2002). Nitrogen dynamics in agricultural soils are very complicated biological and chemical processes. To account for N loss mechanisms and to develop an N loading model, a fundamental understanding of N transformation processes in the soil and N cycle is necessary.

The general N processes in soil are illustrated in Fig. 2. Generally, major forms of N in soils are organic N associated with humus (active and stable in organic pool) and soluble forms of mineral N (mainly NO_3^- and NH_4^+, with low concentration of NO_2^-). Nitrogen cycling and losses consist of the following major processes: mineralization, immobilization, nitrification, denitrification, volatilization, biological N fixation from the atmosphere, decomposition of fresh residue, plant uptake, organic N transport in sediment, and nitrate N losses via leaching, surface runoff and lateral subsurface flow (Fig. 2).

Total N content in the natural soil top one foot ranges from 0.03 to 0.4% (Tisdale et al., 1985). The primary sources of soil N are from fertilizer application, manure application, N fixation from the atmosphere by symbiotic or nonsymbiotic soil bacteria, plant residue and precipitation (Novotny and Olem, 1994). Most soil N is in soil organic matter, which is derived from biological materials such as roots, microflora, fauna, leaf litter and humification processes (Stevenson, 1982). Soil organic and ammonium N are fixed by clays and located in pools that are not immediately available to plants. Parton et al. (1987) reported that the soil organic N is compartmentalized in three pools: the active, slow and recalcitrant pools that have a 1–5 yr, 20–40 yr and 200–1500 yr turnover rates, respectively. Studies from Cambradella et al. (1992) and Delgado et al. (1996a) about soil organic matter (carbon/N studies) compartmentalization support that the microbial biomass, the particulate organic fraction and the mineral associated fraction fit the conceptual compartmentalization pool model described by Parton et al. (1987). The microbial biomass is the most active pool and can have about 5% of the soil N, and the particulate organic matter, while it has a slower turnover time, can also release some N during the growing season of a plant that can be available for uptake as NH_4 or NO_3–N. The mineral associated N pool is the most recalcitrant and has the slower turnover of the three pools as described by Parton et al. (1987). However, some forms of dissolved organic N (DON) are quite mobile (amino acids especially) and can be taken up by some plants. In addition, NH_4 can be transformed into nitrate (NO_3), which is highly mobile. Mobile N can be used by plants, transported by soil water and infiltrated into ground water. N is removed from the soil

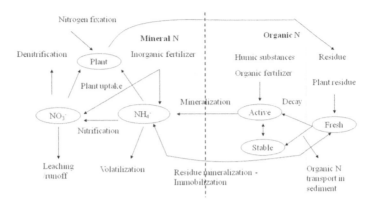

Fig. 2. A simplification of N processes (Havlin et al., 1999).

by plant uptake, surface runoff and subsurface flow (leaching and lateral flow), volatilization, denitrification and erosion and sediment transport.

Mineralization

Mineralization is the process that breaks down organic N compounds in the soil to release ammonium ions, NH_4^+, with the concurrent release of carbon as CO_2 in most cases (Vinten and Smith, 1993). The reverse process of mineralization is immobilization by which ammonium and nitrate are microbially transformed into organic forms. Cropping residues, soil moisture content, soil temperature, and pH are the main factors affecting mineralization and immobilization (Stanford and Epstein, 1974; Haynes, 1986). Immobilization occurs more easily at high C/N ratios (above 30:1). In addition, N fertilizer application stimulates the mineralization process (Haynes, 1986). The promotion of mineralization of soil organic N increases the crop uptake of N (Stevenson, 1982).

The release of N from organic matter is critical to the N cycle and to nitrate leaching in particular. A study done in England (Vinten and Smith, 1993) showed as much as 71 kg N ha^{-1} y^{-1} released from organic N in a field with no manure or N fertilizer application (Burt et al., 1993).

Nitrification and Denitrification

More than 90% of the N fertilizer used in the United States is ammonium salts (Novotny and Olem, 1994). Manure and/or sewage sludge applied to the soil can be quickly decomposed into ammonium. Mineralization converts organic N into NH_4^+. In an aerated, microorganism-rich soil such as farmland, nitrification occurs which converts NH_4^+ to NO_3^- as follows:

$$\text{Organic N} \rightarrow NH_4^{+ \text{ Nitrosomonas}} NO_2^{- \text{ Nitrobacter}} NO_3^- \text{ (Novotny and Olem, 1994).}$$

The reaction from NO_2^- to NO_3^- is much faster than the conversion of NH_4^+ to NO_2^-. Therefore, little nitrite remains in soils. Nitrate is highly soluble and can readily move with soil water. Nitrification occurs between temperatures of 10°C to 45°C with the optimum temperature at 22°C (Stanford and Smith, 1972). Nitrification is also dependent on the soil pH value, which occurs between pH 6 to 10 with the optimum at 8.5 (Dancer et al., 1973). Additionally, nitrification depends on soil moisture content; the nitrification rate decreases with decreasing moisture content (Novotny and Olem, 1994). However, if the soil is saturated for a long period and oxygen is absent or depleted to a point below the oxygen demand, denitrification occurs. In this process, NO_3^- is converted to NO_2, NO, N_2O and N_2 (gaseous forms which return to the atmosphere). This process usually occurs in subsoil with low permeability, and in soils saturated with water for a long period of time, such as a wetland (Carter and Allison, 1960; Havlin et al., 1999).

The phenomenon of denitrification in soil, resulting in a loss of available nitrate has been considered a benign process in reducing the quantities of nitrate loss in surface runoff and subsurface flow such as in tile drainage or aquifers. Therefore, wetland and field ponds and control of drainage in the winter may be useful in reducing leached nitrate in tile drain systems (Hansen et al., 2012; Skaggs et al., 2012). However, if the nitrate reduction does not go entirely to N_2 and N_2O is emitted, another environmental problem is raised because N_2O is a factor in the depletion of the Earth's stratospheric ozone layer and contributes to global warming (Vinten and Smith, 1993; Linn and Doran, 1984; Mosier et al., 2002).

Volatilization

Volatilization (NH_3) refers to the loss of ammonia as a gas into the atmosphere. Because NH_4^+ will more easily convert to NH_3 at high pH, the process is increased at higher pH values. Volatilization also increases with increased wind and temperature (Havlin et al., 1999). Because the nitrification, as discussed above, transforms NH_4^+ to NO_3^- in hours to weeks, volatilization usually happens during a short period after ammonia-based fertilizer application. Once it becomes mobile nitrate, it can no longer volatilize, but it still can be leached out of the root zone. Incorporating fertilizer and applying it right before a rainstorm would push ammonia fertilizer further into the soil profile where it is less available for volatilization (Reddy et al., 1979; Jones and Jacobson, 2002). In addition, applying the ammonia fertilizer on a calm day would also help reduce the volatilization.

Biological N fixation

Biological N fixation is the process by which N_2 gas is converted to bio-available forms of N. Nitrogen fixation is affected by many factors, and N content, soil pH, soil moisture and plant conditions are major factors. Nitrogen fixation supplies N to microorganisms and plants and increases available N level in the soil. In the United States, N fixation produces about one third of the amount of fertilizer applied (Havlin et al., 1999).

Decomposition

Decomposition is the breakdown of fresh organic residue into simpler organic components, and adds organic N to the soil. Factors affecting mineralization also affect the rate of decomposition. In addition to those factors that affect mineralization, the residue characteristics affect the decomposition too.

Nitrate Leaching and Runoff Losses

There are several combined forms of N, including fertilizer added in soils as introduced above (more than 90% of the N fertilizer is ammonium salts), but only the nitrate ion is leached out of soils in appreciable amounts by water passing through the soil profile (Dinnes et al., 2002; Vinten and Smith, 1993). The movement of nitrate in the field is a complex process; Meisinger and Delgado (2002) discussed the principles for managing and reducing nitrate leaching. They reported that one of the main factors that drives leaching is the quantity of nitrate and the water content of the soil during leaching.

Given a quantity of rainfall, the depth of water movement is different for different kinds of soils. Thus, soil structure, pore size, the spatial distribution of pores and their continuity all contribute to the irregular movement of water down the soil profile, which cause the irregular movement of nitrate. The soil moisture front affects the diffusive dispersion of nitrate in the soil solution. The diffusive dispersion of nitrate in the soil solution is the nitrate movement due to the differences in nitrate concentration. Several studies have been done in modeling nitrate transport in the soil (Barraclough, 1989; Addiscott and Whitmore, 1991). The difficulties in modeling nitrate transport are in representing the complex interaction of hydrological and biogeochemical processes controlling the transformation and transport of compounds in the N cycle. A time step of 1 d is usually used for spatially distributed hydrological and water quality models such as AnnAGNPS and SWAT.

Many factors, such as fertilizer application (when, where and how), soil texture, land use, crop rotation, and cultivation can have an effect on the quantity of nitrate leached from a soil. The amount of fertilizer, the timing of fertilizer and the particular type of fertilizer used can affect the fertilizer available for crop uptake and leaching.

Bergstrom and Brink (1986) provided a general relationship between N fertilizer application and N leaching losses. They conducted 10 yr of research on a clay soil in Sweden. They concluded that leaching of nitrate was moderate up to a rate of application of 100 kg N ha^{-1} y^{-1}, increased rapidly thereafter, and reached a rate of 91 kg N ha^{-1} for an application of 200 kg N ha^{-1} y^{-1} in a year in which rainfall was 638 mm.

A similar study was conducted on a Minnesota silt loam soil with subsurface tile drainage (Randall et al., 1993a). Anhydrous ammonia was applied at rates of 0 to 252 kg N ha^{-1} to different plots; they found the nitrate N concentration in soil water increased with increasing application rates. When application rate was above 84 kg N ha^{-1}, the nitrate N concentration was above 10 mg L^{-1}. The optimum application rate for corn production was 168 kg N ha^{-1}. They also concluded that fall application of fertilizer resulted in higher nitrate leaching losses than spring application.

Harris et al. (1984) compared how the timing of fertilizer application affects N leaching losses. They found that half of the N was lost from autumn-applied N and up to 15% of N was lost from spring-applied N. Delgado et al. (1996b) found similar responses using ^{15}N labeled fertilizer, with more ^{15}N recovered by the forage systems that received spring fertilizer applications (42%), than by forage systems that received fall fertilizer applications (22%). The N losses of the 168 kg N ha^{-1} fertilizer application in the fall were higher (50.1%) than the losses from the spring-applied fertilizer (34.5%). Kanwar and Baker (1993) investigated the effects of a single application and split applications of N fertilizer on leaching losses. The nitrate concentration in drainage water was less from split application plots.

On freely drained soils, nitrate leaching can be estimated by an estimation of water flux associated with the soil solution concentration measured. Kolenbrander

(1981) found that for arable soils the leaching of N depends on soil texture, with clay soils losing about half the nitrate of sandy soils as long as N application rate did not exceed 100 to 200 kg N ha^{-1}. Once the N application rate exceeds this range, leaching losses increase rapidly and become less dependent on soil texture. The leaching of N from artificially drained soils is much larger than from freely drained soils depending on the drainage system. For a given site, nitrate leaching was proportional to drain flow.

Several studies showed that arable land was more prone to leaching than grassland (Kolenbrander, 1981; Barraclough et al., 1983; Ryden et al., 1984). The nature of the crop dictates the N requirement and, thus, the nitrate available for leaching. Randall et al. (1993b) investigated the effects of cropping system on nitrate leaching from tile drainage in a Minnesota clay loam soil. They concluded that the N losses from continuous corn systems were much higher than from corn–soybean rotation systems under the fertilizer management treatment recommended to optimize yield. Kanwar and Baker (1993) conducted a similar investigation in Iowa clay soil. They also found that N losses from continuous corn systems were much higher than from corn–soybean rotation systems. However, Melvin et al. (1993) pointed out that the corn–soybean rotation system required less fertilizer application than a continuous corn system; thus, the effects on the quality of tile drainage is from fertilizer application amount, not the crop.

Delgado (1998, 2001) and Delgado et al. (1998, 2001, 2007) found that rooting depth was an important factor that impacted nitrate leaching. Crops with deeper root systems such as cover crops, barley, and winter wheat had a much lower nitrate leaching potential than shallow-rooted crops such as lettuce and potato (Delgado, 1998, 2001; Delgado et al., 1998, 2001, 2007). The deep-rooted crops served as a sieve that can recover nitrate that was applied with the irrigation water, helping to reclaim and improve groundwater since more nitrate was applied with the irrigation water than was leached out of the root zone of the deeper rooted crops (Delgado, 2001; Delgado et al., 2001). The deep-rooted crops, especially the cover crops, served as scavenger crops that recovered nitrate that had already leached out from the root zone of shallow-root crops like potato and lettuce (Delgado, 1998, 2001; Delgado et al., 2001, 2007).

Dowdell et al. (1987) compared leaching losses of N from direct drilled plots and plowed plots over 4 yr. They found that N losses from direct drilled plots were only 48 to 49% of losses from plowed plots. Vinten and Smith (1993) also reported greater leaching losses from plots that have been cultivated (chisel plowed and subsoiled) than from plots left stubble over the winter. The probable reason is that cultivation promotes aeration, and consequently, higher mineralization and lower denitrification losses. However, Harris and Catt (1999) observed greater levels of nitrate from no-tillage plots.

Kanwar and Baker (1993) compared nitrate losses from tile drainage for no–tillage, chisel plow, ridge tillage and moldboard plow. They found that the greatest concentrations were measured in the drainage from moldboard plowed plots. However, the total mass of nitrate in the drainage effluent from moldboard plow was less than that from no-tillage because a larger proportion of water drains through the undisturbed soil, through fairly continuous micropores.

Naveen et al. (1996) compared the effect of no-tillage and conventional tillage on tile drain flow, nitrate concentration and loss in tile effluent in loam soil. They found that flow was significantly higher from no-tillage treatment than that from conventional tillage treatment. The flow-averaged nitrate N concentrations in tile flow were greater from conventional tillage than from no-tillage, but the total loss from these two treatments was not significantly different over the 40-mo study period.

Mitchell et al. (2000) analyzed 5 yr of nitrate N data from the Little Vermilion River watershed and found that concentrations of leached nitrate N followed a seasonal cycle. Nitrate-Nitrogen concentrations varied considerably from the subsurface tile drains between fields depending on the management system. The highest nitrate N loss was from reduced-tillage corn–soybean fields (41 kg N ha^{-1} y^{-1}), followed by reduced tillage white corn–soybean fields (38 kg N ha^{-1} y^{-1}). The lowest nitrate N loss was from grass fields (3.8 kg N ha^{-1} y^{-1}), and from no-tillage corn–soybean and reduced-tillage soybean–corn–corn silage fields (15 kg N ha^{-1} y^{-1}).

Nitrogen Balance

An available N mass balance can be established through summarizing the N gains (mineralization, fixation, fertilizer application) and losses (plant uptake, denitrification, volatilization, and immobilization). Nitrogen can be gained or lost through exchanges with the soil as absorption and desorption. However, this process is usually not simulated in simulations of N processes in models. In addition, precipitation represents another input to the N pool which influences mass balance. The final potential loss is nitrate leaching or loss via surface runoff and organic N transport in sediment. Factors affecting nitrate leaching also affect surface runoff loss.

Because of the complex mechanisms of the N cycle in agricultural soils, long term studies of N balance in agricultural soils are very important to determine the effects of agricultural management practices on leaching of nitrate from agricultural cropland to groundwater and surface water. Such studies are essential for testing the long-term predictive power of models of the agricultural soil–plant N cycle, which should include calculations of mineralization, immobilization, nitrification, denitrification, crop uptake and nitrate leaching. Such models are becoming increasingly important in helping policymakers and land use managers make policy decisions. However, because of the initial condition of the soil organic matter, and uncertainties in measuring mineralization and denitrification rates which cause less accurate estimates of change in organic matter content as well as difficulties in quantifying other N processes, it is very difficult to predict N losses.

Nitrogen Processes IN APEX/EPIC, AnnAGNPS AND SWAT

General Description

There are a large number of N models. A few examples are the Nitrogen Losses and Environmental Assessment Package (Delgado et al., 2010; Shaffer et al., 2010), the LEACHM-N model (Wagenet and Hutson, 1989), the Danish Nitrogen Simulation System (DAISY) (Hansen et al., 1991), and the Root Zone Water Quality model (RZWQM) (Ahuja et al., 2000). This section will be focused on APEX/EPIC, AnnAGNPS and SWAT.

APEX simulates the complete N cycle: atmospheric N inputs, fertilizer/manure N applications, crop N uptake, mineralization, immobilization, nitrification, denitrification, ammonia volatilization, organic N transport on sediment, and nitrate N losses in leaching, surface runoff, lateral subsurface flow, and tile flow. Decomposition and mineralization of fresh organic N are controlled by a decay rate constant. Denitrification occurs only when soil moisture content is above the field capacity. The fertilizer N is considered to dissolve immediately and contribute to the mineral N pool. Plant uptake of N is controlled by plant demand but limited by soil supply of the N. Organic N in each soil layer is partitioned into fresh and stable pools. The organic N loss is estimated using sediment yield, organic N on the soil surface layer and an enrichment ratio; and the soluble N loss is estimated by considering soluble N concentration changes in soil layers (Wang et al., 2012).

In AnnAGNPS, three pools of soil N are considered: stable organic N, active organic N (mineralizable N), and inorganic N. Losses (cell output pathways) include soluble inorganic N in runoff, and sediment-bound organic N from soil erosion. Leaching losses are not considered in the current version. The humic mineralization equation is adapted from the EPIC model (Sharpley and Williams, 1990). Plant uptake of N is modeled with a simple crop growth stage index with adaptations for soil profile nutrient uptake from the TETRANS model (Corwin, 1995). Residue return and decomposition uses equations from RUSLE (Renard et al., 1997).

The N estimation approach in SWAT is more complicated and it simulates five different pools of N in the soil (Neitsch et al., 2002). Two pools of inorganic N are NH_4^+ and NO_3^-, while the other three pools are organic forms of N. Fresh organic N is associated with crop residue and microbial biomass while the active and stable organic N pools are associated with the soil humus. Nitrification simulates the movement of the NH_4^+ pool to NO_3^-; ammonia volatilization simulates the gaseous loss of NH_3 that occurs when ammonium is surface applied. Decomposition and mineralization of fresh organic N are controlled by a decay rate constant as in EPIC. N mineralized from the fresh organic pool is added to the nitrate pool. Decomposition of the residue fresh organic N pool is added to the humus active organic pool. Organic N is allowed to move between the active and stable organic pools. Mineralization from the humus active organic N pool is added to the NO_3^- pool. Crop uptake of N is reduced from the NO_3^- pool.

Initial Nitrogen Contents

To start N simulation, initial content of N has to be defined. This requires either a user input or model defaults. In all models (APEX/EPIC, AnnAGNPS and SWAT), users can define the amount of inorganic and organic N contained in soil layers. If the user does not supply such information, default values for inorganic and organic N concentration (mg kg^{-1}) are used to start the simulation.

As described above, the SWAT model simulates five different pools of N in the soil. Users may define the amount of N in each pool at the beginning of the simulation. Otherwise, SWAT initializes levels of N in different pools. The initial nitrate N level is assumed 7 mg kg^{-1} at the soil surface and exponentially decayed with depth (Neitsch et al., 2002). The organic N for humic materials is assumed to be 1/14 of organic carbon. Two percent of humic organic N is assigned to the active organic N pool and 98% is assigned to the stable organic N pool. Fresh organic N in the residue is initialized as 0.15% of the initial amount of residue in the top 10 mm of the soil surface and is zero for all the other soil layers. The ammonium N pool is initialized to 0 mg kg^{-1}.

In the AnnAGNPS model, similarly, users can define the amount of inorganic and organic N contained in soil layers. If the user does not supply such information, default values for inorganic and organic N concentration (mg kg^{-1} or ppm) are used to start the simulation. In AnnAGNPS, the default values for organic N and inorganic N concentrations are 500 mg kg^{-1} and 5 mg kg^{-1} for the top soil layer and 50 mg kg^{-1} and 0.5 mg kg^{-1} for subsequent soil layers. The amount of active N in the organic N pool is defined based on the cultivation period, which will be introduced later.

The soil input of N level is in concentrations, but AnnAGNPS performs all calculations on a mass basis. To convert a concentration to a mass, AnnAGNPS uses a conversion factor (conv).

$$conv = D \times 10 \times 1000 \times \rho_b \times A_{cell} \qquad [1]$$

The conversion factor represents a weight of soil, calculated by multiplying the volume of soil by its bulk density. It is used to convert soil's nutrient concentration into kilograms, necessary to do mass balances. Similarly, AnnAGNPS allows N levels to be input as concentrations (ppm or g mg^{-1}) and it performs all calculations on a mass basis.

Where:

conv = intensive unit to extensive unit conversion factor (kg),

D = thickness of the composite soil layer (mm),

r_b = bulk density of composite soil layer (g/cc or Metric ton/cubic meter),

A_{cell} = AnnAGNPS cell_area (hectares).

Nitrogen Transformation

Residue Decomposition and Mineralization

The calculations of residue decomposition and mineralization are the same in APEX/EPIC and SWAT, but different in AnnAGNPS. In Chapter 3 of the theoretical document of the SWAT model (Neitsch et al., 2002), the equations of simulating decomposition and mineralization are introduced. In summary, in APEX/EPIC and SWAT, decomposition and mineralization of the fresh organic N pool is allowed only in the first soil layer. Mineralization is calculated from the fresh organic N in crop residue and organic N in humic substances, while decomposition is the breakdown of fresh organic residue into simpler organic components. Mineralization converts organic, plant-unavailable N to inorganic, plant-available N. The decomposition and mineralization processes are controlled by the decay rate constant which is a function of C to N ratio (CNR) and C to P ratio (CPR) in the residue, temperature, soil water content, and composition of crop residue.

$$CNR = \frac{0.58 \times Res}{ON_{fresh} + NO_3} \qquad [2]$$

In both APEX/EPIC and SWAT models, CNR of the residue is calculated as:

Where:

Res = residue in the soil layer (kg ha^{-1}),

ON_{fresh} = fresh organic N in the soil layer (kg ha^{-1}), and

NO_3 = amount of nitrate in the soil layer (kg ha^{-1}).

C/P ratio is calculated as:

$$CPR = \frac{0.58 \times Res}{FOP + P_{sol}} \qquad [3]$$

Where:

Res = residue in the soil layer (kg ha^{-1}),

FOP = fresh organic P in the soil layer (kg ha^{-1}),

P_{sol} = amount of P in the solution in the soil layer (kg ha^{-1}).

In APEX/EPIC, temperature correction factor (TF) for mineralization is calculated as,

$$TF=\frac{T_{soil}}{T_{soil}+\exp(9.93-0.312T_{soil})} \qquad [4]$$

$$T_{soil} \leq 0$$

$$TF=0$$

$$T_{soil} > 0$$

Where:

T_{soil} = the average soil temperature in a day (°C).

$$SWF=\frac{SW}{PO} \qquad [5]$$

Water correction factor (SWF) for mineralization is calculated as:

Where:

SW = the water content of soil layer on a given day,

PO = soil porosity.

In SWAT, the model uses the same algorithm in residue decomposition and mineralization calculation, but the calculation of temperature and water correction factors is slightly different.

The temperature correction factor (TF) for mineralization is:

$$T_{soil} > 0 \qquad [6]$$

$$TF=0.9\times\frac{T_{soil}}{T_{soil}+\exp(9.93-0.312T_{soil})}+0.1$$

$$T_{soil} \leq 0$$

TF = 0

Where:

T_{soil} = the average soil temperature in a day (°C).

In this way, the SWAT-based calculations never allow the temperature factor to fall below 0.1.

The water correction factor (SWF) for mineralization is calculated as:

$$SWF = \frac{SW}{FC} \qquad [7]$$

Where:

   SW = the water content of soil layer on a given day (mm H2O), and
   FC = field capacity, not soil porosity.
   SWAT never allows the water factor to fall below 0.05.
   In both the APEX/EPIC and SWAT model, residue decay rate constant (DCR) is calculated as:

$$DCR = RC \times CNP \times (SWF \times TF)^{0.5} \qquad [8]$$

   Where RC is the rate coefficient for mineralization of the fresh residue, a required user input, and CNP is the minimum value of CNP_a, CNP_b, and 1.0.

$$CNP_a = \exp\left[-0.693 * \frac{(CNR-25)}{25}\right] \qquad [9]$$
$$CNP_b = \exp\left[-0.693 * \frac{(CPR-200)}{200}\right]$$

$$RMN = 0.8 \times DCR \times ON_{fresh} \qquad [10]$$

   Mineralization of fresh organic N in crop residue is estimated for each soil layer by using Eq. [10]:

Where:

   RMN = mineralization rate of fresh organic N (kg ha^{-1} d^{-1}),
   ON_{fresh} = fresh organic N in the soil layer (kg ha^{-1}), and
   DCR = residue decay rate constant (d^{-1}).
   It is assumed that only 80% of fresh organic N in crop residue mineralized into inorganic N. The other 20% was decomposed into humus organic N.
   Three organic N pools are maintained for calculation.
   Decomposition of fresh organic N in crop residue is estimated for each soil layer by using Eq. [11]:

$$RDC = 0.2 \times DCR \times ON_{fresh} \qquad [11]$$

Where:

   RDC = decomposition rate of fresh organic N (kg ha^{-1} d^{-1}),
   ON_{fresh} = fresh organic N in the soil layer (kg ha^{-1}), and
   DCR = residue decay rate constant (d^{-1}).
   N decomposed from fresh organic pool is added to the humus organic pool.
   In the AnnAGNPS model, the residue decomposition and mineralization adapts equations from RUSLE (Renard et al., 1997). They are simulated as follows:

Crop land

   The cell residue N from decomposition is calculated only for the top soil layer for crop land. It is calculated using the following equations:

$$FON = \frac{(Res\_decomp) \times 0.5}{CNR_{harvest}} \qquad [12]$$

Where:

FON = organic N addition from decomposition of crop residue laying on the soil surface on current day (kg),

Res_decomp = crop residue mass decomposition for current day (kg), and

$CNR_{harvest}$ = Ratio of carbon to N for crop at harvest.

Crop residue mass decomposition is calculated as:

$$Res\_decomp=(Res_{Surf})\times 1-exp\{[-temp\_f\times(decomp\_coeff)]\}*A_{cell} \quad [13]$$

Where:

Res_{surf} = surface residue for a cell which is computed from RUSLE module (kg ha^{-1}),

temp_f = RUSLE temperature correction factor (unitless),

decomp_coeff = crop surface residue decomposition coefficient (0–1 unitless, in crop database in RUSLE), and

A_{cell} = AnnAGNPS cell area (hectares).

Temperature correction factor is calculated using the following equation:

$$T_{soil} > 32$$

$$0 < T_{soil} < 32$$

$$T_{soil} < 0$$

$$temp\_f = 0$$

$$temp\_f = \frac{3200\times[(T_{soil}+8)**2]-(T_{soil}+8)**4}{2560000}$$

$$temp\_f = 1 \quad\quad\quad [14]$$

Where:

T_{soil} = the average cell soil temperature (°C).

The above equation is a simplification of the temperature correction factor from RUSLE (Page 151–152, Eq. [5]-7).

Noncropland

The cell residue N from decomposition for non-crop land is calculated for both top soil layer and bottom soil layer. It is calculated using following equations:

$$FON=(Res\_decomp)\times NF \quad\quad\quad [15]$$

Where:

FON = organic N addition from decomposition of non-crop residue (kg),

NF = N fraction of dry total biomass for non-crop field (weight of N/weight of biomass), and

Res_decomp = crop residue mass decomposition for current day (kg).

NF is assumed to be 1% N in dry biomass for grassland, 0.4% for forest systems, and zero for urban or mixed land.

$$\text{Res\_decomp}=(\text{Res}_{sub})\times[1-\exp(-\text{temp\_f}*0.016)]\times A_{cell} \qquad [16\ a]$$

$$\text{Res\_decomp}=(\text{Res}_{surf})\times[1-\exp(-\text{temp\_f}\times0.016)]\times A_{cell} \qquad [16\ b]$$

Non-crop residue decomposition is calculated as:
Where:

Res_{surf-} = surface residue which is computed from RUSLE module (kg ha^{-1}),

Res_{sub-} = subsurface residue which is computed from RUSLE module (kg ha^{-1}),

temp_f = RUSLE temperature correction factor, for non-crop res decomp, used to adjust the calculation of residue decomposition based on a first order rate, constant (unitless).

Residue decomposition coefficient for subsurface is assumed as 0.016.

Equation 16a is for top soil layer and equation 16b is for bottom soil layer.

Humus mineralization

In APEX/EPIC, mineralization of organic N associated with humus is estimated by using Eq. [17]:

$$HMN=CMN\times(AON)\times(SWF\times TF)^{0.5}\left(\frac{BD}{BDP}\right)^2 \qquad [17]$$

Where: HMN = mineralization rate of humus organic N (kg ha^{-1} d^{-1}),

CMN = rate constant for N mineralization from humus. It is a user input and the default value is 0.0003 d^{-1},

AON = active organic N in the soil layer (kg ha^{-1}),

TF = temperature correction factor (Eq. [4]),

SWF = water correction factor (Eq. [5]),

BD = soil bulk density (t m^{-3}), and

BDP = current soil bulk density as affected by tillage (t m^{-3}).

The ratio of BD and BDP takes into account the impact of tillage on mineralization.

The active organic N is calculated as:

$$AON =frac\times ON \qquad [18]$$

Where:

Frac = the fraction of active organic N, and

ON = total organic N in the soil.

The active organic N fraction is calculated based on the following equation:

$$frac=0.4\exp(-0.0277YC)+0.1 \qquad [19]$$

Where YC is the period of cultivation before the simulation starts (year). The concepts expressed in the above equation are based on work by Hobbs and Thompson (1971). For cropland, year of cultivation is set to 50, otherwise, it is set to zero. Below the plow layer, the active pool fraction is set to 40% of plow layer value. This is based on work by Cassman and Munns (1980).

In SWAT, as described in Chapter 3 of the theoretical document (Neitsch et al., 2002), the model maintains three organic N pools: fresh organic N, stable organic N associated with humic substances, and active organic N. The humic organic N level is a user input or initially assumed to be the CNR of 14:1 and humic organic N is partitioned between the active and stable pools. The active N pool is 2% of humic N and the rest is stable pools at the beginning. The fresh organic N pool is set to 0.15% of the initial amount of residue on the soil surface. SWAT uses the same equation as above to calculate the humus mineralization except that the impact of tillage on mineralization is not considered.

In AnnAGNPS, total organic N is maintained at the end of day. The amount of active organic N for mineralization is calculated based on the fraction of active N, which is a function of cultivation years, as it is in the EPIC model (Sharpley and Williams, 1990).

$$HMN=CMN \times (AON) \times (SWF \times TF)^{0.5}$$ [20]

Where:

HMN = mineralization rate of humus organic N (kg ha^{-1} d^{-1}),

CMN = rate constant for N mineralization from humus, user input and default value is 0.0003 d^{-1},

AON = active organic N content of the soil layer (kg ha^{-1}),

SWF = water correction factor (Eq. [6]), and

TF = temperature correction factor (Eq. [7]).

The AnnAGNPS calculates the N mineralization the same way as SWAT (Neitsch et al., 2002), which is adapted from the EPIC model (Sharpley and Williams, 1990). This EPIC mineralization model is a modification of the PAPRAN mineralization model (Seligman and van Keulen, 1980). The model considers two sources of mineralization: the fresh organic N pool associated with crop residue and microbial biomass as introduced in the residue decomposition and mineralization, and the active organic N pool associated with soil humus as Eq. [20].

Temperature, soil moisture, aeration, and pH affect N mineralization rate (Sharpley and Williams, 1990). Mineralization is allowed to occur only if the temperature of the soil layer is above 0°C. Mineralization is also dependent on water availability. A correction factor is used in the mineralization equations to account for the impact of temperature and water on these processes.

N addition from rainfall

To estimate the N contribution from rainfall, APEX/EPIC uses an average rainfall N concentration for a location for all storms; the amount of N in rainfall is estimated as the product of rainfall amount and concentration. SWAT requires user input for the concentration of N in the rainfall for each storm. The amount of N in rainfall is estimated as the product of rainfall amount and concentration. N addition from rainfall is added to the nitrate N pool in both APEX/EPIC and SWAT.

AnnAGNPS does not simulate the N addition from rainfall.

Nitrification and ammonia volatilization

In both APEX/EPIC and SWAT, nitrification and ammonia volatilization are estimated simultaneously. In both models, nitrification and ammonia volatilization are simulated using a combination of the methods developed by Reddy et al. (1979) and Godwin et al. (1984). Nitrification is a function of soil temperature and soil water content, while volatilization is a function of soil temperature, depth and cation exchange capacity. Nitrification and/or volatilization occurs only when temperature of the soil layer exceeds 5°C. SWAT theoretical documentation (Neitsch et al., 2002) provides a detailed description of nitrification and/or volatilization.

Nitrification and volatilization processes are not simulated in the AnnAG-NPS model.

Denitrification

Denitrification converts NO_3^- to NO_2, NO, N_2O and N_2 under anaerobic conditions. Denitrification is a function of temperature, water content, and presence of carbon and nitrate (Sharpley and Williams, 1990; Neitsch et al., 2002). In all models introduced in this book chapter, denitrification is calculated using the following equations:

[21]

$$DN = NO_3 \times \left[1 - \exp(-1.4 \times TF \times orgC) \right]$$

SWF < 0.9

SWF > 0.9

DN=0

Where:

   DN = denitrification rate (kg ha^{-1}),

   NO_3 = amount of nitrate in soil layer (kg ha^{-1})

   TF = nutrient cycling temperature factor, temperature correction factor as used for mineralization (Eq. [4]),

   orgC = organic carbon content (%), and

   SWF = nutrient cycling water factor, as used for mineralization (Eq. [4]).

Fixation

In APEX/EPIC and SWAT, fixation is simulated when the soil does not supply the plant with the amount of N needed for plant growth (Neitsch et al., 2002). The fixed N from atmospheric N_2 is incorporated directly into the plant and never enters the soil. Detailed equations of calculating fixation are described in Chapter 5 of the SWAT theoretical documentation (Neitsch et al., 2002).

Fixation is not simulated in the AnnAGNPS model.

Fertilizer Application

APEX/EPIC provides two options for fertilizer application. For the first option, the user specifies dates, rates, and depths of application of N. The second option is an automated option by the model. In this option, the model decides when and how much fertilizer to apply based on a plant stress level. For user-supplied fertilizer, N fertilizer is considered to dissolve immediately and contribute to the inorganic N pool in the soil.

Similarly to the APEX/EPIC model, in the SWAT model, fertilizer application can be supplied by users or automated by the program. Information on timing of the application, the type of fertilizer or manure applied, amount of fertilizer, and depth distribution of fertilizer have to be supplied. Fertilizer applied partitions into several pools based on the weight fraction of different types of nutrients and bacteria defined by the fertilizer reference database. SWAT maintains the nitrate N and ammonia-N pool; and three other organic N pools. If a user does not supply fertilizer application information, the program decides how much fertilizer to apply and when based on the plant stress level as in APEX/EPIC.

In AnnAGNPS, fertilizer application has to be specified by the user. Information required includes the timing of the application, the type of fertilizer or manure applied, the amount of fertilizer applied and the method of application which refers to whether the application is surface applied or mixed with the soil layer. Fertilizer can be organic or inorganic. There is a built in fertilizer database which is based on the type of fertilizer applied and partitions fertilizer into organic and inorganic pools. Inorganic N from fertilizer application is calculated using the rate of fertilizer applied for current day times the fertilizer inorganic N fraction (from fertilizer reference database). Similarly, organic N from fertilizer application is calculated using the rate of fertilizer applied for current day times the fertilizer organic N fraction (from fertilizer reference database weight by weight).

Plant N Uptake

APEX/EPIC simulates N uptake using a supply and demand approach (Sharpley and Williams, 1990). EPIC calculates the daily crop N uptake based on the difference between the crop N content and the ideal N content for that day. However, soil supply of N which is limited by the mass flow of nitrate N to the roots may limit the daily crop N uptake. Soil supply of N is calculated as:

$$uptN = u \times \frac{NO_3}{SW} \qquad [22]$$

Where:

uptN = rate of N supplied by the soil (kg ha^{-1} d^{-1}),

NO_3 = amount of nitrate in the soil layer (kg ha^{-1}),

SW = soil water content (mm), and

u = water use rate (mm d^{-1}).

The actual amount of N removed from a soil layer is the minimum value of potential N uptake and amount of N supplied by the soil. In other words, the rate of supply from soil to plants cannot exceed the plant demand when mass flow estimates are too large.

The plant growth component of SWAT is a simplified version of the APEX/EPIC plant growth model (Neitsch et al., 2002). The simulation of plant N uptake in SWAT is the same as APEX/EPIC (Neitsch et al., 2002).

In the AnnAGNPS model, the amount of crop nutrient uptake is calculated in a crop growth stage subroutine. This subroutine determines the crop growth stage based on crop data the user entered. Amount of nutrient uptake is calculated based on the crop growth stage. Four growth stages are simulated by AnnAGNPS: initial, development, mature, and senescence. The length of each growth stage can be specified by a user or use the crop data information. At different growth stages, the crop nutrient uptake is different. The crop nutrient uptake is also limited by available nutrients in the composite soil layer. N uptake on current day is calculated as follows:

Where:

uptN = mass of inorganic N taken up by the plant on current day (kg d^{-1}), growth_N_uptake = fraction of N uptake for each growth stage such as initial, development, mature, and senescence,

yield_wt = yield at harvest (kg ha^{-1}),

N_uptake_harvest = N uptake per yield unit at harvest (wt-N/wt-harvest unit, dimensionless), and

stage_length = the number of growing days for current growth stage (days).

$$limited\_uptN = 0.99 \times \frac{inorgN^* conv}{1000000}$$ [23]

Plant N uptake is adjusted based on the availability of nutrients in the soil. If the uptN calculated above is greater than the available inorganic N in the soil layer, then a limited crop N uptake is calculated as:

Where:

Limited_uptN = mass of inorganic N taken up by the plant on current day (kg d^{-1}),

inorgN = inorganic N concentration in the soil layer (ppm), and

conv = conversion factor.

Nitrogen Transport by Runoff

All three models, AnnAGNPS, APEX/EPIC and SWAT assume that surface runoff interacts with the top 10 mm of the soil. Nutrients contained in this surface layer are available for transport to the stream reaches.

APEX/EPIC and SWAT Model

In SWAT model, runoff can be calculated using the SCS curve number procedure and the Green–Ampt method (Neitsch et al., 2009). The potential evapotranspiration can be estimated using the Penman–Monteith method (Monteith, 1965; Allen et al., 1989), Priestley–Taylor method (Priestley and Taylor, 1972) and Hargreaves method (Hargreaves et al., 1985). Erosion caused by rainfall and runoff is computed with the Modified Universal Soil Loss Equation (MUSLE; Williams, 1995). Once the sediment yield is estimated, sediment transport in the channel network is simulated as a function of two processes, deposition and degradation. A detailed description of SWAT can be found in Neitsch et al. (2009).

In APEX/EPIC and SWAT (Neitsch et al., 2002), the nitrate transport is simulated with surface runoff, lateral flow and/or percolation. The concentration of nitrate in the mobile water is calculated, then this concentration is multiplied by the volume of runoff and/or leaching to obtain the mass of nitrate loss through runoff and/or leaching from the soil layer.

Where:

Conc$_{NO3}$ = concentration of nitrate in the mobile water (kg mm^{-1}),

NO$_3$ = amount of nitrate in the soil layer (kg ha^{-1}),

W$_{mobile}$ = amount of mobile water in the soil layer (mm),

q$_e$ = fraction of porosity from which anions are excluded, and

PO = porosity of the soil layer.

Amount of mobile water in the layer is the sum of the amount of water lost by surface runoff, lateral flow and percolation on a given day. Surface runoff is allowed to interact with the top 10 mm of soil and transport nutrients from the top 10 mm of soil. Nutrients removed in surface runoff is calculated as:

Where:

NO_{3surf} = amount of nitrate removed in surface runoff (kg ha^{-1}),

b_{NO3} = nitrate percolation coefficient (kg mm^{-1}), and

Q_{surf} = surface runoff generated on a given day (mm).

The nitrate percolation coefficient is a user input that allows the user to set the concentration of nitrate in surface runoff to a fraction of the concentration in percolation. Similarly, the amount of nitrate lost through lateral flow is calculated as:

$$NO_{3_{lat}} = \beta_{NO_3} {}^* conc_{NO_3} {}^* Q_{lat} \qquad\qquad [27]$$

Where:

NO_{3lat} = amount of nitrate removed in lateral flow (kg ha^{-1}),

b_{NO3} = nitrate percolation coefficient (kg mm^{-1}), and

Q_{lat} = water discharged from the layer by lateral flow on a given day (mm).

Nitrate removed by percolation is calculated as:

$$NO_{3perc} = conc_{NO_3} {}^* W_{perc} \qquad\qquad [28]$$

Where:

NO_{3perc} = amount of nitrate removed to the underlying layer by percolation (kg ha^{-1}), and W_{perc} = amount of water percolating to the underlying layer on a given day (mm).

AnnAGNPS Model

AnnAGNPS uses different methodology to calculate N loss to surface runoff; the total loss to surface runoff (sol_N) includes surface dissolved nitrate N loss (if any from surface applied fertilizers which is surf_sol_N) and loss from the top 10 mm soil surface (cell_soil_sol_N).

When rainfall occurs, runoff interacts with top soil and carries soluble inorganic N in the soil profile away from fields. AnnAGNPS assumes the effective depth of runoff interaction is 10 mm.

When the simulation starts, AnnAGNPS checks if there is an operation on current day, then checks if fertilizer is applied on current day; AnnAGNPS incorporates inorganic N from fertilizer application first, thus fertilizer's impact on soluble N losses is reflected in elevated inorganic N level.

$$cell\_soil\_sol\_N = edi \times \frac{inorgN \times conv}{D \times 1000000} \qquad\qquad [29]$$

Soluble inorganic N removed from soil's top 10 mm by runoff is calculated as:

Where:

cell_soil_sol_N = mass of inorganic N removed from top soil layer through runoff (kg),

edi = effective depth of interaction factor (AnnAGNPS uses 10 mm),

inorgN = inorganic N concentration in the soil layer (ppm),

D = depth of soil layer (mm), and

conv = conversion factor.

When rainfall occurs, AnnAGNPS assumes that rainfall dissolves the inorganic N on the soil surface if there is any from surface application. When the rainfall event is big enough to generate runoff, the runoff carries the dissolved inorganic N away from the field. In this situation, AnnAGNPS assumes that inorganic N on the soil surface is totally dissolved in the water and either carried away by runoff or carried into the soil profile by infiltration. The amount of dissolved inorganic N carried away with runoff or carried into the soil profile with infiltration is determined as follow:

$$surf\_Sol\_N = \frac{Q_{surf}}{(Q_{surf}+inf)} \times surf\_inorgN \qquad [30]$$

If the infiltration is greater than 1.0 mm, the total soluble inorganic N lost to surface runoff is calculated as:

Where:

surf_sol_N = mass of inorganic N in runoff from fertilizer applied on soil surface (kg),

Q_{surf} = surface runoff generated on a given day (mm),

inf = amount of infiltration, calculated in the hydrology module (mm), and

surf_inorgN = inorganic N added through fertilizer application at the soil surface (kg).

Inorganic N lost to infiltration is calculated as:

$$inf\_Sol\_N = surf\_inorgN - surf\_sol\_N \qquad [31]$$

Where:

inf_sol_N = amount of inorganic N infiltrated into soil profile from fertilizer applied on soil surface (kg).

If the infiltration is less than 1.0 mm, the total soluble inorganic N lost to surface runoff is calculated as:

$$surf\_Sol\_N = surf\_inorgN \qquad [32]$$

$$inf\_Sol\_N = 0 \qquad [33]$$

Then, AnnAGNPS resets inorganic N in the soil and on the surface to reflect the impact of the current rainfall event.

Total mass of inorganic N lost in surface runoff includes soil-incorporated and surface-applied N lost.

$$sol\_N = cell\_soil\_sol\_N + surf\_sol\_N \qquad [34]$$

Where:

cell_soil_sol_N = N losses to runoff from composite soil layer (kg),

surf sol_N = N losses to runoff from soil surface (kg), and

sol_N = total soluble N lost to surface runoff (kg) which is equivalent to NO_{3surf} in SWAT model.

Leaching losses are calculated using the updated inorganic N level in the soil.

$$N\_Leaching = \frac{perc\_loss}{SW-Wilting} \times \frac{inorgN \times conv}{1000000} \qquad [35]$$

$$perc\_loss > 0$$

$$N\_Leaching = 0$$

perc_loss<=0

Where:

  N_leaching = leaching loss from soil layer (kg),
  Perc_loss = percolation loss for current day (mm),
  SW = soil water content (mm),
  Wilting = wilting point (mm),
  inorgN = amount of inorganic N in the soil layer (ppm), and
  conv = conversion factor.

  Then, the inorganic N content shall be recalculated to reflect the leaching losses at the end of the day.

  Inorganic N mass balance is maintained for the second soil layer the same way as the top layer except that fertilizer application, rainfall caused runoff and sediment loss are not considered. The leaching from the first layer is added to the second layer and leaching from the second layer is lost to the groundwater system.

Organic N Transport by Sediment

  The APEX/EPIC and SWAT models use a loading function developed by McElroy et al. (1976) and modified by Williams and Hann (1978) to calculate the organic N transported by sediment. It estimates the daily organic N loss by sediment based on the concentration of organic N in the top soil layer, the sediment yield, and the enrichment ratio.

$$sedN = 0.001 \times orgN \times SED \times ER \qquad [36]$$

Where:

  sedN = mass of N attached to sediment and transported to the channel (kg ha^{-1}),

  orgN = concentration of organic N including fresh, stable and active N in the top 10 mm soil layer for current day (ppm),

  SED = the sediment yield produced on current day (t ha^{-1}), and

  ER = enrichment ratio.

  The enrichment ratio occurs because of selective erosion and deposition processes (Knisel, 1980). The smaller particles are more easily transported than coarser particles and the organic N is primarily attached to clay particles. The enrichment ratio is defined as the concentration of nutrients in the eroded material divided by the concentration of nutrients in the soils from the sediment producing area. Both APEX/EPIC and SWAT calculate the enrichment ratio for each storm event, or allow the user to define a particular enrichment ratio for organic N to use for all storms during the simulation.

$$ER = A^* (conc_{sed})^B \qquad [37]$$

  To calculate an enrichment ratio, data analysis showed that the logarithmic relations of enrichment ratio to sediment concentration existed. Enrichment ratio is estimated as the following:

Where:

  A = coefficient of ER,

$conc_{sed}$ = the sediment concentration in surface runoff (Mg m^{-3}), and

B = exponent of ER.

A, and B can be user input. They can be calculated from actual measured values of N in the sediments. Default values are 0.78 for the coefficient and -0.2468 for the exponents in SWAT.

Sediment concentration can be calculated as follows:

$$conc_{sed} = \frac{sed}{10 \times Q_{surf}}$$
[38]

Where:

SED = the sediment produced by the erosion model (t ha^{-1}), and

Q_{surf} = amount of surface runoff on a given day (mm).

Menzel (1980) investigated the relationship of enrichment ratio with sediment discharge with data across the country in Oklahoma, Texas, Michigan, and Georgia. His study showed that slopes, soil texture, land use, tillage practices and the possibly the size of the watershed all affect the value of the enrichment ratio, and sandy soil may have a higher enrichment ratio than other types of soils. However, the logarithmic relations of enrichment ratio to sediment yield are applicable in most cases (Menzel, 1980). Investigation is still needed on the enrichment ratio for predicting P transport by sediment. The complicated factors influencing enrichment ratio and the difficulties in determining it should be kept in mind in estimating nutrient loadings of watersheds for both modelers and model users.

In AnnAGNPS, two assumptions are made for the calculation of mass of N attached to sediment:

a). organic N is comprised of sediment attached N because total N is predominantly organic N in soils.

b). the organic N is associated with clay fraction. This eliminates the need for separate nutrient enrichment ratio (Menzel, 1980).

N attached to sediment is calculated as follows:

$$sedN = frac\_orgN\_clay \times [sed\_part(1,1) + sed\_part(1,2)] \times 1000$$
[39]

Where:

sedN = mass of N attached to sediment (kg),

frac_ orgN_clay = decimal fraction of organic N in clay in soil layer, and

$$frac\_orgN\_clay = \frac{orgN}{(frac\_clay) \times 1000000}$$
[40]

sed_part(1,1) and sed_part(1,2) = Current day's mass of sediment (by particle size and source) at edge of cell in tonnes. Array subscript are: Particle Size (first): 1- clay 2- silt 3- sand 4- small aggregate 5- large aggregate Source (second): 1- irrigation 2- other than irrigation.

Where:

orgN = Concentration of organic_N in the total composite soil layer for current day(ppm), and

frac_clay = fraction of clay to total composite soil.

Nitrogen Balance

Figure 3 shows the soil N pools and flows simulated in APEX/EPIC (Fig. 3a), AnnAGNPS (Fig. 3b) and SWAT (Fig. 3c). No documents were found on APEX/EPIC and SWAT in describing the sequence of calculation for the N balance.

All AnnAGNPS mass balance is based on AnnAGNPS cells and maintained for two composite soil layers.

The mass balance equation for organic N simulation processes is as follows:

$$orgN_t = orgN_{t-1} + \frac{(FON + fer\_orgN - HMN - sedN) \times 10}{Conv} \qquad [41]$$

Where:

1. $orgN_t$ = concentration of organic_N in the total composite soil layer for current day (ppm),

2. $orgN_{t-1}$ = concentration of organic_N in the total composite soil layer for previous day (ppm),

3. FON = organic N addition from decomposition of crop and noncrop residue laying on the soil surface to cell soil layer on current day (kg). Non-crop residue refers to N from litter dry biomass for non-cropland surface residue that is subject to decomposition. Upon decomposition byproducts are considered mixed uniformly in soil layer,

4. Fer_orgN = organic N from fertilizer application such as manure or other sources (kg),

5. HMN = N mineralized from organic N in soil layer on current day (actual argument passed to inorg_N_mass_bal subroutine) (kg), and

6. sedN = current day mass of N attached to sediment loss (kg).

$$orgN_t = orgN_{t-1} - \frac{HMN \times 1000000}{Conv} \qquad [42]$$

Organic N mass balance is maintained for the second soil layer which is the bottom soil layer in AnnAGNPS. Since AnnAGNPS assumes that fertilizer application, rainfall caused runoff and sediment loss are only associated with the top soil layer, the mass balance equation is very simple for the second soil layer. In other words, fertilizer application and rainfall do not interact with the bottom soil layer. Where:

1. $orgN_t$ = concentration of organic_N in the total second composite soil layer for current day (ppm),

2. $orgN_{t-1}$ = concentration of organic_N in the total second composite soil layer for previous day (ppm), and

3. HMN = N mineralized from organic N in soil layer on current day (actual argument passed to inorg_N_mass_bal subroutine) (kg). N mineralization is only calculated for non-crop for the second layer.

For inorganic N, addition from fertilizer application is calculated first, followed by the losses from runoff, denitrification and plant uptake. Then, mass balance was updated for inorganic N that incorporates mineralization of organic N. In other words, mineralization of organic N is not used to calculate losses from runoff, denitrification and plant uptake. At the end of the day, leaching loss is calculated and inorganic N is updated to reflect the leaching loss.

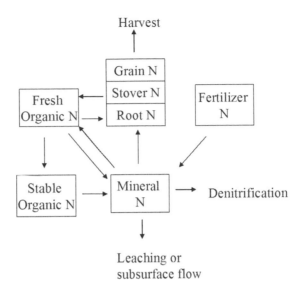

Fig. 3. Soil N pools and processes that move N in and out of pools in APEX/EPIC, AnnAGNPS and SWAT.

Calculation of inorganic N additions to a computational cell

Added fertilizers are considered either well mixed with the top soil layer which is 200 mm, or stay on the soil surface based on the operation effect which is supplied by the user through operation data input. If a soil disturbance exceeds 50% any fertilizer operations are considered as mixed. Otherwise, it assumes the applied fertilizer stays on the soil surface. In addition, when the soil disturbance exceeds 50% it not only incorporates the applied fertilizer from the current operation into the soil, but also incorporates any fertilizer left on the soil surface from previous fertilizer application into the soil.

When a soil disturbance exceeds 50% fertilizers on the soil surface mix with the soil. The amount of fertilizer mixed with soil and the amount of fertilizer left on the soil surface after a soil disturbance is determined by the depth to the impervious layer. If the soil depth to imperious layer is greater than 200 mm (AnnAGNPS sets this layer as the top soil layer, it is also called tillage layer), take all surface fertilizer and incorporate it into the soil.

$$surf\_inorgN = 0 \tag{43}$$

$$mnaN = surf\_inorgN \tag{44}$$

Where:

1. mnaN = mass of inorganic N added to a cell from incorporated inorganic additions such as fertilizers (kg). It is assumed well mixed with soil, and

2. surf_inorgN = inorganic N added through fertilizer application at the soil surface (kg).

When a soil disturbance is less than 50% added fertilizers are considered to stay on the soil surface. For this case, the value of mnaN and surf_inorgN does not change.

$$mnaN = 0 \tag{45}$$

Then, AnnAGNPS checks if there is rainfall on the current day; if there is rainfall on the current day, soil inorganic N will be adjusted to reflect rainfall impact. When rainfall occurs, it dissolves the inorganic N on the soil surface. When the rainfall event is big enough to generate runoff, AnnAGNPS assumes that inorganic N on the soil surface is totally dissolved in the water and either carried away with runoff or carried into the soil profile with infiltration. The amount of dissolved inorganic N carried away with runoff or carried into the soil profile with infiltration is determined based on the amount of rainfall and infiltration as introduced in the surface runoff loss section:

Then, AnnAGNPS resets the mnaN value to reflect the impact of the current rainfall event.

$$mnaN = inf\_sol\_inorgN \qquad\qquad\qquad\qquad\qquad\qquad [46]$$

$$surf\_inorgN = surf\_inorgN - inf\_sol\_N \qquad\qquad\qquad\qquad [47]$$

Where:
1. mnaN = mass of inorganic N added to a cell from incorporated inorganic additions such as fertilizers (kg). It is assumed well mixed with soil,
2. inf_sol_N = amount of inorganic N infiltrated into soil profile from fertilizer applied on soil surface (kg), and
3. surf_inorgN = inorganic N added through fertilizer application at the soil surface (kg).

Calculation of intermediate inorganic N mass balance

The intermediate inorganic N mass balance refers to N pools which includes N additions but prior to N losses as soluble N and plant uptake. Bottom soil layer inorganic N does not change with this operation.

$$inorgN = inorgN + \frac{mnaN*1000000}{conv} \qquad\qquad\qquad\qquad [48]$$

Where:
1. inorgN$_t$ = concentration of inorganic N in the total composite soil layer for current day (ppm),
2. inorgN$_{t-1}$ = concentration of inorganic N in the total composite soil layer for previous day (ppm), and
3. mnaN = inorganic N addition to the soil profile from above calculation.

Calculation of inorganic N losses from a computational cell and reconciliation of inorganic N mass balance at the end of a day

Total mass of inorganic N lost in surface runoff includes soil incorporated and surface applied N lost (Eq. [34]).

$$inorgN_t = inorgN_{t-1} + \frac{(hmnN - uptN - sol\_N - DN\text{-}N\_Leaching)*1000000}{conv} \quad [49]$$

At the end of a day, inorganic mass balance is updated. Mineralized N is added at the end of a day. Therefore, mineralized N is not included in inorganic N available for loss as runoff and/or leaching, plant uptake or denitrification. Where:

1. inorgN $_t$ = concentration of inorganic N in the total composite soil layer for current day (ppm),

2. inorgN$_{t-1}$ = concentration of inorganic N (ppm),

3. hmnN = inorganic N mineralized from organic N in soil layer on current day (kg),

4. uptN = plant uptake N from growth stage subroutine (kg),

5. sol_N = inorganic N lost to runoff (kg),

6. DN = denitrification loss (kg), and

7. N_leaching = leaching loss from soil layer (kg).

$$inorgN_t = inorgN_{t-1} + \frac{(hmnN - uptN - N\_Leaching - DN) * 1000000}{conv} \qquad [50]$$

Inorganic N mass balance is maintained for the second soil layer which is the bottom soil layer. The calculation for the second soil layer is simple because AnnAGNPS assumes that fertilizer application is associated with the top soil layer. In addition, rainfall-caused runoff and sediment loss are only associated with the top soil layer. In other words, fertilizer applications do not interact with the bottom soil layer. Where:

1. inorgN $_t$ = concentration of inorganic N in the total composite soil layer for current day (ppm),

2. inorgN$_{t-1}$ = concentration of inorganic N (ppm),

3. hmnN = inorganic N mineralized from organic N in soil layer on current day (kg),

4. uptN = plant uptake N from growth stage subroutine (kg),

5. DN = denitrification loss (kg), and

6. N_leaching = leaching loss from soil layer (kg).

Table 1 summarizes the processes simulated in each model. SWAT has the most complicated N model. It simulates residue decomposition and mineralization, humus mineralization, nitrification and ammonia volatilization, denitrification, fixation and rainfall N. The SWAT model adapted most of its processes from the APEX/EPIC model except nitrification and ammonia volatilization for which the model uses a combination of the methods developed by Reddy et al. (1979) and Godwin et al. (1984). On the other hand, AnnAGNPS has the least complicated N model. It simulates residue decomposition and mineralization, humus mineralization, and denitrification. However, AnnAGNPS uses the RUSLE technology to evaluate the residue decomposition and mineralization. The following section summarizes studies performed to evaluate the model's performance of N simulations.

Three of the most frequently used statistics recommended by Moriasi et al. (2007), the coefficient of determination (R^2), the Nash–Sutcliffe model efficiency (NSE) and the percent of bias (PBIAS), were used for N simulation evaluation in this book chapter. Values of NSE can range from -∞ to 1 and indicate how accurately simulated values fit the corresponding measured data on a 1:1 line. An NSE value of 1 indicates a perfect fit between the model simulation and the field observation. However, the mean value of the measured data would be considered a more accurate predictor than the simulated output if the NSE value is 0 or lower. The R^2 value measures how well the simulated versus observed regression line approaches an ideal match and ranges from 0 to 1, with a value of 0 indicating no correlation and a value of 1 indicating perfect correlation. The regression slope and intercept also equal 1 and 0, respectively, for a perfect fit. For PBIAS, the closer it is to zero the less bias and better prediction. Description of NSE, R^2 and PBIAS results for specific studies is provided as it is available in the remainder of this discussion.

Model Applications For Nitrogen Simulations

APEX/EPIC Model Application for Nitrogen Simulations

Wang et al. (2012) is the reference case studies for APEX/EPIC around the United States. As documented in Wang et al. (2012) and Gassman et al. (2010), the APEX/EPIC model has been integrated with the SWAT model (Gassman et al., 2010) for watershed simulations. Applications of the model are reviewed in the study by Gassman et al. (2010) and case studies are provided in Wang et al. (2012). APEX/EPIC applications include studies performed for the National Pilot Project for Livestock and the Environment, feedlot, pesticide, forestry, buffer strip, and other conservation practices. The application descriptions in the Gassman et al. (2010) include a summary of calibration and/or validation results obtained for the different studies. The testing of APEX within the Upper North Bosque River (UNBRW) study included comparisons with measured data collected from field plots and indirect validation at the watershed level with APEX simulations embedded within SWAT. In Gassman et al. (2010) study, model results of total N were compared with total N collected from eight field plots ranging in size from 0.01 to 0.52 ha in Erath County, Texas. Those plots were monitored for periods that ranged from roughly 12 to 17 mo between December 1993 and August 1995. Three of the fields received irrigated dairy wastewater applications that ranged between 94 and 586 mm during the monitored periods and the other five field plots received solid dairy manure applications. As shown in the study by Gassman et al. (2010), the percent bias between observed and predicted ranges from 11% to −73% (plus means underprediction and minus means overprediction) for those eight monitoring plots (Gassman et al., 2010). Indirect watershed-level testing of APEX by executing APEX simulations for the dairy waste application fields and then inputting the APEX output into SWAT, which was then used to route surface runoff and pollutant losses from other areas in combination with the APEX inputs to the watershed outlet, was performed by Saleh and Gallego (2007). Model results were compared with field monitoring, and monthly comparison of N (organic, nitrate, and total) losses between observed and simulated had NSE ranging from -0.04 to 0.88 (Saleh and Gallego, 2007).

Table 1. The N transformation processes simulated in each model.

Models	Nitrogen Transformation					
	Residue Decomposition & Mineralization	Humus Mineralization	Nitrification & Ammonia Volatilization	Denitrification	Fixation	Rainfall
AnnAGNPS	Yes	Yes, the same as APEX/EPIC	No	Yes	No	No
APEX/EPIC	Yes	Yes	Yes	Yes	Yes	Yes
SWAT	Yes, the same as APEX/EPIC	Yes, the same as APEX/EPIC	Yes	Yes, the same as APEX/EPIC	Yes, the same as APEX/EPIC	Yes

AnnAGNPS Model Application for Nitrogen Simulations

The AnnAGNPS model was used to assess subsurface drainage management practices for reducing N loadings in the Midwest where subsurface is essential for agricultural production. The model was validated using annual average N loading monitored at the Waterville USGS stream gauge station before it was used to evaluate the impact of drain depth and spacing as well as controlled drainage on N loadings. The simulated 100-yr average annual agricultural N loading was 12.6 kg N ha^{-1} y^{-1}, with 12.2 kg N ha^{-1} y^{-1} dissolved N using those calibrated parameters for runoff and sediment. Average annual N loading (1996–2003) observed at the Waterville USGS stream gauge station was 18.9 kg N ha^{-1} y^{-1} which included point source and nonpoint source N loadings. No additional calibration was performed because it is very difficult to separate agricultural nonpoint source N loading from total N loading which includes point source and nonpoint source at the Waterville USGS stream gauge station (Yuan et al., 2011).

Suttles et al. (2003) applied the AnnAGNPS model to the Little River Research watershed in south central Georgia to simulate sediment and nutrient loadings from the watershed over a seven year period. It was found that, on an annual average basis, N loading was underpredicted in the upper part of the watershed due to underprediction of runoff and overestimation of forested area in the upper part of the watershed. In contrast, N loading was overpredicted in the lower part of the watershed which may have resulted from inadequate simulation of nonpoint-source pollution attenuation by the extensive riparian forests and forested instream wetland areas found in the lower part of the watershed.

Yuan et al. (2003) evaluated AnnAGNPS N simulations on a monthly basis in an agricultural watershed in the Mississippi Delta. The predicted monthly N loading did not match well with the observed N loading (Yuan et al., 2003), and the regression of the monthly predicted with the observed on the line of equal values obtained a R-square of 0.28 and a slope of 0.59. In addition, it was found that the model overpredicted N loadings during the dormant season and underpredicted N loadings during the cropping season. One possible reason for winter (especially during January and December) overprediction is denitrification since denitrification was not simulated in the model. However, it might occur in the Mississippi Delta due to warm winter. A possible reason for model underprediction during the cropping season might be the inaccurate estimation of plant uptake parameter.

Baginska et al. (2003) simulated N loadings from Currency Creek, a small experimental catchment within the Hawkesbury-Nepean drainage basin of the Sydney region. The predicted daily per event N loading was compared with observed daily per event N loading at various temporal scales. It was found that the best simulation results were achieved on an event basis rather than a daily basis. Although the mean values of predicted and measured matched closely, the coefficient of efficiency was usually negative indicating high deviations between predicted daily per event and measured daily per event N loadings. The model slightly underpredicted N loading for the events of smaller magnitude, while overpredicted N loading for the larger events. Despite noticeable inaccuracies in N prediction, the results demonstrated a degree of stability and robustness. The model successfully simulated the trends in N generation and relatively close patterns can be observed between the simulated and observed event N loadings.

The AnnAGNPS model was applied to the Jiulong River watershed (14,700 km^2), an intensive agricultural watershed, located in southeast China (Huang and Hong, 2010). The AnnAGNPS calculated N loadings were evaluated by comparing with the measured values at the subwatershed outlets. The AnnAGNPS simulated total N and dissolved N at all four subwatersheds were close to the measured values with deviation errors (Dv) of less than 5%.

The AnnAGNPS model was applied to the Pipestem Creek Watershed, a typical agricultural watershed, upstream of Pingree, ND, United States for evaluating nonpoint-source pollution (Pease et al., 2010). The predicted N loadings were poorly correlated with the observed values on an event basis. Of the 23 observed events, only four predicted events had N loadings greater than zero.

Shamshad et al. (2008) applied the AnnAGNPS model to the watershed of River Kuala Tasik in Malaysia, a combination of two sub-watersheds, to evaluate the suitability of the AnnAGNPS model in assessing runoff, sediment and nutrient loading under Malaysian conditions. The data for the year 2004 was used to calibrate the model and the data for the year 2005 was used for validation purposes. The study shows that N loading predictions were satisfactory with $R^2 = 0.68$ and E = 0.53, which showed that AnnAGNPS has the potential to be used as a valuable tool for planning and management of watersheds under Malaysian conditions.

SWAT Model Application for Nitrogen Simulations

The SWAT model has been applied to numerous watershed systems throughout the world to evaluate a range of conservation practice effects on water quality (Tuppad et al., 2011; Gassman et al., 2007). Many studies including those summarized by Gassman et al. (2007) have demonstrated SWAT's capability in simulating N losses (Saleh et al., 2000; Santhi et al., 2001; Saleh and Du, 2004; Chu et al., 2004; White and Chaubey, 2005; Arabi et al., 2006; Grunwald and Qi, 2006; Plus et al., 2006; Hu et al., 2007; Jha et al., 2007; Niraula et al., 2012), and results from various studies indicated mixed success of SWAT N simulations. Although many studies reported good performance of SWAT in simulating N losses (Behera and Panda, 2006; Gikas et al., 2006; Jha et al., 2007; Santhi et al., 2001; Stewart et al., 2006), quite a few other studies showed poor performance of SWAT in dissolved N simulations (Chu et al., 2004; Du et al., 2006; Grizzetti et al., 2003; Grunwald and Qi, 2006; Gassman et al., 2007; Hu et al., 2007; White and Chaubey, 2005). The poor performance of simulating dissolved N could be attributed to many factors, including inadequate simulation of landscape processes and/or stream processes. In addition, the poor performance in simulation of different N species, including under- and/or overestimation of individual N species, implies that the good model performance for TN may be a result of smoothing or averaging.

For example, Cerro et al. (2014) evaluated the SWAT model for the Alegria watershed in northern Spain. They reported that SWAT nitrate load simulation results produced Nash-Sutcliffe values of 0.88 for daily loads and concluded that reducing fertilizer amounts by 20% would satisfy European Water Framework Directive standard requirements.

Hoang et al. (2014) applied SWAT to the 622 km² Odense River basin in Denmark. This area represents a lowland system dominated by agricultural production that includes subsurface drains installed in over 50% of the agricultural areas. The SWAT streamflow and nitrate results were consistent with the DAISY-MIKE SHE (DMS) output range for high flow conditions and produced slightly lower results during low-flow periods. They report that SWAT overpredicted surface runoff and underpredicted subsurface drainage flow, resulting in inaccuracies in estimating daily nitrate flux. Several other recent studies propose various SWAT nutrient cycling and/or other in-stream cycling improvements (Vadas and White, 2010; Sen et al., 2012; Kim and Shin, 2011).

Another example is the study performed by Glavan et al. (2010) which provides a critical evaluation of the use of the SWAT model in the Axe catchment, UK, in the context of international performance of the SWAT model. This study concludes that within the constraints of the available data, SWAT's representation of daily nitrate concentration dynamics is poor. Temporal aggregation of model outputs from daily to monthly improved the performance metrics for nitrate.

APEX/EPIC, AnnAGNPS and SWAT Model Application for Evaluating Conservation Practices

Many model applications showed that results from APEX/EPIC, AnnAGNPS and SWAT evaluations can be used with GIS software to assist with identification of hot spots across the field and/or watershed and away from the field and/or watershed, which contributes to precision conservation decisions to maximize environmental benefits and reduce the off-site transport of N in large watersheds (Delgado and Bausch, 2005; Arabi et al., 2006; Gassman et al., 2007; Gassman et al., 2010; Yuan et al., 2011; Santhi et al., 2014). Figure 4 shows GIS maps generated from APEX/EPIC (Fig. 4a), AnnAGNPS (Fig. 4b) and SWAT (Fig. 4c) simulations to show spatial variation of potential N losses across the study area, from which critical areas can be identified to help the implementation of precision conservation practices that will ultimately contribute to increased N use efficiency and reduced N losses to the environment (Delgado and Bausch, 2005; Arabi et al., 2006; Yuan et al., 2011).

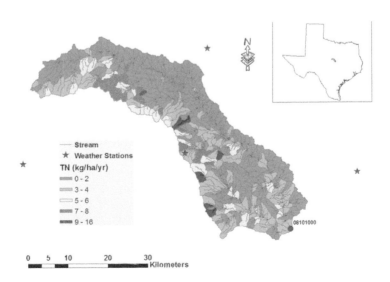

Fig. 4a. APEX simulation of spatial variation of total nitrogen from subareas in the cowhouse watershed located in central Texas (source: Jaehak Jeong from Agriculture and Life Sciences, Texas A & M, Temple, TX).

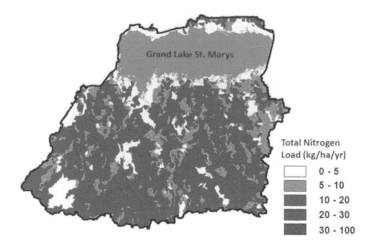

Total Nitrogen
Load (kg/ha/yr)

- [ ] 0 - 5
- 5 - 10
- 10 - 20
- 20 - 30
- 30 - 100

Fig. 4b. AnnAGNPS simulation of spatial variation of total nitrogen load sources to Grand Lake St. Mary's, located in Mercer County, OH.

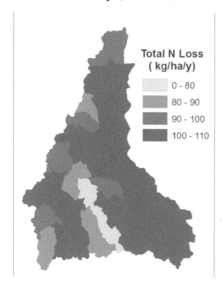

Total N Loss
(kg/ha/y)

- 0 - 80
- 80 - 90
- 90 - 100
- 100 - 110

Fig. 4c. SWAT simulation of spatial variation of total nitrogen from Chicksaw watershed contributing to the Grand Lake St. Mary's, located in Mercer County, OH.

Conclusion

Nitrogen processes simulated in widely utilized hydrologic and water quality models APEX/EPIC, AnnAG-NPS and SWAT were compared in this book chapter. In addition, selected examples of model application were also introduced to illustrate models' performance. Although components of the N cycle included in each model varied, there are a lot of similarities of the models in simulating N processes. Second, considerable variability exists in the prediction of a component such as crop N uptake. Because the many aspects of the N cycle are interconnected and affect soil nitrate concentrations, differences in the prediction of a particular component lead to differences in predicting N losses. The review of many model applications found that all models were able to make reasonable predictions of total N loading in general following model calibration. However, it is common for all models for there to be one application where it performs well, but another application, maybe unique to the model, where it doesn't perform well, regardless of the complexity of the model in representing N processes, particularly for the dissolved N. The comparison between the models suggests our ability to accurately predict N losses from the dominant agroecosystem without calibration is quite uncertain at this time.

Acknowledgments

The United States Environmental Protection Agency through its Office of Research and Development funded and managed the research described here. It has been subjected to Agency review and approved for publication. The authors are grateful for the valuable comments and suggestions provided by many reviewers.

Notice: Although this work was reviewed by the USEPA and approved for publication, it may not necessarily reflect official Agency policy. Mention of trade names or commercial products does not constitute endorsement or recommendation for use.

References

Addiscott, T.M., and A.P. Whitmore. 1991. Simulation of solute leaching in soils of differing permeability. Soil Use Manage. 7:94–102.

Ahuja, L.R., K.W. Rojas, J.D. Hanson, M.J. Shaffer, and L. Ma, editors. 2000. Root zone water quality model. Water Resources Publications, Englewood, CO.

Arabi, M., R.S. Govindaraju, M.M. Hantush, and B.A. Engel. 2006. Role of watershed subdivision on modeling the effectiveness of best management practices with SWAT. J. Am. Water Resour. Assoc. 42:513–528. doi:10.1111/j.1752-1688.2006.tb03854.x

Arnold, J.G., and N. Fohrer. 2005. SWAT2000: Current capabilities and research opportunities in applied watershed modeling. Hydrol. Processes 19:563–572. doi:10.1002/hyp.5611

Arnold, J.G., D.N. Moriasi, P.W. Gassman, K.C. Abbaspour, M.J. White, R. Srinivasan, C. Santhi, R.D. Harmel, A. van Griensven, M.W. Van Liew, N. Kannan, and M.K. Jha. 2012. SWAT: Model use, calibration, and validation. Trans. ASABE 55:1494–1508. doi:10.13031/2013.42256

Baginska, B., W. Milne-Home, and P.S. Cornish. 2003. Modeling nutrient transport in Currency Creek, NSW with AnnAGNPS and PEST. Environ. Model. Softw. 18:801–808. doi:10.1016/S1364-8152(03)00079-3

Barraclough, D. 1989. A usable mechanistic model of nitrates leaching. I. The model. J. Soil Sci. 40:543–554. doi:10.1111/j.1365-2389.1989.tb01295.x

Barraclough, D., M.J. Hyden, and G.P. Davies. 1983. Fate of fertilizer N applied to grassland. I. Field leaching results. J. Soil Sci. 34:483–497. doi:10.1111/j.1365-2389.1983.tb01050.x

Behera, S., and R.K. Panda. 2006. Evaluation of management alternatives for an agricultural watershed in a sub-humid subtropical region using a physical process based model. Agric. Ecosyst. Environ. 113:62–72. doi:10.1016/j.agee.2005.08.032

Bergström, L., and N. Brink. 1986. Effects of differentiated applications of fertilizer N on leaching losses and distribution of inorganic N in the soil. Plant Soil 93:333–345. doi:10.1007/BF02374284

Bingner, R.L., F.D. Theurer, and Y. Yuan. 2003. AnnAGNPS technical processes. http://www. ars.usda.gov/Research/docs.htm?docid=5199 (Accessed 10 March 2013).

Burt, T.P., A.L. Heathwaite, and S.T. Trudgill. 1993. Nitrate: Processes, patterns and management. John Wiley & Sons Ltd., Chichester, UK.

Cambardella, C.A., and E.T. Elliott. 1992. Particulate soil organic matter changes across a grassland cultivation sequence. Soil Sci. Soc. Am. J. 56:777–783. doi:10.2136/sssaj1992.03615995005600030017x

Carter, J.N., and F.E. Allison. 1960. Investigation of denitrification in well-aerated soils. Soil Sci. 90:173–177. doi:10.1097/00010694-196009000-00004

Cassman, K.G., and D.N. Munns. 1980. N mineralization as affected by soil moisture, temperature and depth. Soil Sci. Soc. Am. J. 44:1233–1237. doi:10.2136/sssaj1980.03615995004400060020x

Cerro, I., I. Antiguedad, R. Srinavasan, S. Sauvage, M. Volk, and J.M. Sanchez-Perez. 2014. Simulating land management options to reduce nitrate pollution in an agricultural watershed dominated by an alluvial aquifer. J. Environ. Qual. 43:67–74. doi:10.2134/jeq2011.0393

Chu, T., W. Shirmohammadi, A.H. Montas, and A. Sadeghi. 2004. Evaluation of the SWAT model's sediment and nutrient components in the Piedmont physiographic region of Maryland. Trans. ASAE 47:1523–1538. doi:10.13031/2013.17632

Corwin, L. 1995. Trace element transport - Transient-state solute transport model for the vadose zone. Salinity Laboratory, USDA-ARS, Riverside, CA.

Dancer, W.S., L.A. Peterson, and G. Chesters. 1973. Ammonification and nitrification of N as influenced by soil pH and previous N treatments. Soil Sci. Soc. Am. J. 37:67–69. doi:10.2136/sssaj1973.03615995003700010024x

David, M.B., S.J. Del Grosso, X. Hu, E.P. Marshall, G.F. McIsaac, W.J. Parton, C. Tonitto, and M.A. Youssef. 2009. Modeling denitrification in a tile-drained, corn and soybean agro-ecosystem of Illinois, USA. Biogeochemistry 93:7–30. doi:10.1007/s10533-008-9273-9

Delgado, J.A. 1998. Sequential NLEAP simulations to examine effect of early and late planted winter cover crops on nitrogen dynamics. J. Soil Water Conserv. 53:241–244.

Delgado, J.A. 2001. Use of simulations for evaluation of best management practices on irri-gated cropping systems. In: M.J. Shaffer, L. Ma, and S. Hansen, editors, Modeling carbon and nitrogen dynamics for soil management. Lewis Publishers, Boca Raton, FL. p. 355–381. doi:10.1201/9781420032635.ch10

Delgado, J.A., and W.C. Bausch. 2005. Potential use of precision conservation techniques to reduce nitrate leaching in irrigated crops. J. Soil Water Conserv. 60:379–382.

Delgado, J.A., A.R. Mosier, D.W. Valentine, D.S. Schimel, and W.J. Parton. 1996a. Long term 15N studies in a catena of the shortgrass steppe. Biogeochemistry 32:41–52. doi:10.1007/BF00001531

Delgado, J.A., A.R. Mosier, R.H. Follett, R.F. Follett, D. Westfall, L. Klemedtsson, and J. Ver-meulen. 1996b. Effect of N management on N2O and CH4 fluxes and N recovery in an irrigated mountain meadow. Nutr. Cycling Agroecosyst. 46:127–134. doi:10.1007/BF00704312

Delgado, J.A., M.A. Dillon, R.T. Sparks, and S.Y.C. Essah. 2007. A decade of advances in cover crops: Cover crops with limited irrigation can increase yields, crop quality, and nutrient and water use efficiencies while protecting the environment. J. Soil Water Conserv. 62:110A–117A.

Delgado, J.A., M.J. Shaffer, and M.K. Brodahl. 1998. New NLEAP for shallow and deep rooted rotations. J. Soil Water Conserv. 53:338–340.

Delgado, J.A., P. Gagliardi, M.J. Shaffer, H. Cover, E. Hesketh, J.C. Ascough, and B.M. Dan-iel. 2010. New tools to assess nitrogen management for conservation of our biosphere. In: J.A. Delgado and R.F. Follett, editors, Advances in nitrogen management for water quality. SWCS, Ankeny, IA. p. 373–409.

Delgado, J.A., R.R. Riggenbach, R.T. Sparks, M.A. Dillon, L.M. Kawanabe, and R.J. Ristau. 2001. Evaluation of nitrate-nitrogen transport in a potato-barley rotation. Soil Sci. Soc. Am. J. 65:878–883. doi:10.2136/sssaj2001.653878x

Dinnes, D.L., D.L. Karlen, D.B. Jaynes, T.C. Kaspar, J.L. Hatfield, T.S. Colvin, and C.A. Cambardella. 2002. Nitrogen management strategies to reduce nitrate leaching in tile-drained Midwestern soils. Agron. J. 94:153–171. doi:10.2134/agronj2002.0153

Dowdell, R.J., P. Colbourn, and R.Q. Cannell. 1987. A study of mole drainage with simpli-fied cultivation for autumn–sown crops on a clay soil. Losses of nitrate-N in surface run-off and drain water. Soil Tillage Res. 9:317–331. doi:10.1016/0167-1987(87)90057-2

Du, B., A. Saleh, D.B. Jaynes, and J.G. Arnold. 2006. Evaluation of SWAT in simulating nitrate nitrogen and atrazine fates in a watershed with tiles and potholes. Trans. ASABE 49:949–959. doi:10.13031/2013.21746

Gassman, P.W., M. Reyes, C.H. Green, and J.G. Arnold. 2007. The Soil and Water Assess-ment Tool: Historical development, applications, and future directions. Trans. ASABE 50:1211–1250. doi:10.13031/2013.23637

Gassman, P.W., J.R. Williams, X. Wang, A. Saleh, E. Osei, L.M. Hauck, R.C. Izaurralde, and J.D. Flowers. 2010. The Agricultural Policy/Enviromental EXtender (APEX) model: An emerging tool for landscape and watershed environmental analyses. Trans. ASABE 53:711–740. doi:10.13031/2013.30078

Gikas, G.D., T. Yiannakopoulou, and V.A. Tsihrintzis. 2006. Modeling of non-point source pollution in a Mediterranean drainage basin. Environ. Model. Assess. 11:219–233. doi:10.1007/s10666-005-9017-3

Glavan, M., S. White, and I.P. Holman. 2010. Evaluation of river water quality simulations at a daily time step - Experience with SWAT in the Axe Catchment, UK. Clean: Soil, Air,Water 39:43–54. doi:10.1002/clen.200900298

Godwin, D.C., C.A. Jones, J.T. Ritchie, P.L.G. Vlek, and L.G. Youngdahl. 1984. The water and N components of the CERES models. Proc. Intl. Symp. on Minimum Data Sets for Agrotechnology Transfer, March 1983, Patancheru, India. Intl. Crops Research Institute for the Semi-Arid Tropics, India. p. 95-100.

Grizzetti, B., F. Bouraoui, K. Granlund, S. Rekolainen, and G. Bidoglio. 2003. Modelling diffuse emission and retention of nutrients in the Vantaanjoki Watershed (Finland) using the SWAT model. Ecol. Modell. 169:25–38. doi:10.1016/S0304-3800(03)00198-4

Grunwald, S., and C. Qi. 2006. GIS-based water quality modeling in the Sandusky Watershed, Ohio, USA. J. Am. Water Resour. Assoc. 42:957–973. doi:10.1111/j.1752-1688.2006.tb04507.x

Hansen, L., J.A. Delgado, M. Ribaudo, and W. Crumpton. 2012. Minimizing cost of reducing agricultural nitrogen loadings: Choosing between on- and off-field conservation practices. Environ. Ecol. 3:98–113.

Hansen, S., H.E. Jensen, N.E. Nielsen, and H. Svendsen. 1991. Simulation of nitrogen dynamics and biomass production in winter wheat using the Danish simulation model DAISY. Fert. Res. 27:245–259. doi:10.1007/BF01051131

Harris, G.L., and J.A. Catt. 1999. Overview of the studies on the cracking clay soil at Brimstone Farm, UK. Soil Use Manage. 15:233–239. doi:10.1111/j.1475-2743.1999.tb00094.x

Harris, G.L., M.J. Goss, R.J. Dowdell, K.R. Howse, and P. Morgan. 1984. A study of mole drainage with simplified cultivation for autumn sown crops on a clay soil. 2. Soil water regimes, water balances and nutrient loss in drain waters. J. Agric. Sci. 102:561–581. doi:10.1017/S0021859600042118

Havlin, J.L., J.D. Beaton, S.L. Tisdale, and W.L. Nelson. 1999. Soil fertility and fertilizers. 6th ed. Prentice Hall. Upper Saddle River, NJ.

Haynes, R.J. 1986. The decomposition process: Mineralization, immobilization, humus formation, and degradation. In: R.J. Haynes, editor, Mineral N in the soil-plant system. Academic Press, Orlando, FL. p. 52–126. doi:10.1016/B978-0-12-334910-1.50006-6

Hoang, L., A. van Griensven, P. van der Keur, J.C. Refsgaard, L. Troldborg, B. Nilsson, and A. Mynett. 2014. Comparison and evaluation of model structures for the simulation of pollution fluxes in a tile-drained river basin. J. Environ. Qual. 43:86–99. doi:10.2134/jeq2011.0398

Hobbs, J.A., and C.A. Thompson. 1971. Effects of cultivation on the N and organic carbon content of a Kansas argiustoll (Chernozem). Agron. J. 63:66–68. doi:10.2134/agronj1971.00021962006300010021x

Hu, X., G.E. McIsaac, M.B. David, and C.A.L. Louwers. 2007. Modeling riverine nitrate export from an east-central Illinois watershed using SWAT. J. Environ. Qual. 36:996–1005. doi:10.2134/jeq2006.0228

Huang, J., and H. Hong. 2010. Comparative study of two models to simulate diffuse N and phosphorus pollution in a medium-sized watershed, southeast China. Estuarine Coastal Shelf Sci. 86:387–394. doi:10.1016/j.ecss.2009.04.003

Jha, M.K., P.W. Gassman, and J.G. Arnold. 2007. Water quality modeling for the Raccoon River watershed using SWAT2000. Trans. ASABE 50:479–493. doi:10.13031/2013.22660

Jones, C., and J. Jacobsen. 2002. Nutrient management module No. 3: N cycling, testing and fertilizer recommendations. Montana State University Extension Service. http://landresources.montana.edu/nm/documents/NM3.pdf (Verified on 31 May 2017).

Kanwar, R.S., and J.L. Baker. 1993. Tillage and chemical management effects on groundwater quality. In: Agricultural research to protect water quality: Proceedings of the conference. 21-24 Feb. 1993. Minneapolis, MN. Soil and Water Conservation Society, Ankeny, IA. p. 455-459.

Kim, N.W., and A.H. Shin. 2011. Modification of the channel BOD simulation scheme in SWAT for Korean TMDL application. Trans. ASABE 54:1739–1747. doi:10.13031/2013.39839

Knisel, W.G., editor. 1980. CREAMS: A field-scale model for chemicals, runoff, and erosion from agricultural management systems. U.S. Department Agric. Conserve. Res. Rep. No. 26. USDA Science and Education Administration, Washington, DC.

Knisel, W.G., and K.R. Douglas-Mankin. 2012. CREAMS/GLEAMS: Model use, calibration, and validation. Trans. ASABE 55:1291–1302. doi:10.13031/2013.42241

Kolenbrander, G.J. 1981. Leaching of N in agriculture. In: J.C. Brogan, editor, N losses and surface runoff from land spreading of manures. Netherlands: Martinus Nijhoff/Junk., Dordrecht, the Netherlands.

Linn, D.M., and J.W. Doran. 1984. Effect of water-filled pore space on carbon dioxide and nitrous oxide production in tilled and nontilled soils. Soil Sci. Soc. Am. J. 48:1267–1272. doi:10.2136/sssaj1984.03615995004800060013x

McElroy, A.D., S.Y. Chiu, J.W. Nebgen, A. Aleti, and F.W. Bennett. 1976. Loading functions for assessment of water pollution from nonpoint sources. EPA 600/2-76-151. U.S. Environmental Protection Agency, Washington, DC.

Meisinger, J.J., and J.A. Delgado. 2002. Principles for managing nitrogen leaching. J. Soil Water Conserv. 57:485–498.

Melvin, S.W., J.L. Baker, P.A. Lawlor, B.W. Heinen, and D.W. Lemke. 1993. N management and crop rotation effects on nitrate leaching, crop yield, and N use efficiency. In: Agricultural research to protect water quality: Proceedings of the Conference. 21-24 Feb. 1993. Minneapolis, MN. Soil and Water Conservation Society, Ankeny, IA. p. 411-415.

Menzel, R.G. 1980. Enrichment ratios for water quality modeling. In: W.G. Knisel, editor, CREAM, a field scale model for chemicals, runoff, and erosion from agricultural management systems. U.S. Department Agric. Conserve. Res. Rep. No. 26. USDA Science and Education Administration, Washington, DC. p. 482–492.

Mitchell, J.K., G.F. McIsaac, S.E. Walker, and M.C. Hirschi. 2000. Nitrate in river and subsurface drainage flows from an east central Illinois watershed. Trans. ASABE 43:337–342. doi:10.13031/2013.2709

Moriasi, D.N., J.G. Arnold, M.W. Liew, R.L. Bingner, R.D. Harmel, and T. Veith. 2007. Model evaluation guidelines for systematic quantification of accuracy in watershed simulations. Trans. ASAE 50:885–900. doi:10.13031/2013.23153

Mosier, A.R., J.W. Doran, and J.R. Freney. 2002. Managing soil denitrification. J. Soil Water Conserv. 57:505–513.

Naveen, K.P., L. Masse, and P.Y. Jui. 1996. Tile effluent quality and chemical losses under conventional and no tillage-Part 1: Flow and nitrate. Trans. ASAE 39:1665–1672. doi:10.13031/2013.27683

Neitsch, S.L., J.G. Arnold, J.R. Kiniry, J.R. Williams, and K.W. King. 2002. Soil and Water Assessment Tool. Theoretical Documentation: Version 2000. TWRI Report TR-191, Texas Water Resources Institute, College Station, TX.

Niraula, R., L. Kalin, R. Wang, and P. Srivastava. 2012. Determining nutrient and sediment critical sources areas with SWAT: Effect of lumped calibration. Trans. ASABE 55:137–147. doi:10.13031/2013.41262

Novotny, V., and H. Olem. 1994. Water quality: Prevention, identification, and management of diffuse pollution. Van Nostrand Reinhold, New York.

Parton, W.J., D.S. Schimel, C.V. Cole, and D.S. Ojima. 1987. Analysis of factors controlling soil organic matter levels in great plains grasslands. Soil Sci. Soc. Am. J. 51:1173–1179. doi:10.2136/sssaj1987.03615995005100050015x

Pease, L.M., P. Oduor, and G. Padmanabhan. 2010. Estimating sediment, nitrogen, and phosphorous loads from the Pipestem Creek watershed, North Dakota, using AnnAGNPS. Comput. Geosci. 36:282–291. doi:10.1016/j.cageo.2009.07.004

Plus, M., I. La Jeunesse, F. Bouraoui, J.M. Zaldívar, A. Chapelle, and P. Lazure. 2006. Modelling water discharges and nitrogen inputs into a Mediterranean lagoon: Impact on the primary production. Ecol. Modell. 193:69–89. doi:10.1016/j.ecolmodel.2005.07.037

Randall, G.W., J.L. Anderson, G.L. Malzer, and B.W. Anderson. 1993a. Impact of N and tillage management practices on corn production and potential nitrate contamination of groundwater in Southern Minnesota. In: Agricultural research to protect water quality: Proceedings of the conference. 21-24 Feb. 1993. Minneapolis, MN. Ankeny, IA: Soil and Water Conservation Society. 172-175.

Randall, G.W., D.J. Fuchs, W.W. Nelson, D.D. Buhler, M.P. Russelle, W.C. Koskinen, and J.L. Anderson. 1993b. Nitrate and pesticide losses to tile drainage, residual soil N, and N uptake as affected by cropping systems. In: Agricultural research to protect water quality: Proceedings of the conference. 21-24 Feb. 1993. Minneapolis, MN.Soil and Water Conservation Society, Ankeny, IA. p. 468-470

Reddy, K.R., R. Khaleel, M.R. Overcash, and P.W. Westerman. 1979. A nonpoint source model for land areas receiving animal wastes: II. Ammonia volatilization. Trans. ASAE 22:1398–1404. doi:10.13031/2013.35219

Renard, K.G., G.R. Foster, G.A. Weesies, D.K. McCool, and D.C. Yoder. 1997. Predicting soil erosion by water: A guide to conservation planning with the Revised Universal Soil Loss Equation (RUSLE). USDA Agriculture Handbook No. 703. USDA Science and Education Administration, Washington, DC.

Ryden, J.C., P.R. Ball, and A.E. Garwood. 1984. Nitrate leaching from grassland. Nature 311:50–53. doi:10.1038/311050a0

Saleh, A., J.G. Arnold, P.W. Gassman, L.M. Hauck, W.D. Rosenthal, J.R. Williams, and A.M.S. McFarland. 2000. Application of SWAT for the Upper North Bosque River Watershed. Trans. ASAE 43:1077–1087. doi:10.13031/2013.3000

Saleh, A., and B. Du. 2004. Evaluation of SWAT and HSPF within BASINS program for the Upper North Bosque River watershed in Central Texas. Trans. ASAE 47:1039–1049. doi:10.13031/2013.16577

Saleh, A., and O. Gallego. 2007. Application of SWAT and APEX using the SWAPP (SWAT-APEX) program for the Upper North Bosque River watershed in Texas. Trans. ASABE 50:1177–1187. doi:10.13031/2013.23632

Santhi, C., J.G. Arnold, J.R. Williams, W.A. Dugas, R. Srinivasan, and L.M. Hauck. 2001. Validation of the SWAT model on a large river basin with point and nonpoint sources. J. Am. Water Resour. Assoc. 37:1169–1188.

Santhi, C., N. Kannan, M.J. White, M. Di Luzio, J.G. Arnold, X. Wang, and J.R. Williams. 2014. An integrated modeling approach for estimating the water quality benefits of conservation practices at the river basin scale. J. Environ. Qual. 43:177–198. doi:10.2134/jeq2011.0460

Seligman, N.G., and H. van Keulen. 1980. PAPRAN: A simulation model of annual pasture production limited by rainfall and N. In: M.J. Frissel and J.A. van Veeds, editors, Simulation of N behaviour of soil-plant systems, Proc. Workshop. Centre for Agricultural Publishing and Documentation. Wageningen, Netherlands. p. 192-221.

Sen, S., P. Srivastava, P.A. Vadas, and L. Kalin. 2012. Watershed-level comparison of predictability and sensitivity of two phosphorus models. J. Environ. Qual. 41:1642–1652. doi:10.2134/jeq2011.0242

Shaffer, M.J., J.A. Delgado, C.M. Gross, R.F. Follett, and P. Gagliardi. 2010. Simulation processes for the Nitrogen Loss and Environmental Assessment Package (NLEAP). In: J.A. Delgado and R.F. Follett, editors, Advances in nitrogen management for water quality. SWCS, Ankeny, IA. p. 361–372.

Shamshad, A., C.S. Leow, A. Ramlah, W.M.A. Wan Hussin, and S.A. Mohd Sanusi. 2008. Applications of AnnAGNPS model for soil loss estimation and nutrient loading for Malaysian conditions. Int. J. Appl. Earth Obs. Geoinf. 10:239–252. doi:10.1016/j.jag.2007.10.006

Sharpley, A.N., and J.R. Williams. 1990. EPIC—Erosion/Productivity Impact Calculator: 1. Model Documentation. U.S. Department of Agriculture Technical Bulletin No. 1768. USDA Science and Education Administration, Washington, DC.

Skaggs, R.W., N.R. Fausey, and R.O. Evans. 2012. Drainage water management. J. Soil Water Conserv. 67:167A–172A. doi:10.2489/jswc.67.6.167A

Stanford, G., and S.J. Smith. 1972. N mineralization potentials of soils. Soil Sci. Soc. Am. Proc. 36:456–472.

Stanford, G., and E. Epstein. 1974. N mineralization-water relations in soils. Soil Sci. Soc. Am. Proc. 38:103–107. doi:10.2136/sssaj1974.03615995003800010032x

Stevenson, F.J. 1982. Organic forms of soil N. In: F.J. Stevenson, editor, N in agricultural soils. American Society of Agronomy, Madison, WI. p. 67–122. doi:10.2134/agronmonogr22.c3

Stewart, G.R., C.L. Munster, D.M. Vietor, J.G. Arnold, A.M.S. McFarland, R. White, and T. Provin. 2006. Simulating water quality improvements in the upper North Bosque River watershed due to phosphorus export through turfgrass sod. Trans. ASABE 49:357–366. doi:10.13031/2013.20410

Sullivan, P.T., and Y. Gao. 2016. Assessment of nitrogen inputs and yields in the Cibolo and Dry Comal Creek watersheds using the SWAT model, Texas, USA 1996–2010. Environ. Earth Sci. 75:725. doi:10.1007/s12665-016-5546-0

Suttles, J.B., G. Vellidis, D. Bosch, R. Lowrance, J.M. Sheridan, and E.L. Usery. 2003. Watershed-scale simulation of sediment and nutrient loads in Georgia Coastal Plain streams using the Annualized AGNPS model. Trans. ASAE 46:1325–1335. doi:10.13031/2013.15443

Tisdale, S.L., W.L. Nelson, and J.D. Beaton. 1985. Soil fertility and fertilizers. Macmillan Publishing, London.

Tuppad, P., K.R. Douglas-Mankin, T. Lee, R. Srinivasan, and J.G. Arnold. 2011. Soil and Water Assessment Tool (SWAT) hydrologic/water quality model: Extended capability and wider adoption. Trans. ASABE 54:1677–1684. doi:10.13031/2013.39856

Vadas, P.A., and M.J. White. 2010. Validating soil phosphorus routines in the SWAT model. Trans. ASABE 53:1469–1476. doi:10.13031/2013.34897

Vinten, A.J.A., and K.A. Smith. 1993. N cycling in agricultural soils. In: T.P. Burt, A.L. Heathewaite, and S.T. Trudgill, editors, Nitrate processes, patterns and management. John Wiley & Sons Ltd, England. p. 39–75.

Wagenet, R.J., and J.L. Hutson. 1989. LEACHM: Leaching estimation and chemistry model: A process based model of water and solute movement, transformations, plant uptake and chemical reactions in the unsaturated zone. Continuum Vol. 2. Water Resource Institute, Cornell University, Ithaca, NY.

Wang, X., J.R. Williams, P.W. Gassman, C. Baffaut, R.C. Izaurralde, J. Jeong, and J.R. Kiniry. 2012. EPIC and APEX: Model use, calibration, and validation. Trans. ASABE 55:1447–1462. doi:10.13031/2013.42253

White, K.L., and I. Chaubey. 2005. Sensitivity analysis, calibration, and validations for a multisite and multivariable SWAT model. J. Am. Water Resour. Assoc. 41:1077–1089. doi:10.1111/j.1752-1688.2005.tb03786.x

Williams, J.R. 1990. The erosion productivity impact calculator (EPIC) model: A case history. Philos. Trans. R. Soc. Lond., B 329:421–428. doi:10.1098/rstb.1990.0184

Williams, J.R. 1995. The EPIC model. In: V.P. Singh, editor, Computer models of watershed hydrology. Water Resources Publications, Highlands Ranch, CO. p. 909–1000.

Williams, J.R., and R.W. Hann. 1978. Optimal operation of large agricultural watersheds with water quality constraints. Tech. Report No. 96. Texas A&M University, Texas Water Resources Institute, College Station, TX.

Yuan, Y., R.L. Bingner, M.A. Locke, F.D. Theurer, and J. Stafford. 2011. Assessment of subsurface drainage management practices to reduce N loadings using AnnAGNPS. Appl. Eng. Agric. 27:335–344. doi:10.13031/2013.37075

Yuan, Y., R.L. Bingner, and R.A. Rebich. 2001. Evaluation of AnnAGNPS on Mississippi Delta MSEA watersheds. Trans. ASAE 44:1183–1190. doi:10.13031/2013.6448

Yuan, Y., R.L. Bingner, and R.A. Rebich. 2003. Evaluation of AnnAGNPS N loading in an agricultural watershed. J. Am. Water Resour. Assoc. 39:457–466. doi:10.1111/j.1752-1688.2003.tb04398.x

Yuan, Y., R.L. Bingner, F.D. Theurer, R.A. Rebich, and P.A. Moore. 2005. Phosphorus component in AnnAGNPS. Trans. ASAE 48:2145–2154. doi:10.13031/2013.20100

Yuan, Y., R.L. Bingner, and F.D. Theurer. 2006. Subsurface flow component for AnnAGNPS. Appl. Eng. Agric. 22:231–241. doi:10.13031/2013.20284

Yuan, Y., and L. Chiang. 2015. Sensitivity analysis of SWAT nitrogen simulations with and without in-stream processes. Arch. Agron. Soil Sci. 61:969–987. doi:10.1080/03650340.2014.965694

GPS, GIS, Guidance, and Variable-rate Technologies for Conservation Management

John Fulton* and Matthew Darr

Abstract

The concept of precision agriculture can best be described as the spatial application of practices and technologies to agriculture to better capture and understand the precise interactions between the crop and the land. Precision agriculture enhances current farming techniques not only by improving crop yields, but also by endorsing a better stewardship of the environment by demonstrating how tillage, fertilizers, pesticides, and even water usage can be minimized. Data-driven decisions are also producing information to help farmers and agronomists improve field level decisions and ultimately profitability. Digital tools are now available to help manage inputs and provide information to support in-season decisions and provide benchmarking capabilities. With these advanced tools and technologies, farmers can experiment with new and innovative ways of growing, maintaining, and harvesting their crops to maximize production and profits.

Technology has become a mainstay in agriculture as it has provided a means to become more efficient with inputs and machinery management. Similarly, it enables farmers to better manage land and enhances implementation of conservation management strategies. Precision agriculture, or precision farming, is the term used to describe the adoption of technology and site-specific management practices in agriculture. The advent of the US Global Positioning System (GPS) and its availability to civilians in the early 90's initiated the advancement of precision agriculture worldwide. Technologies such as GPS-based (or today Global Navigation Satellite System [GNSS]-based) autoguidance technology are at minimum a standard option on farm equipment if not an embedded technology. In reality, technology enables adopters to implement and manage farmland in a much improved manner while ensuring preservation of conservation structures such as grassed waterways, buffer strips and other set aside areas. Upon adoption

Abbreviations: ASC, automatic section control; GIS, geographic information system; GLONASS, Global Orbiting Navigation Satellite System; GNSS, Global Navigation Satellite System; GPS, Global Positioning System; RTK, real-time kinematic; VRT, variable rate technology.

J. Fulton, Agricultural Engineering Department, Ohio State University, Columbus, OH 43210; M. Darr, Agriculture and Biosystems Engineering Department, Iowa State University, Ames, IA 50011. *Corresponding author (fulton.20@osu.edu).

doi:10.2134/agronmonogr59.2014.0016

of technology, agriculturalists and farmers in particular have realized tangible benefits such as savings of inputs and costs, reduced fatigued over long days, and better matching of application rates to crop or soil requirements. This chapter covers how precision agriculture technologies not only provide noticeable benefits to farmers but also how they concurrently enhances conservation management to meet environmental and sustainability requirements. These modern technologies can be used by nutrient managers and conservation practitioners to apply precision conservation on the ground to increase the effectiveness of conservation practices, improve management of natural resources, and increase sustainability.

Sound and effective conservation planning has become more important than ever due to the growing population and limited resources on earth. The increasing global population increases the demand for food while requiring preservation of our natural resources, especially water and soil. However, advancement of geospatial technologies and the site-specific management practices they enable have provided farmers, managers and land owners tools to help improve and conserve soil and water. These modern technologies include satellite based positioning systems such as the global positioning system (GPS), GPS-based guidance, variable-rate technology (VRT), geographic information systems (GIS), online mapping and visualization tools, remote sensing, and analytic tools for modeling of erosion, terrain, profit and environmental monitoring.

A wide range of precision agriculture technologies are available on the market and have been adopted by US farmers. The intention of these technologies has been to increase farm productivity while providing farmers the ability to adapt their crop management strategy to in-field variability. While the adoption rate for precision agriculture technologies continues to increase, past cropping patterns and practices tend to determine current practices that farmers use, even when new equipment and technology is purchased. However, it is conceivable that farmers can re-think the use of such technologies to increase equipment efficiency and productivity. This idea includes implementing new or revising current management strategies.

A major benefit of precision agriculture technologies and site-specific management is that they are able to improve environmental stewardship to sustain farm ecosystems. These technologies can mitigate adverse impacts on the environment from farming practices while creating sustainable agroecosystems. Precision agriculture also permits implementation of nontraditional and innovative management techniques. Practices such as strip tillage and the use of cover crops can help address soil compaction issues plus conserve water during dry periods. Additionally, more accurate placement of inputs (pesticides, nutrients, and fungicides) including manure and litter results in a more sustainable farm operation. While the implementation of technology and new practices is mostly driven by economics, there are environmental, conservation stewardship and sustainability components that influence adoption as well. Modern-day technologies such as remote sensing, geographic information systems (GIS), and modeling can be applied to precision conservation to improve management of natural resources and increase effectiveness of conservation efforts (Berry et al., 2003; 2005).

GPS and GNSS Positioning Technology

The Global Positioning System (GPS) was the key enabling sensor that permitted precision agriculture to evolve. GPS receivers were first introduced into agriculture in the early 1990s and provided the ability to locate a user's position on earth. GPS positions without correction could pinpoint a receiver's position to within 20 to 30 m. The development of differential GPS, or DPGS, greatly improved positioning resulting in accuracies to within several meters. Today, more refined DGPS options exist to correct raw GPS positions accurate at the meter level using corrected services such as WAAS down to centimeter horizontal and vertical accuracies when implementing real-time kinematic (RTK) positioning. Accurate DGPS positions allow for the development of precision agriculture technologies that collect spatial data across farmland, control equipment, and automate specific on-board functions during field operation. GPS is also being used to track animal movements and other farm assets.

One of the important outcomes of GPS has been the ability to create accurate field boundaries to define management zones for which decisions are made. Conservation practices (i.e., grassed waterways) can be mapped along with other areas designated as nonmanagement regions of the farm. Additional benefits of GPS mapping are:

1. Accurately defining boundaries of fields and conservation practice areas.

2. Calculating area of cropable land, set aside areas such CRP ground, buffer strips, etc.

3. Calculating distances, including heights

4. Identifying land use and area calculations for FSA or other agency reporting

5. Determining field boundaries and within-field management zones for:

    a. Management decisions and maps for equipment operators

    b. Development of prescription maps for VRA

    c. Input for Auto-Swath and other precision ag technology

    d. Field boundary locations for grass waterways and other conservation structures.

6. Marking buildings and other structures

7. New mapping units permit users to take images that are tagged with GPS locations during field and crop scouting (location + image)

Boundary definition using heads-up digitization within GIS and similar packages will be addressed along with utilizing DGPS to map boundaries. The impact of different levels of GPS and/or GNSS correction services and how they impact the accuracy of spatially defining boundaries will also be discussed. The development of buffers around conservation practice areas based on GPS and/or GNSS accuracy to reduce potential damage and application of inputs of existing conservation structures will be covered.

Overview of GPS

GPS satellites broadcast continuous independent radio signals to Earth, containing specialized data messages describing each satellite's identity, operational health, predicted position, and a timestamp. Knowing the precision position of the GPS satellites, a GPS receiver can calculate how far away multiple satellites are based on the amount of time needed to receive each satellite's data message. The distance traveled (D) by the data message is computed by knowing the rate of travel (R) and time of travel (T); $D = RT$. This is possible because the satellites are in precise orbits (approximately 20,000 km from Earth) and because the radio waves travel at the speed of light; approximately 300,000 km s^{-1}. With the assumption that the GPS satellites are in fixed orbits, with Earth at the center of those orbits, a geometric coordinate system can be derived to locate any point within the interior sphere of the constellation; this geometric coordinate system can then be converted into a geodetic coordinate system, consisting of latitude, longitude, and height. A GPS receiver must process the signals from a minimum of four GPS satellites before the receiver can accurately derive its location within the coordinate system, but the architecture of the satellite constellation is designed such that at least six satellites should be available to the user at any time of day throughout the world (except for the polar regions; Hurn, 1989).

GNSS

Global Navigation Satellite System, commonly called GNSS, is a term being used more frequently today. GNSS refers to the use of multiple satellite constellations to determine a geographic position anywhere in the world. The two most notable satellite based navigation systems are the United States' Global Positioning System (GPS) and the Russian Federation's Global Orbiting Navigation Satellite System (GLONASS). Others that are under development or are not fully operational include Europe's Galileo, China's COMPASS, and India's IRNSS.

These satellite navigation systems use a constellation of orbiting satellites that work in combination with a network of ground stations placed strategically around the world. More importantly, GPS or receiver manufacturers' have started to track both the GPS and GLONASS systems simultaneously to increase the coverage (number of satellites in view by a receiver) to compute a position solution. Once Galileo becomes operational, expect all three systems to be tracked simultaneously. In these cases, the term GNSS will be used on the manufacturer's specifications sheet for the receiver. Therefore the term GNSS is used to describe a receiver that has the ability to track multiple satellite navigation systems; GNSS = GPS + GLONASS + Galileo + future systems that come online. The result is improved accuracy, integrity, and coverage by tracking both systems. When purchasing a "GPS" receiver, one might consider a receiver which tracks systems beyond just GPS and enables better positioning capabilities in areas where part of the sky is obscured by trees, buildings, hills or other items.

Real-time Kinematic (RTK)

Real-Time Kinematic (RTK) represents a differential correction procedure that provides centimeter level or better positional accuracies. RTK has become common within agriculture, surveying and construction industries due to its ability to provide accurate and repeatable positions on the earth; especially application requiring this level of positioning. Simply, RTK employs a process called carrier-phase enhancement to estimate correctional errors that are transmitted in real-time to a rover receiver from an individual base station (e.g., single-base solution) or collection of base stations (e.g., network solution) serving as references to compute real-time positional errors.

Applications of RTK are used for navigation and machinery automation especially within agriculture and construction. Machinery automation includes automatic machine guidance where the machine drives itself and blade control on equipment like bulldozers and graders. Advantages of RTK and machinery automation are the increase in machine productivity and accuracy of job execution. Currently, over 80% of the GPS in the US are RTK systems.

More specific to agriculture, RTK systems are being used to implement
- Controlled traffic scenarios– equipment following the same path during field operations
- Precision Strip Tillage (and Strip Fertilization)
- Automatic Section Control (ASC) on sprayers, planters and other application equipment
- Precise Seed and Fertilization applications
- Elevation data for drainage and terrain modeling

Guidance and Autocontrol

GPS-based guidance evolved shortly after GPS receivers were introduced in agriculture. Guidance technology was adopted quickly by farmers as it helped operators maintain desired paths. Two levels of GPS-guidance exist and are typically termed (i) lightbar or (ii) autoguidance. Lightbars are similar to navigation systems in cars in which the equipment operator still drives the machine while the lightbar technology indicates the desired path. In this case, the driver must maneuver the machine to maintain the desired path. Autoguidance describes the ability of the machine to drive the prescribed path without the operator interacting with the steering wheel. In either case, the operator must define the desired path (e.g., straight, curve, etc.) called the AB line which can be saved and recalled for future field operations. The uses of guidance include:
- Spraying
- Tillage
- Planting
- Spreaders or applicators
- Harvesting
- Forage mowing and conditioning

The following provides a list of benefits.

1. Extends operational hours by minimizing operator fatigue

2. Able to operate at night and in adverse conditions (e.g., dust)

3. Improves field efficiency and capacity

   a. Cover more acres with fewer hours of operation

   b. Eliminate guess rows

   c. Minimize driver errors such as skips and overlaps

   d. Reduce overlap on implements and sprayers

4. Reduces per acre fuel consumption

5. Able to operate at faster field speeds

Fig. 1. Illustration of autoguidance technology implemented during planting within a conservation management system. The autoguidance permits the tractor and planter to operate accurately on planned paths in a high-residue cover crop conservation system.

6. Allows for more accurate placement of agronomic inputs such as fertilizer and herbicides

7. Enhances the adoption of controlled traffic

8. Allows nontraditional management

a. Strip tillage followed by planting

b. Specialized harvesting, such as peanuts grown in split-rows

9. Able to implement banded spraying–focused application of products whereby less product is used versus full-coverage

Autoguidance (Fig. 1) has been one of the leading precision agriculture technologies adopted in the United States. It has become a standard feature on larger equipment, with capabilities built into smaller machinery. All equipment manufacturers provide autoguidance as a feature today. In many cases, RTK is being embraced with autoguidance affording the ability to accurately drive the same paths (AB lines) over time. Guidance, especially RTK autoguidance, provides a benefit of reducing overlap and missed areas during field operations. One study estimated overlap as 9% of the implement width, reducing efficiency (Backman et al., 2009). Most importantly, autoguidance enables farmers to change their traditional practices or cropping patterns (Fig. 2; Taïx et al., 2006). Guidance technology now permits alternate cropping patterns to be considered. Most cropping patterns (e.g., row alignment at planting) are based on the previously established management patterns. In many cases these patterns are the same as those that have been used over several decades. These traditional patterns are developed over growing seasons by experience, and are often

Fig. 2. Example of a nontraditional cropping system whereby crop strips are altered across a field to maximize light interception (e.g. leaf area index) and thereby crop yields.

determined by where machinery enters the field as well as by where the equipment operator perceives the longest and straightest side of the boundary. In some cases, the establishment of row patterns can be intuitive (e.g., long, narrow field) but in irregular shaped fields evidence exists that these established paths may not be optimal for equipment productivity.

Controlled Traffic

Considerable research has shown controlled traffic to be an effective means of reducing compaction in agricultural operations (Buckingham, 1975; Dumas et al., 1973). Equipment is confined to predetermined paths to decrease the area of soil affected and restrict traffic to drier soils. RTK technology is being used to guide equipment with centimeter accuracy. Manufacturers including John Deere and Trimble Inc. have agricultural guidance systems on the market that automatically steer the tractor along a predetermined path. These systems are often used for installation of in-furrow irrigation systems where accuracy is crucial. Guidance systems also allow for easy implementation of controlled traffic. Gan-Mor and Clark (2001) indicated that controlled traffic can lessen and in some cases eliminate the need for deep tillage operations. Raper and Bergtold (2007) reported a 6% fuel savings and 9% draft force reduction could be achieved with controlled traffic subsoiling.

Variable-rate Technologies

Variable-rate technology (VRT) enables equipment to vary the application rates of inputs such as seed and fertilizer across fields. This capability to varying the rate of inputs allows practitioners to implement site-specific management in an attempt to match inputs with field variability. A variable rate strategy can provide a reduction of inputs through minimizing over-application, thereby maximizing profitability and reducing environmental risks. VRT was the "original" precision agriculture technology but struggled to gain adoption by farmers due to the lack of easy-to-use systems and the field level knowledge required to effectively generate prescriptions (Rx) maps.

Examples of common VRT practices include:
• Planter and seeder for varying population and hybrids (Fig. 3)
• Nutrient and lime application
• Fungicide applications
• Minimal or reduced tillage

Variable-rate technology represents a means to improve field execution of planting and application of inputs. However, an important step to ensure success of VRT implementation is to properly calibrate the technology and equipment along with making sure operators have the proper training. VRT can introduce more complexity to a field operation requiring additional education of operators to ensure they have the knowledge and skills to utilize it.

Variable-rate technology includes a combination of components to work and apply inputs at a precise time and/or location (Fig. 4). Two strategies exist to implement VRT: (i) *map-based* and (ii) *sensor-based*. In the map-based strategy, a DGPS is used for specifying field locations, plant or soil samples are collected using a field grid map, and subsequently a map is generated for site-specific applications. The sensor-based method is an "on-the-go" method because it uses real-time

Fig. 3. A. Example of a planter equipped with variable-rate technology (VRT). B. An in-cab display. C. Hydraulic drive.

sensors for controlling and varying application as the sensing and application equipment traverses the field. The sensor-based method provides an immediate, direct measure of the crop or soil with an immediate decision that alters the rate of application based on the measurement. A third strategy to vary rates can be *manual* control where the machine operator manually changes the rates using the in-cab display (Fig. 3) during field operation. While manual control is not as elegant or accurate as map- or sensor-based VRT, it is a simple, less-expensive method that can allow a farmer to become familiar with the extent of in-field variability, and the potential of precision management.

A VRT system includes integrated components, such as a DGPS receiver, in-cab display, software, and rate controller. Figure 4 presents common VRT components to execute a map-based strategy.

Advantages of Variable-rate Technology

Variable rate technology offers several benefits including economic and environmental savings. By tailoring the application rates to the needs of the crop or soil, the efficiency of inputs is improved by reducing over- or under-application and potential farm profitability increases. Environmental benefits with adopting a variable-rate strategy can include but are not limited to (i) eliminate over-application of inputs, (ii) improve plant health and thereby uptake of applied fertilizers, and (iii) reduce the risk of off-site transport of nutrients or chemicals through surface runoff and leaching into water sources.

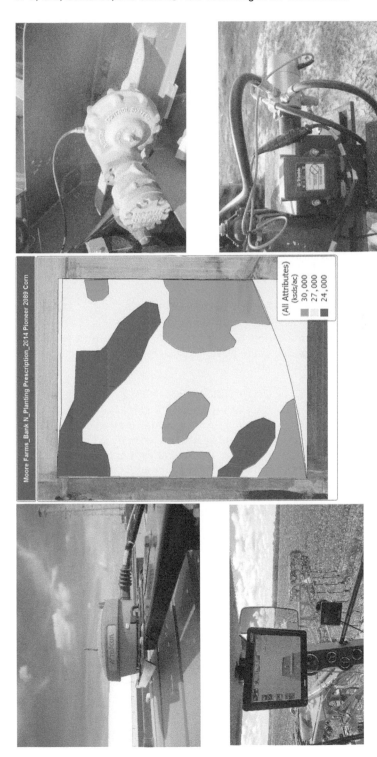

Fig. 4. DGPS enabled, automatic section control can provide more accurate application of pesticides and nutrients within conservation structures than manual control (e.g., operator decision), reducing over- or underapplication (Fulton et al., 2011).

Variable-rate Technology Challenges

To improve profitability of precision agriculture technology and improve its performance, the operation of agriculture equipment must be precise (Mowitz, 2003). Proper calibration by following the manufacturers' recommendations and standards is a necessity. The American Society of Agricultural and Biological Engineers (ASABE) has developed procedures for calibrating equipment which is a useful tool for producers utilizing VRT (ASABE, 2009; ASABE, 2017). Potential challenges to the implementation of VRT is the expense of implementation, especially since many VRT technologies cannot be retrofitted to older machinery, and the increased technical knowledge needed to run the equipment. Time delays between sampling or sensing, data processing, and application rate as well as system error from GPS signal loss near trees and buildings can also be concerns.

Variable-rate Technology Implementation

Manufacturers are starting to make VRT components a standard item on new equipment. In this case, it is an operator's decision to implement the use of VRT. It is also important to check with manufacturers of aftermarket VRT systems to ensure that they are compatible with existing equipment.

Future of Variable-rate Technology

It appears that variable-rate technology will continue to rapidly advance as electronic controls and communication technologies advance within agriculture. Future technological advances include planting, tillage, pest management, crop diagnosis, and water irrigation (Table 1).

Automatic Section Control Technology

Automatic section control (ASC) technology is a common technology on sprayer, fertilizer applicators and planters. ASC provides the capability to automatically turn on and off sections of the applicator or planter based on a no-apply area (e.g., conservation structure) or an area that had already been covered by the machine. This technology works in conjunction with the application map, substantially reducing over-application of pesticides, fertilizers and seed in conservation

Table 1. Potential new methods of variable-rate technology (VRT).

Field operation	Variable-rate practice
Planting	Seed placement at a depth dependent on soil moisture
	Varying seeding rate as a function of soil type and/or other field characteristics
	Planting varieties as a function of soil type or other field characteristics
Tillage	Residue management dependent on soil organic matter or biomass levels
	Till to the depth of compaction layer
	Till only areas with high field traffic
Pest management	Sensor-based insect or weed control
	Apply chemical in identified moderate to high risk areas only
	Apply chemical dependent on insect or weed type
Crop diagnosis	Apply chemical dependent on plant stress and nutrient deficiency
Water and irrigation	Apply water and chemicals in a site specific stress-based manner

practice areas or areas that have previously been applied or covered. ASC is a precision agriculture technology that has been readily adopted by farmers and custom applicators since a standard technology option on agriculture equipment.

Automatic section control technology turns boom-sections or rows off in previously covered areas, or automatic on or off when operating on headland turns, within point rows, or over terraces and grassed waterways. Figures 5 and 6 present manual operation versus the use of ASC on sprayers and planters, respectively.

The key benefit to planting using ASC technology is input savings by minimizing double planted or over-sprayed areas of the field (Fig. 6). An Auburn University study (Runge et al., 2014) outlined seed savings from using ASC between 1 to 12% from field to field with an overall average savings of 4.3%. These reported seed savings were only contributed to the ASC technology and did not consider the influence of guidance technology. Including both ASC and guidance technology as a means to reduce double planted areas could increase overall seed savings ranging from 3% to 35% for individual fields (Runge et al., 2014). The seed savings per field is directly dependent on size and shape. Higher savings are realized in large, odd-shaped fields or fields having conservation structures such as terraces, buffers, and grassed waterways (Fig. 5). Other benefits of ASC include:

1. Increased application accuracy and input use efficiency

2. Reduced overlap at headlands and point rows

    a. Over-application due to operator error will be reduced

    b. Reduced crop damage from over-application

3. Improved environmental stewardship

    a. Pesticide and nutrient applications in environmentally sensitive areas (e.g., riparian buffers) will be minimized

    b. The integrity of conservation structures will be maintained because pesticide spray-outs are unlikely to occur

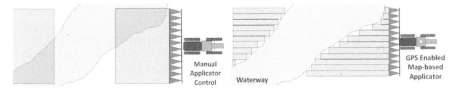

Fig. 5. DGPS enabled, automatic section control can provide more accurate application of pesticides and nutrients within conservation structures than manual control (e.g. operator decision), reducing over- or under-application (Fulton et al., 2011).

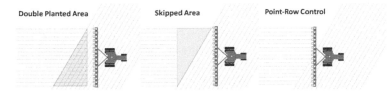

Fig. 6. Illustration of doubled planted and skipped areas of a field where the tractor operator decides when to turn on and off the planter versus DGPS-based automatic section control (Runge et al., 2014).

    c. Double or triple rate applications resulting from overlap will be virtually elimi-
nated thereby minimizing run-off and leaching of pesticides and nutrients.

4. Over-applications of chemicals resulting from equipment malfunctions will
be more rapidly detected by an operator who focuses on overall system performance rather than on navigating and turning on and off sprayer sections

5. Greater fuel efficiency is ensured because automated control allows fields to
be navigated more efficiently with less tortuous paths

6. Decreased overall inputs thereby increasing savings

    Automatic section control technology on sprayers has been able to eliminate
herbicide misapplication to environmentally sensitive areas. The benefits of ASC
technologies extend beyond pesticide applicators (Jaynes et al., 2001). Nitrogen
leaching into ground water has been shown to be dependent on both application
rate and soil characteristics of the application area. Similarly, overapplication of
nitrogen commonly occurs in the end-of-row zones of fields. The application of
map-based ASC alleviates these problems through the elimination of off-target
application and application to environmentally sensitive areas.

Controller Area Networks

Controller Area Networks or CAN Bus systems began to propagate on agricultural machinery in the mid-1990s as a network for sensors and machine controllers
to communicate. Early applications enabled engines, transmissions, and cab displays to communicate information between controllers and also to the machine
operator. Fundamentally, this is analogous to Ethernet networks that link computers within private and public networks today. By the late 1990s new standards
were under development to create uniformity in the communication protocol and
signal definitions which today are the backbone of the majority of precision agriculture application. This new standard, known commonly as ISOBUS, has been
an enabling tool in supporting flexible and brand independent choices for producers related to technology adoption.

    From the perspective of an agricultural producer, ISOBUS is a behind-the-
scenes technology that is empowering the growth of data driven agricultural
applications and advanced control systems. These are now common in agricultural machinery. ISOBUS provides the basic architecture for yield data to be
recorded and mapped in real time, for control signals to vary the drive speed of
individual planter row units based on specific machine settings, and for new sensors, such as those used to actively measure crop health or soil conditions, to be
directly integrated into existing ag machinery. Additionally, Virtual Terminals,
which are a core component of ISOBUS technology, allow for a single display
to perform multiple functions and control multiple implements simultaneously.
This technology has reduced the cost of adoption as well as the complexity of
operation of precision agriculture equipment by eliminating the requirement to
purchase a new control display for each new precision agriculture feature.

    ISOBUS by nature of its common communication platform is also enabling
new advances in data collection, cloud data integration, and in-cab data visualization. Recent advances in data communication have led to innovations that extract
specific ISOBUS information and relay this information directly to cloud storage or
onto mobile devices. This enhances the visualized quality of data and improves the
probability of actionable outcomes for producers. By leveraging the existing cloud

and mobile computing infrastructure agricultural producers are able to benefit from large technology investments within the consumer and industrial marketplaces.

Yield Monitoring and Mapping

Yield monitors (Fig. 7) provide producers the ability to collect crop performance information to generate yield maps. These maps can be used to diagnose problems and serve as the basis of site-specific management. Calibration of yield monitors is required for each crop type and any significant changes in the physical character- istics of the crop (e.g density, surface properties, etc.). Calibration is performed by weighing the harvested grain or cotton then entering the known weights into the yield monitor display to determine the calibration constants over a flow range. This calibration data then adjusts the yield estimates to minimize errors.

Yield maps (Fig. 8) provide an illustration of variations in crop yields across fields plus offering spatial data for postharvest analyses. These maps have the ability to indicate where some areas could be micro-managed over areas as small as a few square feet to either (i) increase yield or (ii) reduce input costs. Further, temporal yield maps can be developed to determine yield stability from year-to-year. While yield maps indicated low to high yielding areas of a field, they can also be used for end-of-year performance and diagnose issues influencing yield during the growing season such as:

• Soil type differences
• Fertility
• Weed pressure or control
• Insect infestations
• Drainage
• Soil compaction
• Tillage
• Equipment malfunction

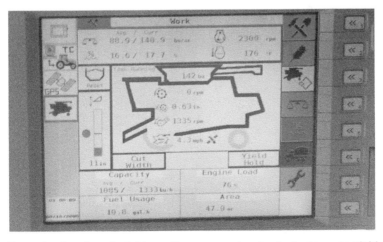

Fig. 7. Example of an in-cab yield monitor screen showing instantaneous yield and grain moisture content.

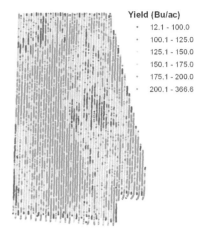

Yield (Bu/ac)
· 12.1 - 100.0
· 100.1 - 125.0
125.1 - 150.0
150.1 - 175.0
175.1 - 200.0
· 200.1 - 366.6

Fig. 8. Example of a corn yield map illustrating low (red) and high (green) yielding areas.

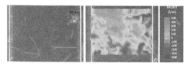

Fig. 9. Visible aerial image (left) of a field along with a profit map (right) illustrating areas of profit (green) and loss (red). While the whole analyses indicated a new profit each year, persistent areas of loss (red) exist requiring evaluation for alternate practices including possible implementation of conservation structures (Courtesy of Iowa Soybean Association).

Profitability maps can be generated from the yield map by incorporating input costs of production. A yield map converted to a profit map (Fig. 9) indicates areas within a field that have poor return on investment. These areas along the eastern and southern boundaries of the larger field may be eligible for enrollment in a conservation practice which may help offset profits lost by low yields. Using yield maps over many years can help better plan for and implement conservation practices such as buffer strips or grassed waterways. The key factor required to successfully implement precision agriculture is the ability to effectively collect yield and similar data spatially and then analyze the data to identify areas of risk or areas to place conservation structures or implement conservation practices.

Benefits of Yield Maps

1. Produce a final report card of crop production

2. Provide instantaneous yield

3. Quickly identify high and low yielding areas

4. Guidance for tailoring future management practices

5. Help evaluate varieties, nutrients, and other inputs

6. Lead to maximized returns

7. Can be used to aid in landlord negotiations to perform improvements, split on crop shares, or modify rental agreements.

Summary

Farm equipment across the United States has increased in size over the years to improve field efficiency and timeliness of operations. Unfortunately, this trend has made it difficult to manage NRCS conservation structures (e.g., grassed waterways, filter strips, herbaceous and wooded riparian buffers; examples in Fig. 10). Often these structures are "sprayed-out" during routine herbicide applications (Fig. 11). Further, due to irregular shapes of crop fields, double and triple application occurs frequently. Precision agriculture technologies enable application operators to avoid sensitive areas or not allow application to occur within these areas.

Grassed waterways suffer greatly from unintentional application of pesticides directly to waterway areas (Donald, 2002). These applications cause direct and irreversible destruction of vegetation and will require reseeding to once again establish these conservation practice areas (Fig. 3). Avoiding damage to these conservation areas has become increasingly difficult for agricultural producers because of a trend toward larger, faster equipment for improving production efficiency. For example, it is common to see seeding equipment extend beyond 12 m (40 ft) in width, sprayers that exceed 27 m (90 ft), and fertilizer application equipment exceeding 18 m (60 ft). Preserving waterways and field boundaries with large equipment is particularly difficult with conservation tillage management systems because burn-down herbicides are used so extensively. Furthermore, the size of modern day agricultural equipment makes it difficult for operators to consistently avoid or manage around different conservation and environmental management areas within and bordering fields and pastures. Typically, the shape, width, and length of these areas pose a difficult scenario since management is

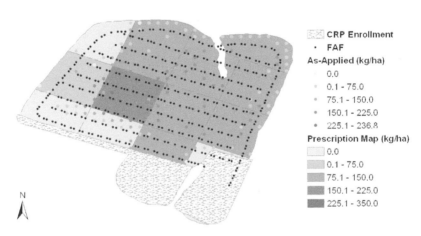

Fig. 10. Example map illustrating a field boundary with CRP or buffer strip installed to protect and reduce non-point source pollution.

Fig. 11. A sprayed-out grassed waterway with a small portion of a buffer strip impacted by a herbicide treatment.

limited to the equipment width (Fig. 3). When a 27-m sprayer (common width across the United States) intersects a grassed waterway, it becomes hard for an operator to manually turn the boom ON and OFF while delivering product so as not to spray the waterway but still guarantee the cropland receives the applied product. An additional difficulty is the ability of applicator operators to identify the boundary of vegetated conservation and environmental management areas during early spring burn down of field vegetation prior to planting. Many times, the vegetation has not come out of its dormancy, making it difficult to distinguish between cover crops and the vegetated conservation areas. Therefore, the ability to map the actual managed field boundary using GPS and the subsequent automated ability of equipment to turn ON or OFF application based on this map greatly reduces this issue and preserves these areas. The use of large application equipment leads to common deficiencies in conservation management. Specifically, these include skipped areas, double applications, unintentional applications, or application to environmentally sensitive areas. In addition to destroying conservation areas, double pesticide or nutrient applications can result in contamination of ground and/or surface water.

Future of Precision Agriculture

Data driven information for agriculture is an evolution occurring in agriculture today. The use of data to understand and improve decision-making at the farm is referred to as digital agriculture. Online tools and mobile applications (APPs) are becoming standard for communicating information, storing data and providing digital tools for farmers and their trusted advisors. Weather and geospatial data layers such as soils maps are more readily available to farmers through these digital tools and precision agriculture software. This information is being connected to farm or farmer-specific data to help analyze the farm operation and field-by-field performance. Today, tractors, combines and now sprayers are connected to the internet through embedded modems and telemetry technology allowing for easier transfer of data to and from machines. Collection of the data can provide insights on how to improve productivity, efficiency and farm performance. The ability to share data is being enabled today so farmers can benchmark themselves against other farmers. All facets of the industry including seed, fertilizer and chemical companies are offering a variety of precision agriculture services that can help farmers implement new practices such as the four Rs of nutrient stewardship: right rate, right source, right time, and in the right place [www.nutrientstewardship.com/4rs/(verified 23 May 2017)].

Small unmanned aircraft systems (sUAS) are affordable technology that farmers can use to conduct scouting or collect data and information from across their fields. The future of sUAS is promising to collect imagery and samples across fields with small on-the-ground robots also collecting data and information to implement more informed decisions during the season. Robots and automated machinery will be disruptive technologies bringing dramatic change to improve crop production, optimize inputs and placement, sense crop and soil conditions, and other farm management practices. Integrating these emerging technologies into farming practices offers opportunities to greatly improve the economic and environmental sustainability of agronomic production systems.

References

ASABE. 2017. ASABE Standard X573: Procedures for evaluating variable-rate granular material application accuracy. ASABE, St. Joseph, MI.

ASABE. 2009. ASABE Standard S341.4: Procedure for measuring distribution uniformity and calibrating granular broadcast spreaders. ASABE, St. Joseph, MI.

Backman, J., T. Oksanen, and A. Visala. 2009. Parallel guidance system for tractor-trailer system with active joint. In: E.J. van Henten, D. Goense, and C. Lokhorst, Precision agriculture '09: Papers presented at the 7th European Conference on Precision Agriculture. Wageningen, the Netherlands. 6-8 July 2009. Wageningen Academic Publishers, Wageningen, the Netherlands. p. 615-622.

Berry, J.K., J.A. Delgado, R. Khosla, and F.J. Pierce. 2003. Precision conservation for environmental sustainability. J. Soil Water Conserv. 58:332–339.

Berry, J.K., J.A. Delgado, F.J. Pierce, and R. Khosla. 2005. Applying spatial analysis for precision conservation across the landscape. J. Soil Water Conserv. 60:363–370.

Buckingham, F. 1975. Controlled traffic can stop compaction of agricultural soils. Crops and soils Magazine. June-July. pp. 13-14.

Donald, W.W. 2002. Glyphosate effects on ground cover of tall fescue waterways and estimated soil erosion. J. Soil Water Conserv. 57:237–243.

Dumas, W.T., A.C. Trouse, L.A. Smith, F.A. Kummer, and W.R. Gill. 1973. Development and evaluation of tillage and other cultural practices in a controlled traffic system for cotton in the Southern Coastal Plains. Trans. ASAE 16:0872–0876. doi:10.13031/2013.37648

Fulton, J.P., D. Mullenix, A. Brooke, A. Winstead, B. Ortiz, and A. Sharda. 2011, Automatic Section Control (ASC) Technology for Sprayers. Timely Information Web publication, Alabama Cooperative Extension System. https://sites.aces.edu/group/crops/precisionag/Publications/Timely%20Information/Automatic%20Section%20Control%20(ASC)%20Technology%20for%20Sprayers.pdf (verified 19 May 2017).

Gan-Mor, S., and R.L. Clark. 2001, July. DGPS-based automatic guidance–implementation and economical analysis. In: Proceedings of 2001 ASAE Annual Meeting. July 2001. American Society of Agricultural Engineers. St. Joseph, MI.

Hurn, J. 1989. GPS: A guide to the next utility. Trimble Navigation, Inc. Sunnyvale, CA.

Jaynes, D.B., T.S. Colvin, D.L. Karlen, C.A. Cambardella, and D.W. Meek. 2001. Nitrate loss in subsurface drainage as affected by nitrogen fertilizer rate. J. Environ. Qual. 30:1305–1314. doi:10.2134/jeq2001.3041305x

Mowitz, D. 2003. Precision planting pays off. Successful Farming 101:33–42.

Raper, R.L., and J.S. Bergtold. 2007. In-row subsoiling: A review and suggestions for reducing cost of this conservation tillage operation. Appl. Eng. Agric. 23:463–471. doi:10.13031/2013.23485

Runge, M., J.P. Fulton, T. Griffin, S. Virk, and A. Brooke. 2014. ANR-2217: Automatic section Control technology for row crop planters. Alabama Cooperative Extension System, Auburn University, Auburn. AL.

Taïx, M., P. Souères, H. Frayssinet, and L. Cordesses. 2006. Path planning for complete coverage with agricultural machines. In: S. Yuta, H. Asama, E. Prassler, T. Tsubouchi, and S. Thrun, Field and service robotics: Recent advances in research and applications, Springer Tracts in Advanced Robotics Vol. 24. Springer Berlin Heidelberg. Heidelberg, Germany. p. 549–558.

Identifying Riparian Zones Best Suited to Installation of Saturated Buffers: A Preliminary Multi-Watershed Assessment

M.D. Tomer,* D.B. Jaynes, S.A. Porter, D.E. James, and T.M. Isenhart

Abstract

Saturated riparian buffers are a new type of conservation practice that divert subsurface tile drainage water from direct discharge to surface water into distribution pipes that discharge the tile water into riparian soils. This enables natural processes of biological uptake and denitrification to decrease nutrient loads that are being lost from croplands via tile drains, reducing water quality impacts from agriculture at relatively little cost. This chapter suggests and evaluates draft criteria that identify riparian zones within a watershed that are suited to installation of saturated buffers. Soils criteria, evaluated using soil survey information, include subsurface accumulations of soil organic matter (SOM) (> 1% SOM at 0.75–1.2 m depth), relatively fine-textured subsoils (< 50% sand at 0.75–1.2 m depth), and a shallow water table (< 1 m depth) April through June. These criteria highlight riparian locations where soil conditions should enhance nutrient removal. Criteria are also proposed to avoid locations where streambank failure and/or inundation of crops adjacent to the buffer may occur, which are evaluated using high-resolution digital elevation models, now widely available through LiDAR (light detection and ranging) surveys. The criteria were evaluated in three Midwestern HUC-12 watersheds dominated by fine-grained glacial deposits. Results showed topographic criteria were more restrictive than soils criteria, especially in the flattest landscapes, but 30 to 60% of streambank lengths in the test watersheds were deemed suitable to installation of saturated buffers. This evaluation contributed to inclusion of a saturated buffer siting tool in the Agricultural Conservation Planning Framework (ACPF). Local information is needed to design this practice to fit site conditions.

In the U.S. Midwest, agricultural nutrient losses that are carried by artificial subsurface (tile) drainage water have been associated with impacts on water quality

Abbreviations: ACPF, Agricultural Conservation Planning Framework; LiDar, light detection and ranging; MLRA, Major Land Resource Area; RPAs, Riparian Assessment Polygons.

M.D. Tomer, D.B. Jaynes, S.A. Porter, and D.E. James, USDA-ARS, National Laboratory for Agriculture and the Environment, Ames, IA; T.M. Isenhart, Department of Natural Resource Ecology and Management, Iowa State University, Ames, IA. *Corresponding author (mark.tomer@ars.usda.gov)

doi:10.2134/agronmonogr59.2013.0018

(David et al., 2010; Raymond et al., 2012). A number of measures can be put in place to reduce nutrient losses from tile drainage, including cover crops (Kaspar et al., 2012), drainage water management (Adeuya et al., 2012), denitrifying bio-reactors (Schipper et al., 2010), and nutrient removal wetlands (Tomer et al., 2013). Riparian buffers are also known to be capable of removing subsurface nutrients, particularly nitrate, but from groundwater that passes beneath riparian buffers rather than tile drainage (Mayer et al., 2007). Tile drainage water typically bypasses riparian zones via installed drainage pipes, preventing any interaction of drainage water with riparian soils or vegetation. A recent advance to overcome this bypass issue has been the saturated buffer (Jaynes and Isenhart, 2014). This riparian practice enables a subsurface discharge of drainage water along distribution pipes laid at shallow depth (0.3–0.6 m) and installed along the upper edge of a riparian zone. A water level control gate is installed to direct normal tile flows toward these subsurface discharge pipes. Larger flow rates that occur during times of increased precipitation can pass through the gate structure to ensure drainage rates from farmed fields are not impeded by the gate diversion. The saturated buffer practice has been evaluated in several settings, and removal rates may be (but are not always) equivalent to the proportion of tile drainage volume that is diverted to subsurface discharge (D.B. Jaynes, 2016, unpublished data).

The purpose of this chapter is to propose a mapping technique that identifies locations that are most appropriate for installation of saturated riparian buffers in Midwestern tile drained watersheds. Prediction of actual N removal rates from tile drainage water using this practice is difficult on a site by site basis, but we believe it possible to identify where high rates of N removal should most readily be achievable, and identify where this practice can be installed with minimal risks of unintended consequences. Saturated buffers are inexpensive to install compared to other types of practices that require more land area and/or significant construction costs (Jaynes and Isenhart, 2014); this is especially true where riparian buffers are already established. Therefore, even modest N removal rates (i.e., < 30%) may achieve N removal at acceptable costs compared to other types of N-removal practices that require more land area. Where high rates of N removal can be achieved, the saturated buffer practice will be among the most preferred practice options on grounds of cost efficiency. However, more information is needed to identify the extent of sites that are suitable for saturated riparian buffer installations before the potential role of this practice in reducing watershed-scale nutrient loads can be fully elucidated.

Context

We are developing an approach that identifies riparian sites suited for saturated buffers for inclusion in the Agricultural Conservation Planning Framework (ACPF) toolbox for ArcGIS (ESRI, 2014), and to be consistent with the riparian assessment tools that are part of that toolbox (Tomer et al., 2015a, 2015b). The riparian assessment includes an approach to discretize all of a watershed's riparian corridors into a series of 250- by 90-m polygons that are each evaluated and ranked in terms of upland runoff-contributing area and apparent width of shallow water table zones, as determined through analyses of high resolution (1- to 5-m grid) digital elevation models. The rankings convey the relative importance of opportunities at a riparian site to use buffer vegetation for slowing or filtering surface runoff and for interacting with shallow groundwater, as compared to other riparian sites across that watershed. Once cross-classified according to these rankings, buffer designs are suggested that identify buffer widths and types of vegetation appropriate to riparian settings throughout the watershed. See Tomer et al. (2015a) for further details. The approach used to classify opportunities to install saturated buffers employs the same spatial discretization of the riparian corridor into Riparian Assessment Polygons (RAPs). However, our current intent, in the context of the ACPF toolbox, is to allow the saturated buffer siting tool to be run independent of (i.e., with or without) the riparian assessment.

We begin with two disclaimers. First, locations of tile drainage outfalls and expected flow rates are not readily available for most watersheds. A full application of this tool toward saturated buffer installations requires local knowledge, which we did not access in the watersheds used for demonstration in this chapter. Runoff flow paths that can be identified through the ACPF toolbox (or other terrain analyses programs) will often provide a good indicator of where tile outfalls can be found. However, major outfalls may carry greater flows than this practice can readily accommodate. Initial field trials of this practice have focused on field-scale drainage systems less than 160 acres in size (D.B. Jaynes, 2016, unpublished data). Second, the reader is advised that the saturated buffer tool described here is in draft form, and subject to revision on inclusion in the ACPF toolbox (Porter et al., 2015).

The saturated buffer siting tool, as conceived and demonstrated herein, should highlight riparian soils and slope conditions that are most conducive to installation and successful performance of this practice, provided field-scale drainage systems discharge in those same vicinities. Detailed knowledge of the local drainage system will be necessary for saturated buffer installation. One critical limitation to the design and performance of this system is to ensure sediment does not accumulate and clog the distribution pipes. Surface intakes, if part of the contributing drainage system, pose a risk for reducing subsurface discharge. Outletting the distribution pipes to the stream should reduce this risk, and should be included in saturated buffers receiving drainage from surface intakes. Note this means the system will have more than one discharge outlet, which will make performance monitoring more difficult.

Approach and Methods

Criteria for site selection

Our approach comprises criteria intended to identify suitable soil conditions, presence of shallow groundwater, and appropriate slope conditions. These criteria are tested using soils data extracted from the gSSURGO (soil survey) (Natural Resources Conservation Service, 2014) database, and high-resolution topographic data (see Tomer et al., 2015a, 2015b) obtained through Light Detection and Ranging (LiDAR) surveys.

Suitable soil conditions are associated with a sufficient residence time and a source of organic carbon at depth in the soil to encourage nitrate removal through denitrification. A sufficient residence time should occur where soils are fine textured, such that water will take at least several hours to drain away and that the water table will become mounded around (or just below) the distribution pipes. An environment conducive to denitrification will encourage nitrate concentrations to be halved every 6 to 12 h of residence time (Moorman et al., 2015); during which lateral flows through the soil (rather than vertical flows to depths where soil organic matter decreases) should dominate. Coarse-textured, sandy soils may not provide the desired flow rate (not rapid) and direction (lateral), especially where coarse-textured soils are found at depth. We set a criterion to only include soils that have < 50% sand at 0.75- to 1.2-m depth, as an average calculated by weighting textural data by horizon thickness. The second soil criterion ensures the presence of organic carbon to enable denitrification to occur at depth in the soil. The criterion we chose was > 1% soil organic matter (SOM) at the 0.75-to 1.2-m depth interval. Note that while high SOM contents are common in Midwestern riparian subsoils, deep sandy soils are also found in riparian areas where glacial outwash deposits are present.

The final soils criterion was set to ensure the presence of a shallow water table within the riparian zone, to identify and prioritize sites where lateral water flows that encourage denitrification should occur. Soil survey information includes seasonal (April–June) water table depth information to support land use suitability interpretations. Soils exhibiting a seasonal water table within 1.0 m of the surface were considered suitable. While sites with shallower (i.e., < 0.5 m) water table depths may exhibit the greatest denitrification rates because greater (e.g., > 4%) SOM concentrations occur in the upper profile of many riparian soils, there is also a risk of seepage flows and rainfall-runoff erosion where the saturated buffer practice brings the water table to, or very near, the surface. This consideration led us to choose an intermediate (< 1.0 m) seasonal water table depth criteria. Riparian assessment polygons were considered to have soils suitable for the saturated buffer practice where soil map units meeting these three conditions (SOM and sand contents at 0.75–1.2 m, and shallow seasonal water table) occupied at least a 20-m width along the 250-m long RAP.

While soils criteria were selected to prioritize soil conditions suitable for nitrate removal, topographic criteria were selected to minimize the risk of unintended consequences from saturated buffer installation. First, we sought to eliminate areas with steep banks, where raising the water table could increase the risk of bank sloughing. Provisionally, a 10-m minimum width of land area within a RAP having a surface elevation within 1.5 m of the channel (Tomer et al., 2015a) was selected for this criterion. The width of this "low-lying" land area along the channel was determined using tools in the ACPF riparian assessment (Porter et al., 2015; Tomer et al., 2015a). Field reviews of several watersheds have led us to believe this criterion will successfully avoid steep banks. The NRCS standard for the saturated buffer (Natural Resources Conservation Service, 2016) specifies a maximum bank height of 2.4 m.

The third and final criterion is intended to avoid areas with a flat riparian zone, where crops planted just above the riparian buffer could be inundated by buffer saturation. However, in addition, sloping riparian terrain may not be suited to optimal performance of the saturated buffer practice because of increased flow rates including preferential flows that can carry water rapidly downslope; return flows (Kirkby, 1988) could lead to runoff and erosion across a steep riparian zone with a mounded water table, and could be a problem in some settings. Noting that wider riparian buffers may be necessary to adequately filter runoff where slopes are > 10% (Liu et al., 2008), we selected a slope range of 2 to 8% as being in the range that would avoid risks of saturated buffers being either too flat or too steep. Those RAPs where 2 to 8% slopes occupy at least 35% of the RAP area were deemed suitable for saturated buffers for the purpose of this study. Our intent with the final development of this tool for the ACPF toolbox is to provide user options under most if not all criteria; that is, the minimum SOM at depth could be selected at 0.5%, enabling flexibility for Midwest areas where greater SOM concentrations are seldom found deep in the profile, considering that roots of buffer vegetation may provide a reasonable carbon source to facilitate denitrification (Dosskey et al., 2010). Allowing the user to select the fraction of the RAP that must have 2 to 8% slopes to meet the slope criterion is another option being considered.

A key point is that our intent is not to dictate where saturated buffers should or should not be placed, but simply to identify locations where the practice should perform well in terms of nitrate removal efficiency, and to help avoid areas where some risk of unintended consequences is apparent. As we learn more about saturated buffers and how well they perform in different settings, the prudence of each of these criteria will become clearer.

Description of test watersheds

We applied this demonstration analysis to three headwater watersheds in Iowa and Illinois (Fig. 1). Bear Creek in north central Iowa has been the subject of several studies on riparian buffers (Schultz et al., 2004), and comprises hummocky terrain or recent glacial origin (Wisconsinan age; 12,000–14,000 yr ago) that is drained through a valley carved by glacial meltwater. Lime Creek, located in northern Illinois, was identified as watershed 6 in an initial assessment of the ACPF riparian assessment tools (Tomer et al., 2015a), and was described in some

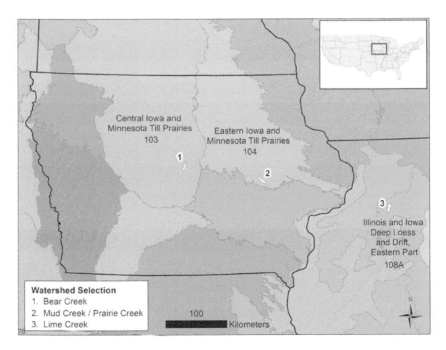

Fig. 1. Map showing locations of three test watersheds and the Major Land Resources Areas (MLRA) in which they are found. See http://www.nrcs.usda.gov/wps/portal/nrcs/ detail/soils/survey/?cid_=_nrcs142p2_053624 for further details on MLRAs.

detail by Tomer et al. (2015b, 2013). This watershed is bounded to the north by a terminal moraine from a Wisconsinan age glacial advance and its southern half is dominated by a glacial-lacustrine plain. Bear and Lime Creek watersheds are of similar age in terms of glacial origin, but are contrasted in terms of type and extent of fluvial deposits. Mud Creek and Prairie Creek are found on a landscape that originated as Illinoian till, which has been more dissected by stream development given the older age (approx. 500,000 yr).

Digital elevation models for these three watersheds were derived from LiDAR survey data at 1- to 3-m grids, and preprocessed to eliminate false impoundments such as at bridges and culverts (discussed by Tomer et al., 2013) using a 'cutter' tool that is available in the ACPF toolbox (Porter et al., 2015). Following flow routing analysis that maps upslope contributing areas and channelized flows across the watersheds, the extents of perennial streams were identified using hillshade images of topographic data and aerial photographs. The perennial stream network and adjoining riparian lands were then discretized using 250- by 180-m RAPs as described by Tomer et al. (2015a). This method uses the 'strip-map index' feature of ArcGIS (Ver. 10.3; ESRI, 2014) to create the series of polygons; the stream polyline is then used to split each RAP enabling results to be tabulated separately for each side of the stream. Data that were used to evaluate soils criteria (SOM, texture, seasonal water table) were extracted from the gSSURGO database (Natural Resources Conservation Service, 2014); data used to evaluate topographic criteria (i.e., area of 2–8% slopes, and minimum 10-m width of area within 1.5 m of channel for bank height interpretation) were obtained from terrain analysis tools that calculate slope and flow accumulation rasters as described by ESRI, (2014) and Porter et al. (2015).

Results and Discussion

Three test watersheds showed a range of riparian conditions meeting the above-described provisional criteria to identify potential sites for saturated buffer installations (Table 1). In general, however, soil survey information suggests suitable soil conditions are common, and may be nearly ubiquitous in some watersheds. Many Midwest watersheds exhibit riparian soil conditions with SOM > 1% to depth, fine subsoil textures, and seasonal water table depths within 1.0 m of the surface. In Bear and Lime Creeks, of these three soils criteria, riparian soils failed to meet the SOM criteria (> 1% SOM at 0.75–1.2 m) most frequently, but 79 to 84% of the RAPs still passed this criteria. Coarse-textured outwash materials have been identified in the lower part of Bear Creek watershed and were found to reduce buffer performance for nitrate removal by Simpkins et al. (2002). However, these coarse textures occur below fine sediments that have accumulated during and since the Holocene, and/or are not extensive enough to be identified as separate map units in the gSSURGO database. This emphasizes the need for on-site investigation to confirm soil conditions are suitable when designing and installing saturated buffers. In general, soil survey information indicates most riparian soils in these watersheds are conducive to conditions that would be sought for saturated buffer installations that should achieve substantial reductions in tile-drainage nitrate loads (Fig. 2, top row).

Topographic criteria may be more important than soil criteria because they are aimed at preventing unintended consequences that could lead to stream bank

Table 1. Summary of results indicating extent of conditions that are favorable for riparian buffer installations in three test watersheds.

Attribute	Bear Creek	Lime Creek	Mud Creek and Prairie Creek
Watershed area, ha	7489	6965	6720
Total stream length, km	50.1	51.4	52.3
Stream length captured by RAPs†, km	44.6	49.8	46.7
Stream bank length captured by RAPs†, km; equal to twice stream length	89.2	99.7	93.4
Number of RAPs†	274	350	298
Number of RAPs† meeting soils criterion			
Soil texture (< 50% sand at 0.75–1.2 m depth)	266	350	298
SOM‡ (> 1% at 0.75–1.2 m depth)	217	296	297
Seasonal water table < 1.0 m	268	337	297
Number of RAPs† meeting topographic criteria			
Bank height (> 10 m width is < 1.5 m above channel)	268	194	221
Slope (> 35% of RAP† is 2–8% slopes)	213	211	209
Number of RAPs† meeting all five criteria	164	103	162
Stream bank length meeting all five criteria, km	54.7 (61%)	30.6 (31%)	52.3 (56%)

†RAP, Riparian Assessment Polygon.
‡SOM, Soil Organic Matter.

failure or crop inundation. Damage of either type can increase economic costs associated with this practice beyond those anticipated during design, which could lead to this practice becoming unpopular among landowners. Results for these three watersheds indeed suggest that topographic criteria may limit the extent of suitable saturated buffer sites more frequently than do soil criteria, at least in areas of the Midwest dominated by finer-textured glacial deposits. In all three watersheds, the number of RAPs meeting any individual criterion was least for one of the two topographic criteria, compared to the number of RAPs meeting any of the soils criteria (Table 1, Fig. 2).

In Bear Creek watershed, soils with low SOM at depth in the upper part of the watershed, and steep slopes near the stream in the middle lengths of the watershed (see Fig. 3, top) were the two most common apparent limitations on the potential extent of saturated

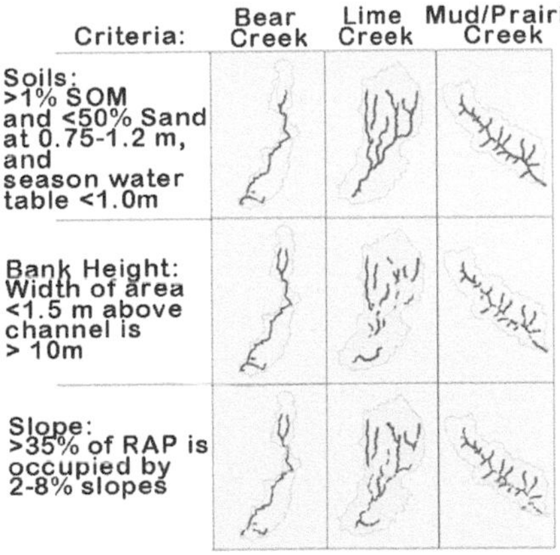

Fig. 2. Map matrix illustrating the extent of Riparian Assessment Polygons that met soil and topographic conditions in three test watersheds.

Criteria:	Bear Creek	Lime Creek	Mud/Prairie Creek
Soils: >1% SOM and <50% Sand at 0.75-1.2 m, and season water table <1.0m			
Bank Height: Width of area <1.5 m above channel is > 10m			
Slope: >35% of RAP is occupied by 2-8% slopes			

Fig. 3. Maps of three test watersheds illustrating the spatial distribution of riparian soil and topographic conditions meeting provisional criteria for siting saturated buffers. Zoomed in maps show areas where RAPs (Riparian Assessment Polygons) were excluded due to being too steep (> 8%) along Bear Creek (top), too flat (< 2% slopes) along Lime Creek (middle), and where bank heights may be too great along Mud Creek and Prairie Creek (lower).

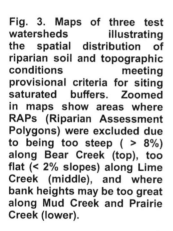
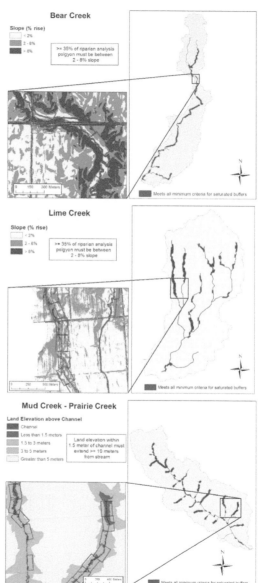

buffer installations. Nevertheless, 61% of the RAPs in this watershed met all five provisional criteria for saturated buffer installation. One of the research sites being used to monitor and evaluate this practice is in Bear Creek (Jaynes and Isenhart, 2014), and this site met all five criteria (not shown).

Lime Creek watershed had the least extent of RAPs indicated as suitable for saturated buffer placements. Low SOM contents at depth occurred along some upper stream lengths, while deeply dug ditches and flat landscapes in the lower parts of the watershed led to topographic criteria being failed in lower parts of the watershed. Areas that were too flat to meet the slope criteria for Lime Creek watershed are shown in Fig. 3 (center panel). There were 31% of the RAPs in Lime Creek that met all five criteria, however, most of these were found along head-water reaches in the northwest part of the watershed. The upper (most northern) reaches of Lime Creek watershed drain farmlands dominated by a terminal moraine, which are sloping and may not have extensive tile drainage. Again, on-site investigation would be required to ascertain this.

The Mud Creek and Prairie Creek watershed has the most favorable soil conditions for saturated buffer sites, compared to Bear and Lime Creek water-sheds (Table 1). Topographic criteria essentially provide the only limitations to siting saturated buffers along Mud Creek and Prairie Creek, with flat slopes and/or incised banks providing the main limitations along the lower main channel, and steep slopes and or incised streambank conditions occurring along upper reaches. Upper reaches with incised conditions are depicted in Fig. 3 (lower panel). More than half the RAPs (56%) nevertheless met all five criteria for potential saturated buffers.

Conclusions

This study has illustrated a draft tool for identifying potential sites for installation of saturated buffers (Jaynes and Isenhart, 2014), which on being finalized will be added to the ACPF toolbox (Porter et al., 2015; Tomer et al., 2015a, 2015b). Results from three watersheds located in separate landscape regions (MLRAs) suggest that suitable sites should be relatively common in many Midwest watersheds with tile-drained croplands. Topographic criteria should be regarded as more important than soil criteria to minimize the risk of unintended consequences from the saturated buffer practice. Ongoing research should help clarify soils criteria that will achieve optimal N removal in saturated buffers. Results from this study suggest that topographic conditions will most commonly limit site suitability in most Midwestern watersheds, but also suggest 30 to 60% of riparian zones may be suited to installation. Limitations we found were consistent with expectations based on distributions of slopes and incised channels that are common in these watersheds. Per the advice in the NRCS standard for this practice (Natural Resources Conservation Service, 2016), on-site soil investigations should be undertaken to confirm site and soil suitability as an early step in the saturated buffer design and installation process.

References

Adeuya, R., N. Utt, J. Frankenberger, L. Bowling, E. Kladivko, S. Brouder, and B. Carter. 2012. Impacts of drainage water management on subsurface drain flow, nitrate concentration, and nitrate loads in Indiana. J. Soil Water Conserv. 67:474–484. doi:10.2489/jswc.67.6.474

David, M.B., L.E. Drinkwater, and G.F. McIsaac. 2010. Sources of nitrate yields in the Mississippi River basin. J. Environ. Qual. 39:1657–1667. doi:10.2134/jeq2010.0115

Dosskey, M.G., P. Vidon, N.P. Gurwick, C.J. Allen, T.P. Duval, and R. Lowrance. 2010. The role of riparian vegetation in protecting and improving chemical quality in streams. J. Am. Water Resour. Assoc. 46:261–277. doi:10.1111/j.1752-1688.2010.00419.x

ESRI. 2014. ArcGIS 10.3.1 for Desktop, quick start guide. ESRI, Redlands CA. http://desktop.arcgis.com/en/arcmap/10.3/get-started/quick-start-guides/arcgis-desktop-quick-start-guide.htm (accessed 21 April 2017).

Jaynes, D.B., and T.M. Isenhart. 2014. Reconnecting tile drainage to riparian buffer hydrology for enhanced nitrate removal. J. Environ. Qual. 43:631–638. doi:10.2134/jeq2013.08.0331

Kaspar, T.C., D.B. Jaynes, T.B. Parkin, T.B. Moorman, and J.W. Singer. 2012. Effectiveness of oat and rye cover crops in reducing nitrate losses in drainage. Agric. Water Manage. 110:25–33. doi:10.1016/j.agwat.2012.03.010

Kirkby, M. 1988. Hillslope runoff processes and models. J. Hydrol. 100:315–339. doi:10.1016/0022-1694(88)90190-4

Liu, X., X. Zhang, and M. Zhang. 2008. Major factors influencing the efficacy of vegetated buffers on sediment trapping: A review and analysis. J. Environ. Qual. 37:1667–1674. doi:10.2134/jeq2007.0437

Mayer, P.M., S.K. Reynolds, M.D. McCutchen, and T.J. Canfield. 2007. Meta-analysis of nitrogen removal in riparian buffers. J. Environ. Qual. 36:1172–1180. doi:10.2134/jeq2006.0462

Moorman, T.B., M.D. Tomer, D.R. Smith, and D.B. Jaynes. 2015. Evaluating the potential role in denitrifying bioreactors in reducing watershed-scale nitrate loads: A case study comparing three Midwestern (USA) watersheds. Ecol. Eng. 75:441–448. doi:10.1016/j.ecoleng.2014.11.062

Natural Resources Conservation Service. 2014. Gridded Soil Survey Geographic (gSSURGO) Database User Guide. USDA-NRCS. https://www.nrcs.usda.gov/wps/portal/nrcs/detail/soils/survey/geo/?cid%20=%20nrcs142p2_053628. (Accessed 30 Mar. 2016).

Natural Resources Conservation Service. 2016. Conservation Practice Standard, Saturated Buffer, Code 604. https://www.nrcs.usda.gov/wps/portal/nrcs/detailfull/national/technical/cp/ncps/?cid_=_nrcs143_026849. (Accessed 3 Apr. 2017).

Porter, S.A., M.D. Tomer, D.E. James, and K.M.B. Boomer. 2015 Agricultural Conservation Planning Framework ArcGIS® user's manual. www.northcentralwater.org/acpf/ (Verified 30 March 2016).

Raymond, P.A., M.B. David, and J.E. Saiers. 2012. The impact of fertilization and hydrology on nitrat fluxes from Mississippi watersheds. Curr. Opin. Environ. Sustain. 4:212–218. doi:10.1016/j.cosust.2012.04.001

Schipper, L.A., W.D. Robertson, A.J. Gold, D.B. Jaynes, and S.C. Cameron. 2010. Denitrifying bioreactors– an approach for reducing nitrate loads to receiving waters. Ecol. Eng. 36:1532–1543. doi:10.1016/j.ecoleng.2010.04.008

Schultz, R.C., T.M. Isenhart, W.W. Simpkins, and J.P. Colletti. 2004. Riparian forest buffers in agroecosystems–lessons learned from the Bear Creek Watershed, central Iowa, USA. Agrofor. Syst. 61:35–50.

Simpkins, W.W., T.R. Wineland, R.J. Andress, D.A. Johnston, G.C. Caron, T.M. Isenhart, and R.C. Schultz. 2002. Hydrogeological constraints on riparian buffers for reduction of diffuse pollution: Examples from the Bear Creek watershed in Iowa, USA. Water Sci. Technol. 45:61–68.

Tomer, M.D., K.M.B. Boomer, S.A. Porter, B.K. Gelder, D.E. James, and E. McLellan. 2015a. Agricultural Conservation Planning Framework: 2. Classification of riparian buffer design-types with application to assess and map stream corridors. J. Environ. Qual. 44:768–779. doi:10.2134/jeq2014.09.0387

Tomer, M.D., W.G. Crumpton, R.L. Bingner, J.A. Kostel, and D.E. James. 2013. Estimating nitrate load reductions from placing constructed wetlands in a HUC-12 watershed using LiDAR data. Ecol. Eng. 56:69–78. doi:10.1016/j.ecoleng.2012.04.040

Tomer, M.D., S.A. Porter, K.M.B. Boomer, D.E. James, J.A. Kostel, M.J. Helmers, T.M. Isenhart, and E. McLellan. 2015b. Agricultural Conservation Planning Framework: 1. Developing multi-practice watershed planning scenarios and assessing nutrient reduction potential. J. Environ. Qual. 44:754–767. doi:10.2134/jeq2014.09.0386

Vegetative Filters

M.G. Dosskey,* S. Neelakantan, T. Mueller, Z. Qiu

Abstract

Vegetative filters, such as filter strips and riparian buffers, are widely applied in agricultural landscapes to mitigate sediment and chemical runoff from fields to waterways. They are typically installed along field margins and stream banks in strips having constant width. Nonuniform or concentrated flow of runoff from agricultural fields, however, can limit the pollutant trapping effectiveness of vegetative filters having this landscape configuration. Using precision conservation technologies, alternative configurations can be identified that may substantially improve their water quality performance. The key technologies include digital elevation models (DEM) coupled with algorithms for determining in a geographic information system (GIS) precisely where field runoff would flow. Several physically-based empirical models capitalize on these technologies to identify where filtering capabilities are more favorable. The models are used to identify more effective locations for vegetative filters to be placed in landscapes and watersheds and to create variable-width configurations along field margins.

Vegetative filters, such as filter strips and riparian buffers, are widely installed in agricultural landscapes to mitigate pollutant movement from fields to waterways. Runoff of excess rainfall from fields carries with it sediments, nutrients, and other chemicals that degrade water quality in streams. Vegetative filters are placed along field margins and stream banks to intercept the runoff flow and promote deposition and retention of the sediment and chemicals. Several reviews of research on processes and performance indicate they can be very effective, particularly for sediment and sediment-attached pollutants (Dosskey, 2001; Baker et al., 2006; Mayer et al., 2007; Helmers et al., 2008).

Vegetative filters are conventionally designed to have a constant width along a field margin or stream bank. They retain a greater fraction of the pollutants in runoff when the runoff is uniformly distributed through the entire filter area (Dillaha et al., 1989; Franklin et al., 1992; Dosskey et al., 2002; Blanco-Canqui et

Abbreviations: AOI, area of interest; DEM, digital elevation models; GIS, geographic information systems; SRI, Sediment Retention Index; TI, Topographic Index; VFSMOD-W, Vegetative Filter Strip Modeling System; WII, Water Inflow Index; WNI, Wetness Index.

M.G. Dosskey, USDA Forest Service, National Agroforestry Center, 1945 North 38th Street, Lincoln, NE, USA 68583; S. Neelakantan, Washington State Department of Natural Resources, MS 47020, Olympia, WA 98504; T. Mueller, Decision Science and Modeling Team, Deere & Company, Mercury Building, 4140 NW 114th Street, Urbandale, IA 50263; Z. Qiu, Department of Chemistry and Environmental Science, New Jersey Institute of Technology, University Heights, Newark, NJ 07102. *Corresponding author (mdosskey@ fs.fed.us)

doi:10.2134/agronmonogr59.2013.0019

Precision Conservation: Geospatial Techniques for Agricultural and Natural Resources Conservation
J.A. Delgado, G.F. Sassenrath, and T. Mueller, editors. Agronomy Monograph 59.

al., 2006). However, runoff will most often be nonuniform, diverging from subtle ridges and converging into swales that deliver more runoff to some portions of field margin or riparian zones than to others (Dillaha et al., 1986; Franklin et al., 1992; Fabis et al., 1993; Ludwig et al., 1995; Dosskey et al., 2002; Bereswill et al., 2012; Pankau et al., 2012; Hancock et al., 2015). Where this occurs, performance of a constant-width filter is limited because effectiveness is reduced along portions receiving additional load while portions that contact little or no runoff do not contribute materially to pollutant retention (Dillaha et al., 1986; Misra et al., 1996; Dosskey et al., 2002; Berry et al., 2003; Delgado and Berry, 2008). Performance of constant-width filters may be boosted by additional practices that redistribute runoff throughout the filter area, such as by grading the field, constructing spreaders, or installing grass barriers (Franklin et al., 1992; Blanco-Canqui et al., 2006). Alternatively, performance may be boosted by simply redistributing the filter area: installing more filter area where there is greater load and less filter area where the load is smaller (Dosskey et al., 2005).

The concept of placing more filter area where runoff load is relatively greater can be applied at whole landscape and watershed scales as well. While vegetative filters are conventionally placed along field margins and stream banks, only some portions of these general locations may intercept significant runoff loads and have site conditions that favor effective pollutant retention (e.g., Agnew et al., 2006; Dosskey et al., 2011b). Furthermore, there may be locations elsewhere in landscapes and fields where a vegetative filter could yield even greater water quality performance (Dosskey et al., 2013).

Geospatial technologies present an opportunity to use precision conservation to enhance the water quality performance of vegetative filters by matching filter location and design to the magnitude of the pollutant runoff load and filtering conditions at the site (Berry et al., 2003; 2005; Delgado and Berry, 2008). The key technology is the availability of digital elevation models (DEM) that can be analyzed in a geographic information system (GIS) to determine runoff flow directions and land slopes at very high spatial resolution. One method employs this information to develop variable-width filter designs around field margins. Other methods use it to scope whole landscapes and watersheds for more effective locations to install filters. These are referred to as terrain-based models because they all rely, at least in part, on elevation data that represents an area's terrain and the terrain's control on the distribution of water flow across the landscape. Application of these methods promises to improve the water quality performance of vegetative filter installations.

Field-Scale Design

AgBufferBuilder

A model was developed for designing the size and configuration of filter area along a field margin to match nonuniform patterns of overland runoff from an agricultural field (Dosskey et al., 2011a). The model was subsequently automated by adapting it to terrain analysis in a GIS (Dosskey et al., 2015). Its utility was further enhanced by modifying it to enable estimation of the performance of existing and hypothetical filter designs. The automated tool, called AgBufferBuilder, was developed to run with ArcGIS (Esri, Redlands, CA). The key feature of AgBufferBuilder is that it sizes the filter in segments along a field margin in proportion to the size of field area that drains to each segment, that is, buffer area ratio. This approach can account for varying sizes and irregular shapes of run-off-contributing areas that produce nonuniform runoff. This computer program, user's guide, practice data sets, and other documentation can be downloaded from https://nac.unl.edu/tools/AgBufferBuilder.htm.

Data Inputs and Operations

In the GIS, a DEM is overlaid on an aerial orthophotograph of the field and its grid cell structure is used to divide the margin of the field into many short segments (one grid cell each), determine contributing area and average slope to each segment, and provide a grid structure for mapping filter area within each contributing area. Contributing area is determined using the ArcGIS Flow Accumulation function which employs the D8 algorithm of Jenson and Domingue (1988).

For determining the appropriate buffer area ratio to apply in each contributing area, the design model employs seven discrete relationships between buffer area ratio and trapping efficiency (i.e., percent of input load retained by a filter) (Fig. 1). These relationships were developed by simulation modeling using the process-based Vegetative Filter Strip Modeling System (VFSMOD-W; Muñoz-Carpena and Parsons, 2014). Each relationship corresponds to a different combination of site conditions (slope, soil texture, field soil cover and management condition) and type of pollutant (sediment or sediment-attached) for a design storm of 61 mm (2.4 in) in 1 h generating runoff to a well-established grass filter. From information about site conditions and pollutant type, the GIS program selects the

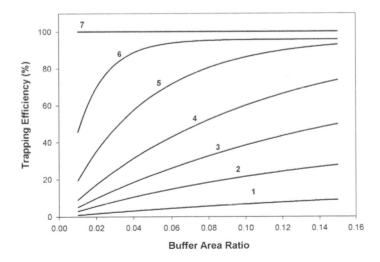

Fig. 1. Relationships between pollutant trapping efficiency and buffer area ratio for a well-established grass filter during a rainfall event of 61 mm in 1 h under seven different combinations of land slope, soil texture class, soil cover and management, and pollutant type. For a given buffer area ratio, higher trapping efficiencies are associated with flatter slopes, coarser soils, greater soil cover, and sediment as opposed to sediment-bound pollutants. [Reproduced from Dosskey et al. (2011a) by permission of the Soil and Water Conservation Society.]

relationship that best describes each contributing area (method details provided in Dosskey et al., 2011a; 2015) and then applies a user-supplied value for trapping efficiency to determine the buffer area ratio to apply in each contributing area. To the extent that the DEM and flow algorithm indicate runoff would be nonuniform, the resulting design will vary in size along the field margin but provide a nearly constant level of pollutant trapping efficiency at all segments of the margin.

AgBufferBuilder also contains an assessment procedure that enables the user to estimate the trapping efficiency of alternative designs by applying the relationships in Fig. 1, but in reverse. On the aerial orthophotograph, the user draws the filter polygons, either by outlining existing filters or by drawing hypothetical designs. With the DEM and site information, the AgBufferBuilder program then computes the buffer area ratio associated with each contributing area, and calculates the associated trapping efficiency. Finally, AgBufferBuilder calculates the contributing area-weighted average trapping efficiency of all segments along the field margin to derive an estimate of performance on a whole-field basis.

Examples of Design and Assessment

In the following example, a digital aerial orthophotograph of a 15.2-ha (37.6-acre) field was obtained from the USDA Natural Resources Conservation Service (NRCS) Geospatial Data Gateway website (https://gdg.sc.egov.usda.gov/). A DEM having approximately 10-m grid spacing was obtained for that field from the U.S. Geological Survey (USGS) National Map (https://nationalmap.gov/) and was resampled to a 5-m grid.

Design results from AgBufferBuilder for this crop field are shown in Fig. 2A. The sinuous contours and variable slope revealed by the DEM suggest that runoff load does not distribute uniformly to the margin around this field. Consequently, the AgBufferBuilder-designed filter (in red) has a highly variable configuration. Despite the variable configuration, this design is expected to provide a nearly constant level of sediment trapping efficiency along the entire field margin. In this scenario, the designed filter covers 0.80 ha (2.0 acres) and is estimated to provide a sediment trapping efficiency of 64% under the design storm condition at all segments of field margin through which field runoff drains and for the field as a whole.

Assessment of a hypothetical filter using AgBufferBuilder is shown in Fig. 2B. In this scenario, a more practical alternative to the AgBufferBuilder-designed filter was drawn along the field margin (in yellow). This design is more continuous and rectangular in shape making it easier to install and farm around, but it generally preserves larger size where the design procedure indicates that greater sediment load would be trapped. In this scenario, the filter covers 0.83 ha (2.1 acres) and is estimated to provide a sediment trapping efficiency of 44%. The subtle change in design from Fig. 2A to 2B involved shifting some filter area from locations (grid cells) that would intercept more sediment to other locations that would intercept less sediment, resulting in lower overall sediment trapping efficiency for the field. This example illustrates the potential sensitivity of filter performance to a change in configuration where runoff is nonuniform.

Uses and Limitations of AgBufferBuilder

AgBufferBuilder is intended to be used in field planning situations. The field professional would run the design procedure to provide the farmer a reference design and the opportunity to propose alternative, more adoptable designs and to compare estimates of size and performance.

Several comparisons of estimated performance between an AgBufferBuilder design and a constant-width filter of the same overall size (hectares), like the example in Fig. 2, consistently indicate better performance from the variable-width AgBufferBuilder design (Dosskey et al., 2015). These results suggest that substantial enhancement of filter performance could be gained from utilizing precision design methods where field runoff is nonuniform. By installing more filter area where runoff load is greater and less filter where the load is smaller, a higher level of performance can be achieved without the additional expense of adding more width to a constant-width filter and/or installing additional practices that redistribute concentrated flow uniformly over a constant-width filter.

Reliability of modeled results using AgBufferBuilder derive from tests of VFSMOD-W (Muñoz-Carpena et al., 1999, 2007; Abu-Zreig et al., 2001; Sadeghi et al., 2004) and comparison of AgBufferBuilder results to field measurements (Dosskey et al., 2015). However, accuracy of the DEM database can also affect reliability of the results. For example, land shaping and drainage modifications that are more recent than when the source elevation data was collected can alter runoff patterns from what would be predicted by the DEM and flow algorithm. If those alterations are substantial, AgBufferBuilder results will not accurately represent performance in the field. Therefore, design maps and assessments created with AgBufferBuilder should be used only after some form of field inspection.

Landscape-Scale Placement

Several different physically-based empirical models have been advanced that employ DEMs for identifying where in landscapes and watersheds runoff flow is more likely to concentrate and offer relatively better conditions for reducing pollutant loading with vegetative filters. These models are used for prioritizing

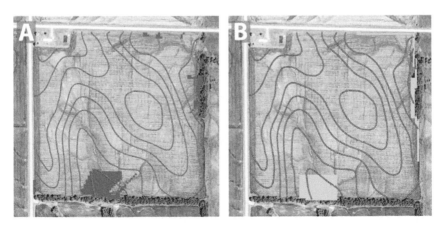

Fig. 2. Digital aerial orthophotograph of a 15.2 ha⁻¹ (37.6 acres) agricultural field showing 1-m contours and (A) AgBufferBuilder-designed vegetative filter (in red), and (B) an alternative configuration that is more continuous and rectangular (in yellow). These designs were produced using a 10-m digital elevation model resampled to a 5-m grid.

locations for filter installation within landscapes and watersheds. They are not restricted to specific landscape positions such as field margins or riparian zones and, instead, assess all portions of a landscape.

Landscape placement models are simple equations that enumerate relative potential for filter effectiveness. An index value is calculated for every DEM grid cell in the area of interest (AOI) so comparisons can be made among sites across multiple fields and throughout watersheds. To keep them simple, they generally have few variables, but information derived from a DEM is central to each one. The DEM and a flow direction algorithm are used in a GIS to indicate where filters would intercept relatively greater runoff flow. Additional geospatial data, including soils and land cover, may be used to refine these results for pollutant mitigation potential and for particular pollutants and/or transport mechanisms. Index values gauge the relative capability of a vegetative filter placed in a given location to mitigate pollutant load; a higher value is better. Because these indexes employ the same fine-scale DEMs as the design tool, these models may also be used to distinguish relatively better locations for filters within individual fields.

Index Models

Four index models are: Wetness Index, Topographic Index, Water Inflow Index, and Sediment Retention Index.

· Wetness Index (WNI; Beven and Kirkby, 1979; Moore et al., 1991) is calculated for each grid cell in the AOI by

$$WNI = \ln\left(A_s \div \tan\beta\right) \quad [1]$$

where A_s is the contributing area draining to the grid cell per unit length of a side of the grid cell (m^2 m^{-1}) and β is the slope angle of the cell (in degrees). Larger values of WNI identify where runoff flow converges from larger source areas to flatter slope locations. At these locations relatively more runoff is likely to accumulate and infiltrate along with its dissolved pollutants, deposit its sediment more easily, and raise the water table closer toward interaction with the rooting zone (Tomer et al., 2003; Burkart et al., 2004).

· Topographic Index (TI; Walter et al., 2002; Lyon et al., 2004) is calculated for each grid cell by

$$TI = \ln\left[A_s \div (\tan\beta \times K_{sat} \times D)\right] \quad [2]$$

where A_s and β are the same as for the Wetness Index, and K_{sat} is the saturated hydraulic conductivity of the soil profile (m d^{-1}) having depth D (m) above a layer that restricts percolation such as a bedrock or dense soil layer. Where there are multiple soil layers above a restrictive layer, K_{sat} is the thickness-weighted mean conductivity of all soil layers above the restrictive layer (Qiu, 2009). The TI is a modification of the WNI that accounts for depth and permeability of the soil in the grid cell for more accurately gauging probability of the water table rising above the soil surface and generating erosive saturation-excess overland flow. It has been field validated in areas having shallow, permeable soils where runoff from source areas is transported toward streams primarily through subsurface

lateral flow (Schneiderman et al., 2007; Easton et al., 2008). The TI has been inter-
preted to assess the relative risk for each cell to become a source of pollutants to
overland runoff, rather than as a filter, and would be used to identify precise locations where high-loading land uses should be avoided (Walter et al., 2009). While
development of this index has focused on identifying pollutant source areas, it
may also be useful for identifying locations where subsurface lateral flow is more
likely to flow through root zones where dissolved pollutants could be immobi-
lized and transformed.

The following two index models, Water Inflow Index and Sediment Retention Index (Dosskey et al., 2011b) are companion indexes that gauge two different
aspects of runoff to, and pollutant retention by, vegetative filters under the same
conditions. They were derived by simulation modeling using the Vegetative Filter Strip Modeling System (VFSMOD-W version 5; Muñoz-Carpena and Parsons,
2003). The VFSMOD-W is a process-based model which describes Hortonian, or
infiltration-excess, overland runoff of water and sediment from cultivated source
areas to and through vegetative filters. These indexes are quantitative rather than
relative, having been scaled to describe freshly-tilled landscapes generating runoff to a well-established 12 m-wide grass filter during a design storm event having
a 2-yr, 24-h return frequency.

· Water Inflow Index (WII) is calculated for each grid cell by

$$\text{WII} = 0.81 \times \left(A \times R \div K_{sat}^{0.5} \right)^{0.8076} \qquad [3]$$

where A is the contributing area (acres), R is the rainfall-runoff erosivity
factor (in·ft·tf·in·[ac·hr·yr]$^{-1}$) of the Revised Universal Soil Loss Equation (RUSLE;
Renard et al., 1997), and K_{sat} is the saturated hydraulic conductivity of the surface
soil layer (in h^{-1}). The WII combines size of source area with infiltration properties
of its soil to gauge the volume of overland runoff (m^3 of water per m of land contour) that would flow into a given grid cell during a design storm event.

· Sediment Retention Index (SRI) is calculated for each grid cell by

$$\text{SRI} = 18.6 \times \left[A \times R \times K \times (LS) \div D_{50} \right]^{0.4333} \qquad [4]$$

where

$$[LS] = \left(A_s \div 22.13 \right)^{0.4} \times \left(\sin \beta \div 0.0896 \right)^{1.3} \qquad [5]$$

A and R are the same as for the Water Inflow Index, A_s and β are the same
as for the Wetness and Topographic Indexes, K is the soil erodibility factor [t (ac
EI)$^{-1}$] of RUSLE, and D_{50} is the median particle diameter (mm) of the surface soil
layer which is assigned to its soil texture class according to Rawls and Brakensiek
(1983). The SRI gauges the amount of sediment (kg retained per m of land contour) that would be deposited in the vegetative filter.

Data Inputs and Operations

For all of these landscape placement indexes, the DEM provides a spatial grid
for calculating index values throughout an area of interest. Contributing area (A)
and specific contributing area (A_s) to each grid cell can be determined from the

DEM by using the ArcGIS program's built-in Flow Accumulation function which employs the D8 algorithm of Jenson and Domingue (1988). Potentially more precise results can be obtained using the ArcGIS plugin program TauDEM (Tarboton, 1997, 2013) that employs the D∞ algorithm. The slope of grid cells can be determined from the DEM using the ArcGIS Slope function. Among other sources, DEMs can be obtained from the USGS National Map (https://nationalmap.gov/).

These indexes, except for Wetness Index, also require soils data. The SSURGO soils database can provide soil attributes K, K_{sat}, D, and texture class for determining D_{50} of the soil map unit associated with each grid cell. The SSURGO database can be obtained from the USDA-NRCS Web Soil Survey accessible through the USDA-NRCS Geospatial Data Gateway (https://gdg.sc.egov.usda.gov/).

If the models are to be applied only to cultivated lands, as would be appropriate for WII and SRI, additional land cover data would be required to identify these portions of the overall area of interest. Land cover data can be obtained from the Multi-Resolution Land Characteristics Consortium website (https://www.mrlc.gov/). Values for R can be estimated from physical maps provided in Renard et al. (1997).

Example Results of Landscape Placement Indexes

Examples of results produced by the Water Inflow, Sediment Retention, Wetness, and Topographic indexes are shown in Fig. 3. These results were generated for a 67 km^2 agricultural watershed in Clinton County, MO, of which only a 2.6 km^2 (1 mi^2) area is shown. Projected in red onto an aerial photo of this area are the grid cells having the largest index values by each index, that is, the top 10% of all grid cells in the larger watershed that are on land classified as agricultural, that is, not classified as forest or water in the land cover database, so that large streams and forested riparian areas are not included.

For all of the indexes, spatial patterns of high-valued cells in Fig. 3 generally look like an extension of the stream network into agricultural land. The WII provides the best representation of the four for where runoff flow concentrates into swales, gullies, and stream channels. This general pattern in all four of the indexes reflects the dominance of the contributing area variable in these indexes (Dosskey et al., 2013). The highest index values in an area of interest generally indicate location of the largest stream courses because they have the largest contributing area. Values become smaller as one moves to grid cells upstream in the channel and decline abruptly as major tributaries are passed, and the pattern progresses up ever-smaller stream channels, ephemeral waterways, gullies, and swales and eventually to the tops of ridges that have no contributing area. Departures from this general pattern will reflect variation in soil and/or slope, depending on the particular index.

Uses and limitations of landscape placement indexes

These indexes can be used to help establish priorities for installing vegetative filters within large planning areas (Qiu, 2003; 2009; Qiu et al., 2009). Priority would be assigned to locations having higher index values, which typically point to swales where runoff converges. While farming convenience typically dictates placing filters along field margins and riparian zones, these indexes may identify more effective portions of field margins and riparian zones, and even locations within fields, where installation would likely yield greater water quality impact and cost-effectiveness.

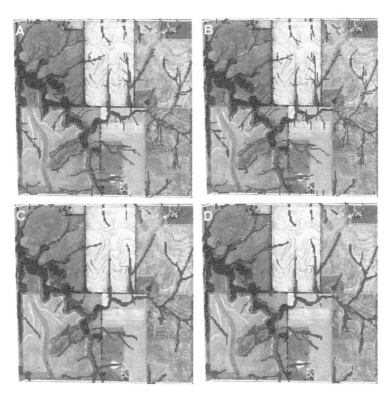

Fig. 3. Aerial view of a 2.6 km² portion of an agricultural watershed showing locations identified in the top 10% of index values in that watershed (in red) using: (A) Water Inflow Index, (B) Sediment Retention Index, (C) Wetness Index, and (D) Topographic Index. A 10-m digital elevation model and grid was used in these analyses. (Reproduced from Dosskey et al. [2013] by permission of John Wiley & Sons)

Agricultural grid cells having the largest values; however, they may not be appropriate locations for installing vegetative filters because they may indicate where runoff too frequently would concentrate to a degree that would overtop and/or erode a vegetative filter and render it ineffective. Dosskey et al. (2011b) used an empirical method to estimate maximum threshold values of WII and SRI by assigning to them values associated with the upper end of grassed waterways. As index values become larger than this, the highly-engineered grassed waterways are needed for conveying concentrated runoff flows without causing gully erosion. More work is needed to refine this aspect of using landscape placement indexes for prioritizing locations for vegetative filters.

Choosing a Landscape Placement Index

Although these indexes may produce generally similar results, there can be important differences that make one index preferable over another (Dosskey et al., 2013). In choosing the best index, one must consider both accuracy and ease of use. Index models are designed to integrate numerous complex biophysical processes and relationships into simple empirical models for the purpose of making them easy for planners to compute and apply. Utility of these approximations, however, is gained at the expense of some accuracy. The following considerations may help in determining how to make this tradeoff.

Greater accuracy can be gained by choosing an index that is tailored more closely to the specific pollutant type of concern and its primary transport and mitigation mechanisms. For example, the SRI is tailored for depositing sediment from infiltration-excess (Hortonian) overland runoff from cultivated fields, while TI is tailored to prevent mobilization of sediments and nutrients from the soil surface of sites prone to generating saturation-excess overland runoff. Furthermore, SRI is the only index that directly gauges level of pollutant mitigation. The others gauge only the level of water movement, and a level of pollutant mitigation must be inferred. To make an informed choice, some type of prior assessment must be made to determine the nature of the pollution problem in a given planning area.

Greater accuracy may also be gained by using a scaled index, that is, index values that are directly proportional to impact. For example, values for the WRI and SRI are given directly in quantity of water or sediment retained by the filter. A scaled index makes it easier to judge how much better one site would be than another than if an unscaled index, such as WNI, is used.

Planners may find results easier to generate and interpret using an index that requires fewer GIS data layers and input parameters. The WNI would be easiest of the five indexes to apply because it employs only a DEM. The other indexes additionally require spatially-explicit soil information that must be obtained from the SSURGO database and TI requires the most parameters from SSURGO. Since the SRI is applicable only to cultivated cropland, a land cover database may also be needed. In fact, the application of any of these indexes may be assisted by land cover data, especially in mixed land-use planning areas.

Limitations of Terrain-based Models

The use of DEMs enables identifying runoff patterns at a fine scale of resolution necessary for treating agricultural runoff effectively with vegetative filters. The terrain-based models described here emphasize the importance of intercepting runoff from larger source areas to the mitigation potential of vegetative buffers.

Reliability of terrain-based tools, however, depends strongly on the accuracy and resolution of the DEM. In agricultural landscapes, terraces, ditch systems, subsurface drain tiles, tillage furrows, and road berms can be extensive and substantially influence spatial patterns of runoff flow (Souchere et al., 1998; Frankenberger et al., 1999; Dosskey et al., 2003; Hösl et al., 2012; Buchanan et al., 2012). Recent, high-resolution imaging data may be required to detect these features and accurately assess flow directions, particularly on relatively flatter terrain (Dosskey et al., 2005). Most publicly-available DEMs have horizontal resolution of 10 m or 30 m and vertical accuracy of about 1.6 m (Gesch, 2007). Finer horizontal grid scales can be produced by resampling, but the accuracy of the topographic information remains unchanged. Many areas are being updated with high-resolution LiDAR (light detection and ranging) or recent photogrammetric data that produce DEMs having much better accuracy (Xhardé et al., 2006). There are also techniques for manually modifying DEMs to account for observed drainage modifications (e.g., Duke et al., 2003; Gironás et al., 2010). Regardless, the reliability of results produced by DEM-based tools should be assessed by some form of field validation.

Conclusion

Several physically-based empirical models capitalize on DEM and GIS technologies to identify where conditions are more favorable for mitigating pollutant runoff with vegetative filters. Despite some limitations, these placement and design models capture a concept that is largely absent from current conventional placement (riparian and field margin) and design (constant width) strategies: to place relatively more filter where more load can be intercepted. Employing this concept through the use of terrain-based tools promises to greatly enhance the water quality benefits derived from vegetative filters.

References

Abu-Zreig, M., R.P. Rudra, and H.R. Whiteclay. 2001. Validation of a vegetative filter strip model (VFSMOD). Hydrol. Processes 15:729–742. doi:10.1002/hyp.101

Agnew, L., S. Lyon, P. Gérard-Marchant, V.B. Collins, A.J. Lembo, T.S. Steenhuis, and M.T. Walter. 2006. Identifying hydrologically sensitive areas: Bridging the gap between science and application. J. Environ. Manage. 78:63–76. doi:10.1016/j.jenvman.2005.04.021

Baker, J.L., M.J. Helmers, and J.M. Laflen. 2006. Water management practices: Rain-fed cropland. In: M. Schnepf and C. Cox, editors, Environmental benefits of conservation on cropland: Status of our knowledge. Soil and Water Conservation Society, Ankeny, IA. p. 89–130.

Bereswill, R., B. Golla, M. Streloke, and R. Schulz. 2012. Entry and toxicity of organic pesticides and copper in vineyard streams: Erosion rills jeopardize the efficiency of riparian buffer strips. Agric. Ecosyst. Environ. 146:81–92. doi:10.1016/j.agee.2011.10.010

Berry, J.R., J.A. Delgado, R. Khosla, and F.J. Pierce. 2003. Precision conservation for environmental sustainability. J. Soil Water Conserv. 58:332–339.

Berry, J.R., J.A. Delgado, F.J. Pierce, and R. Khosla. 2005. Applying spatial analysis for precision conservation across the landscape. J. Soil Water Conserv. 60:363–370.

Beven, K.J., and M.J. Kirkby. 1979. A physically based, variable contributing area model of basin hydrology. Hydrol. Sci. Bull. 24:43–69. doi:10.1080/02626667909491834

Blanco-Canqui, H., C.J. Ganzer, and S.H. Anderson. 2006. Performance of grass barriers and filter strips under interrill and concentrated flow. J. Environ. Qual. 35:1969–1974. doi:10.2134/jeq2006.0073

Buchanan, B.P., K. Falbo, R.L. Schneider, Z.M. Easton, and M.T. Walter. 2012. Hydrological impact of roadside ditches in an agricultural watershed in central New York: Implications for non-point source pollutant transport. Hydrol. Processes 27:2422–2437. doi:10.1002/hyp.9305

Burkart, M.R., D.E. James, and M.D. Tomer. 2004. Hydrologic and terrain variables to aid strategic location of riparian buffers. J. Soil Water Conserv. 59:216–223.

Delgado, J.A., and J.R. Berry. 2008. Advances in precision conservation. Adv. Agron. 98:1–44. doi:10.1016/S0065-2113(08)00201-0

Dillaha, T.A., R.B. Reneau, S. Mostaghimi, and D. Lee. 1989. Vegetative filter strips for agricultural nonpoint source pollution control. Trans. ASAE 32:0513–0519. doi:10.13031/2013.31033

Dillaha, T.A., J.H. Sherrard, and D. Lee. 1986. Long-term effectiveness and maintenance of vegetative filter strips. Bulletin 153. Virginia Water Resources Research Center, Virginia Polytech. Inst. and State Univ., Blacksburg, VA.

Dosskey, M.G. 2001. Toward quantifying water pollution abatement in response to installing buffers on crop land. Environ. Manage. 28:577–598. doi:10.1007/s002670010245

Dosskey, M.G., D.E. Eisenhauer, and M.J. Helmers. 2005. Establishing conservation buffers using precision information. J. Soil Water Conserv. 60:349–354.

Dosskey, M.G., M.J. Helmers, and D.E. Eisenhauer. 2011a. A design aid for sizing filter strips using buffer area ratio. J. Soil Water Conserv. 66:29–39. doi:10.2489/jswc.66.1.29

Dosskey, M.G., M.J. Helmers, D.E. Eisenhauer, T.G. Franti, and K.D. Hoagland. 2002. Assessment of concentrated flow through riparian buffers. J. Soil Water Conserv. 57:336–343.

Dosskey, M., M. Helmers, D. Eisenhauer, T. Franti, and K. Hoagland. 2003. Hydrologic routing of farm runoff and implications for riparian buffers. In: J.D. Williams and D. Kolpin, editors, Agricultural hydrology and water quality. TPS-03-1. American Water Resources Association, Middleburg, VA. p. 1–4.

Dosskey, M.G., S. Neelakantan, T.G. Mueller, T. Kellerman, M.J. Helmers, and E. Rienzi. 2015. AgBufferBuilder: A geographic information system (GIS) tool for precision design and performance assessment of filter strips. J. Soil Water Conserv. 70:209–217. doi:10.2489/jswc.70.4.209

Dosskey, M.G., Z. Qiu, M.J. Helmers, and D.E. Eisenhauer. 2011b. Improved indexes for targeting placement of buffers of Hortonian runoff. J. Soil Water Conserv. 66:362–372. doi:10.2489/jswc.66.6.362

Dosskey, M.G., Z. Qiu, and Y. Kang. 2013. Comparison of DEM-based indexes for targeting the placement of vegetative buffers in agricultural watersheds. J. Am. Water Resour. Assoc. 49:1270–1283. doi:10.1111/jawr.12083

Duke, G.D., S.W. Kienzie, D.L. Johnson, and J.M. Byrne. 2003. Improving overland flow routing by incorporating ancillary road data into digital elevation models. J. Spatial Hydrology.

Easton, Z.M., M.T. Walter, and T.S. Steenhuis. 2008. Combined monitoring and modeling indicates the most effective agricultural best management practices. J. Environ. Qual. 37:1798–1809. doi:10.2134/jeq2007.0522

Fabis, J., M. Bach, and H.G. Frede. 1993. Vegetative filter strips in hilly areas of Germany. In: J.K. Mitchell, editor, Proceedings of the International Symposium on Integrated Resource Management and Landscape Modification for Environmental Protection, Chicago, IL. 13 December 1993. American Society of Agricultural Engineers, St. Joseph, MI. p. 81–88.

Franklin, E.C., J.D. Gregory, and M.D. Smolen. 1992. Enhancement of the effectiveness of forested filter zones by dispersion of agricultural runoff. Report No. 270. Water Resources Research Institute of the University of North Carolina System, Raleigh, NC.

Frankenberger, J.R., E.S. Brooks, M.T. Walter, M.F. Walter, and T.S. Steenhuis. 1999. A GIS-based variable source area hydrology model. Hydrol. Processes 13:805–822. doi:10.1002/(SICI)1099-1085(19990430)13:6<805::AID-HYP754>3.0.CO;2-M

Gesch, D.B. 2007. The National Elevation Dataset. In: D. Maune, editor, Digital Elevation Model technologies and applications: The DEM user's manual. American Society for Photogrammetry and Remote Sensing, Bethesda, MD. p. 61–82.

Gironás, J., J.D. Niemann, L.A. Roesner, F. Rodriguez, and H. Andrieu. 2010. Evaluation of methods for representing urban terrain in storm-water modeling. J. Hydrol. Eng. 15:1–14. doi:10.1061/(ASCE)HE.1943-5584.0000142

Hancock, G., S.E. Hamilton, M. Stone, J. Kaste, and J. Lovette. 2015. A geospatial methodology to identify locations of concentrated runoff from agricultural fields. J. Am. Water Resour. Assoc. 51:1613–1625. doi:10.1111/1752-1688.12345

Helmers, M.J., T.M. Isenhart, M.G. Dosskey, S.M. Dabney, and J.S. Strock. 2008. Buffers and vegetative filter strips. In: Upper Mississippi River Sub-basin Hypoxia Nutrient Committee, editor, Final report: Gulf hypoxia and local water quality concerns workshop, Ames, IA. 26-28 September 2005. American Society of Agricultural and Biological Engineers, St. Joseph, MI. p. 43–58.

Hösl, R., P. Strauss, and T. Glade. 2012. Man-made linear flow paths at catchment scale: Identification, factors and consequences for the efficiency of vegetated filter strips. Landsc. Urban Plan. 104:245–252. doi:10.1016/j.landurbplan.2011.10.017

Jenson, S.K., and J.O. Domingue. 1988. Extracting topographic structure from digital elevation data for geographic information system analysis. Photogramm. Eng. Remote Sens. 54:1593–1600.

Ludwig, B., J. Boiffin, J. Chadœuf, and A.V. Auzet. 1995. Hydrologic structure and erosion damage caused by concentrated flow in cultivated fields. Catena 25:227–252.
doi:10.1016/0341-8162(95)00012-H

Lyon, S.W., P. Gérard-Marchant, M.T. Walter, and T.S. Steenhuis. 2004. Using a topographic
index to distribute variable source area runoff predicted with the SCS-curve number
equation. Hydrol. Processes 18:2757–2771. doi:10.1002/hyp.1494

Mayer, P.M., S.K. Reynolds, M.D. McCutchen, and T.J. Canfield. 2007. Meta-analysis of
nitrogen removal in riparian buffers. J. Environ. Qual. 36:1172–1180. doi:10.2134/
jeq2006.0462

Misra, A.K., J.L. Baker, S.K. Mickelson, and H. Shang. 1996. Contributing area and concentration effects on herbicide removal by vegetative buffer strip. Trans. ASAE
39:2105–2111. doi:10.13031/2013.27713

Moore, I.D., R.B. Grayson, and A.R. Ladson. 1991. Digital terrain modelling: A review of
hydrological, geomorphological, and biological applications. Hydrol. Processes 5:3–
30. doi:10.1002/hyp.3360050103

Muñoz-Carpena, R., and J.E. Parsons. 2003. VFSMOD-W: Vegetative Filter Strip Modeling System v.2.x. Univ. of Florida, Gainesville, FL. http://abe.ufl.edu/carpena/vfsmod/
index.shtml (accessed 3 April 2017).

Muñoz-Carpena, R., and J.E. Parsons. 2014. VFSMOD: Vegetative Filter Strips Mod-
eling System Model Documentation and User's Manual, version 6.x, Univ. of
Florida, Gainesville, FL. http://abe.ufl.edu/carpena/files/pdf/software/vfsmod/VFS-
MOD_UsersManual_v6.pdf (accessed 3 April 2017).

Muñoz-Carpena, R., J.E. Parsons, and J.W. Gilliam. 1999. Modeling hydrology and
sediment transport in vegetative filter strips. J. Hydrol. 214:111–129. doi:10.1016/
S0022-1694(98)00272-8

Muñoz-Carpena, R., Z. Zajac, and Y.M. Kuo. 2007. Global sensitivity and uncertainty
analysis of the water quality model VFSMOD-W. Trans. ASABE 50:1719–1732.
doi:10.13031/2013.23967

Pankau, R.C., J.E. Schoonover, K.W.J. Williard, and P.J. Edwards. 2012. Concentrated
flow paths in riparian buffer zones of southern Illinois. Agrofor. Syst. 84:191–205.
doi:10.1007/s10457-011-9457-5

Qiu, Z. 2003. A VSA-based strategy for placing conservation buffers in agricultural watersheds. Environ. Manage. 32:299–311. doi:10.1007/s00267-003-2910-0

Qiu, Z. 2009. Assessing critical source area in watersheds for conservation buffer planning
and riparian restoration. Environ. Manage. 44:968–980. doi:10.1007/s00267-009-9380-y

Qiu, Z., C. Hall, and K. Hale. 2009. Evaluation of cost-effectiveness of conservation buffer
placement strategies in a river basin. J. Soil Water Conserv. 64:293–302. doi:10.2489/
jswc.64.5.293

Rawls, W.J., and D.L. Brakensiek. 1983. A procedure to predict Green-Ampt infiltration
parameters. In: American Society of Agricultural Engineers, editor, Advances in infiltration: Proceedings of the National Conference on Advances in Infiltration. Publ. No.
11-83. Chicago, IL. 12–13 December 1983: ASAE, St. Joseph, MI. p. 102–112.

Renard, K.G., G.R. Foster, G.A. Weesies, D.K. McCool, and D.C. Yoder. 1997. Predicting soil
erosion by water: A guide to conservation planning with the Revised Universal Soil
Loss Equation (RUSLE). USDA Agric. Handb. 703. USDA, Washington, DC.

Sadeghi, A.M., R. Muñoz-Carpena, and C. Graff. 2004. Optimization of buffer-grass criteria as
part of national P-index evaluation using VFSMOD-W model. In: Abstracts of the 2004 ASA-
CSSA-SSSA International Annual Meetings, Seattle, WA. 31 Oct.–4 Nov., 2004. American
Society of Agronomy, Madison, WI.

Schneiderman, E.M., T.S. Steenhuis, D.J. Thongs, Z.M. Easton, M.S. Zion, A.L. Neal, G.F.
Mendoza, and M.T. Walter. 2007. Incorporating variable source area hydrology into a
curve-number-based watershed model. Hydrol. Processes 21:3420–3430. doi:10.1002/
hyp.6556

Souchere, V., D. King, J. Daroussin, F. Papy, and A. Capillon. 1998. Effects of tillage on
runoff directions: Consequences on runoff contributing area within agricultural
catchments. J. Hydrol. 206:256–267. doi:10.1016/S0022-1694(98)00103-6

Tarboton, D.G. 1997. A new method for the determination of flow directions and upslope areas in grid digital elevation models. Water Resour. Res. 33:309–319.

Tarboton, D.G. 2013. Terrain analysis using digital elevation models (TauDEM). Utah State Univ., Logan, UT. http://hydrology.usu.edu/taudem/taudem5/index.html (Accessed 3 April 2017).

Tomer, M.D., D.E. James, and T.M. Isenhart. 2003. Optimizing the placement of riparian practices in a watershed using terrain analysis. J. Soil Water Conserv. 58:198–206.

Walter, M.T., J.A. Archibald, B. Buchanan, H. Dahlke, Z.M. Easton, R.D. Marjerison, A.N. Sharma, and S.B. Shaw. 2009. New paradigm for sizing riparian buffers to reduce risks of polluted storm water: Practical synthesis. J. Irrig. Drain. Eng. 135:200–209. doi:10.1061/(ASCE)0733-9437(2009)135:2(200)

Walter, M.T., T.S. Steenhuis, V.K. Mehta, D. Thongs, M. Zion, and E. Schneiderman. 2002. A refined conceptualization of TOPMODEL for shallow-subsurface flows. Hydrol. Processes 16:2041–2046. doi:10.1002/hyp.5030

Xhardé, R., B.F. Long, and D.L. Forbes. 2006. Accuracy and limitations of airborne LiDAR surveys in coastal environments. In: Proceedings of the 2006 IEEE International Geoscience and Remote Sensing Symposium, Denver, CO. 31 July–4 August 2006. The Institute of Electrical and Electronics Engineers, Inc., New York. p. 2412–2415. doi:10.1109/IGARSS.2006.625

Identifying and Characterizing Ravines with GIS Terrain Attributes for Precision Conservation

David J. Mulla* and Shannon Belmont

Abstract

Ravines are a source of sediment loading into surface waters of the Minnesota River Basin (MRB). Ravines formed as the natural product of a landscape adjusting to disequilibrium in the main channel of the MRB caused by a massive glacial flood 11,500 yr ago that lowered the base level of the river channel. Precision conservation techniques are needed to locate and correct ravines that contribute large amounts of sediment. This study uses a geographic information system (GIS) to identify the location of all ravines in the MRB and quantify their spatial distribution, area extent, and connectivity to the mainstem MRB. An analysis of test ravines with 3-m light detection and ranging (LiDAR) digital elevation model (DEM) was conducted to quantify uncertainty in ravine aerial extent estimates. Ravines could be located with an accuracy of 90% using a GIS algorithm involving slope steepness, flow accumulation, and standard deviation of aspect. Calculations from the GIS algorithm to delineate ravines show that ravines compose a total of 197,830,000 m^2 (0.45%) of the basin landscape. Watersheds and agroecoregions along the main channel of the MRB had a greater incidence of ravines than in other locations due to their proximity to the lower base level of the main channel. Statistical and GIS-based analyses of ravine morphometrics showed that the elevation change from ravine to the main channel of the MRB was strongly correlated with ravine volume ($r = 0.64$) and relief ($r = 0.8$); both are characteristics that lead to greater sediment loading from ravines. Thus, ravines located near the main channel tended to be larger and steeper than ravines located farther from the channel. With the techniques developed in this study, conservationists can identify, for the first time, all ravines in the MRB and quantify features that are strongly related to sediment loading, such as volume, area, and relief.

The Minnesota River is listed by American Rivers (1997) as one of the 20 most polluted rivers in the United States, and the Minnesota Pollution Control Agency has listed 18 reaches of the Minnesota River impaired for turbidity (MPCA, 2005) under section 303d of the Clean Water Act (USEPA, 2009). In addition, the Minnesota

Abbreviations: DEM, digital elevation model; DOQ, digital orthoquad; GIS, geographic information system; HUC, hydrologic unit code; LiDAR, light detection and ranging; MRB, Minnesota River Basin.

D.J. Mulla, Dep. of Soil, Water & Climate, 1991 Upper Buford Circle, University of Minnesota, St. Paul, MN 55108; S. Belmont, Dep. of Environment & Society, 5215 Old Main Hill, Utah State Univ., Logan, UT 84322-5215 (swb.in.ut@gmail.com). *Corresponding author (mulla003@umn.edu).

doi:10.2134/agronmonogr59.2013.0020

Precision Conservation: Geospatial Techniques for Agricultural and Natural Resources Conservation
J.A. Delgado, G.F. Sassenrath, and T. Mueller, editors. Agronomy Monograph 59.

River contributes over 80% of the sediment entering Lake Pepin (Kelley and Nater, 2000), a natural lake in the Mississippi River downstream of Minneapolis and St. Paul, which is impaired for both sediment and phosphorus. These identified impairments have launched a multifaceted effort to understand the hydrologic and geomorphic processes at work in the MRB to identify the sediment sources within the watershed with a particular focus on near-channel sources such as ravines, streambanks, and stream bluffs (Sekely et al., 2002). Mitigation of the impairment is not only required by the Federal Clean Water Act but is also important to improving habitat for fish and aquatic organisms, protecting recreational and aesthetic values, and slowing the infilling of Lake Pepin on the Mississippi River (Rada et al., 1990; Sekely et al., 2002; Sheeder and Evans, 2004; Engstrom et al., 2009).

Landscape History

The Minnesota River flows 539 km from its source, Big Stone Lake (in west-central Minnesota on the South Dakota border), southeast to Mankato, where it makes an abrupt turn and travels northeast to its confluence with the Mississippi River at Fort Snelling, just south of St. Paul, MN (Fig. 1).

The modern Minnesota River Valley is a relic channel of the Glacial River Warren. This relic river channel was carved initially 11,500 yr ago, when meltwater from the Laurentide Ice Sheet caused Glacial Lake Agassiz to overtop Big Stone Moraine, a terminal moraine near the present-day town of Browns Valley, MN. Huge volumes of water poured out during this time, carving a valley 45 m deep at the mouth and up to 70 m deep where it takes a sharp turn to the northeast in Mankato (Thorleifson, 1996). The Minnesota River Valley is up to 8 km across in places, yet the "underfit" modern Minnesota River is a mere 60 to 100 m wide.

Since settlement, sediment loading to Lake Pepin has risen sevenfold from the upper Mississippi and St Croix River watersheds (Engstrom et al., 2009). Intensive

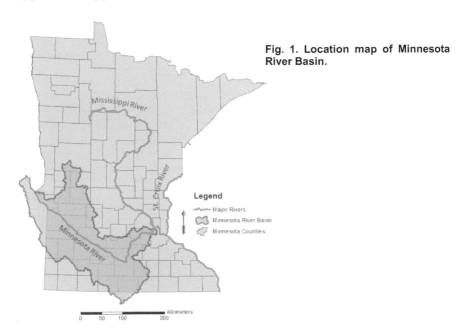

Fig. 1. Location map of Minnesota River Basin.

agricultural production combined with an increasingly wetter climate, especially in the MRB, is one cause of this increase (Kelley and Nater, 2000; Simon and Rinaldi, 2000; Knox, 2006; Mulla and Sekely, 2009). Much of the sediment loading into Lake Pepin is driven by near channel sources, such as ravines, streambanks, and stream bluffs, rather than by upland farmer fields (Engstrom et al., 2009).

History of Ravine Evolution

A ravine is defined as a small, narrow, deep depression that is smaller than a valley but larger than a gully (Bates and Jackson, 1984). Poesen et al. (2003) differentiate ravines from ephemeral gullies, which are defined as channels eroded by overland flow that "can easily be filled by normal tillage." The transitions between rill, gully, and ravine lie on a continuum and their definitions are subjective (Poesen et al., 2003). Ravines represent an advanced stage of soil erosion by water and are related to erosional processes such as concentrated overland flow, seepage erosion, and landslides (Montgomery and Dietrich, 1988; Faulkner et al., 2008).

Gully or ravine initiation is the result of nonequilibrium conditions in a landscape (Simon and Darby, 1999). It is known that gully and ravine initiation is related to a critical threshold involving slope and catchment area (Vandaele et al., 1996) and begins when velocity, shear stress, power, or some other flow characteristic exceeds a critical threshold (Sidorchuk, 2006). The main external drivers of gully or ravine formation include tectonic uplift, base-level lowering, changes in hydrology, loss of vegetation, and/or anthropogenic disturbances to the watershed (increased impervious surfaces or infiltration reductions due to soil compaction from machinery; Schumm, 1999; CCMR, 2004). Base-level lowering at the end of the last glaciation is the primary factor controlling ravine initiation in the MRB. More recent anthropogenic alterations to the hydrology, including installation of tile drain outlets in ravines, may be driving a new era of dynamic geomorphic instability, which may be impacting erosion and incision rates and the correlated increase in sedimentation in Lake Pepin.

The incised channel left in the wake of the receding Glacial River Warren left the Minnesota River tributaries stranded 40 to 70 m above the new channel bottom. It is probable that this steep gradient, known as a knick point, formed at the confluence of the tributaries with the Minnesota River and quickly began to migrate upstream, cutting down through the glacial substrate. As the post-glacial landscape evolved and patterns of precipitation fluctuated, drainage networks adapted and shifted, capturing streams in some cases and abandoning channels and ravines in others, leading to the pattern of ravines ranging in length from 500 to 2000 m (Fig. 2) that occur along the border of the Minnesota River Valley and are distributed throughout the basin.

There is a pressing need to identify the locations of ravines and to control the sediment delivered from them to the MRB. This study is the first to map all ravines in the MRB. This need is consistent with precision conservation, which is an emerging concept to identify and mitigate small areas that have a disproportionately large impact on water quality. The advent of high-resolution DEMs using LiDAR imagery has greatly enhanced the ability of GIS terrain analysis to identify small critical areas for implementation of precision conservation practices (Galzki et al., 2011; Belmont et al., 2015). Once ravine locations and their

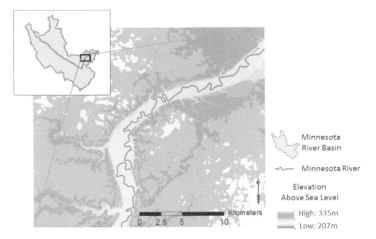

Fig. 2. Ravine drainage pattern along a section of the Minnesota River Valley between Henderson and Belle Plaine, MN. Lower elevations are shown with dark colors, higher elevations with yellow colors.

magnitudes are identified, the largest and steepest of these can be targeted for implementation of practices to mitigate their impact on water quality.

Objectives

Ravines are common features within the MRB, due to the base-level fall and the resulting disequilibrium of the landscape. To mitigate sediment loading into the Minnesota River and focus remediation strategies, the main sediment sources to the river must be understood and quantitatively constrained.

This research involves GIS-based terrain analysis of physical characteristics of ravines. To better understand the contribution of ravines to sediment losses in the MRB, this study sets out to accomplish the following objectives:

- Develop a terrain attribute-based algorithm using ArcGIS (ESRI, 1994) to identify the locations, area, frequency, and spatial distribution of ravines on the landscape of the MRB; and
- Identify morphometric characteristics of ravines that can be used to identify which ravines will most probably to deliver large amounts of sediment to the MRB.

The information from this study could be combined to identify ravines that have the greatest potential for delivering sediment to the MRB and to identify ravines that require precision conservation practices to mitigate sediment loss.

Methods

Ravine Morphometrics and Classification
Ravine Selection

The accuracy of the GIS-based ravine identification technique described below was evaluated by identifying 65 ravines for field survey. The initial selection process consisted of creating an evenly spaced grid across the entire MRB, with each grid cell representing 1/64 of a USGS quadrangle (spatial extent about 2 km²). Where possible, one ravine was selected per grid cell. Field accessibility and landowner permission limited the original ravine set to 65 used for field verification.

Field validation was done in June 2007 by a team that verified the existence of the 65 GIS-identified ravines; measured ravine width, depth, and angle of side slope; and took notes on incoming tile lines, culverts, and overall vegetative state in upland contributing areas.

Data Resolution: Error Analysis

The highest-resolution data available for the entire Minnesota River Basin at the time of this study were 30-m DEM tiles from the USGS (http://seamless.usgs.gov) and 3-m digital orthoquad (DOQ) photographs available from the Minnesota Geospatial Information Office's Land Management Information Center (http://www.mngeo.state.mn.us/chouse/wms/geo_image_server.html). High-resolution LiDAR data were also available from spring 2005 for Blue Earth County. These high-resolution data were used to quantify the uncertainty associated with using the 30-m DEM- and DOQ-derived data sets to identify ravines so that basin-wide ravine identification could be made using the coarser resolution data. Nine Blue Earth County ravines were selected at random from the larger group of 65 ravines for evaluation of 30-m DEM accuracy using high-resolution LiDAR data.

The nine ravines were manually digitized from the 3-m LiDAR on the basis of boundaries where the slope changed significantly from surrounding non-ravine cropland areas, and the area, mean slope, and total relief were calculated for each (Fig. 3; Belmont et al., 2015). Typically, ravines were wide and long (at least 30 m wide and 500 m long) and could be easily differentiated from surrounding cropland not only by slope but also by land cover, because trees grow in ravines but not on adjacent cropland. The same nine ravines were digitized a second time using only the 30-m DEM as a guide and a third time using only the boundaries of the forested areas on the DOQ as a guide. The ravine polygon areas digitized from the DOQ were calculated from the 30-m DEM. The assumption behind the notion that forested land can be used as an indicator of ravine locations is

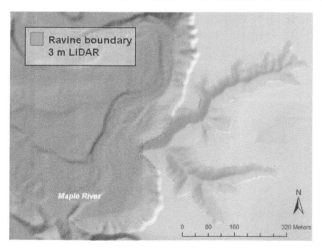

Fig. 3. Manually digitized ravine planform area from 3-m light detection and ranging (LiDAR) digital elevation model.

that ravines are too steep to safely farm using mechanized equipment. The 3-m LiDAR data was considered to be the best and most comprehensive data available and was used as the standard against which the other data sets were compared. The area, mean slope, and relief of each 30-m and DOQ ravine polygon were then compared to the LiDAR polygon area, slope, and relief, and a percentage difference was calculated for all nine ravines combined.

Digitizing Ravines and Delineating Watersheds

From the results of the data-set resolution error analysis, it was determined that the most reliable visual source for defining ravine boundaries in the MRB were DOQs based on 3-m LiDAR and supplemented by the 30-m USGS DEM as a semitransparent drape over the landscape to help define specific areas where forested bluffs obscured the direction and terminus of ravine features. All 65 study ravines were digitized with this methodology.

Pour-point outlets for each ravine were chosen manually using the DOQ overlain with a semitransparent flow-accumulation grid calculated from the 30-m DEM. The point file was required as an input for the ArcInfo watershed delineation process described by Balco (2001). That process was used to delineate the catchment area or upland contributing area for each ravine (Belmont et al., 2015). The script also required a DEM of the region and a flow direction grid, which was calculated from the 30-m DEM using the D8 algorithm utilized by ArcGIS (Barnes et al., 2013, 2014). The resulting delineated watersheds included the ravine area. The digitized ravine polygon shapefile was used to clip out the ravine portion of the delineated watershed so that attribute extraction and analysis could be performed on the ravines and their contributing areas separately.

Terrain Attribute Extraction

Ravine size is one of the most important attributes that should be quantified for the purposes of understanding ravine sediment production (Belmont et al., 2015). Larger ravines tend to have steeper gradients and greater sediment delivery than smaller ravines. There are many ways to evaluate ravine size, including planform area (2-D), actual surface area (3-D), volume, perimeter, extent of branching, longest channel length, and total channel length (Belmont et al., 2015). There are also relationships that combine attributes to quantify ravine size, such as drainage density (total channel length divided by area) and longest channel length divided by area (which estimates width).

The potential energy gradient, or severity of incision in the longest downhill direction of a ravine, is a second characteristic of high interest (Gran et al., 2009). In general, severity of incision increases when steepness of the ravine main channel increases. The incision extent may relate to the age of the ravine, the location within the MRB, the proximity to the base-level fall, the current erosional forces at work within the ravine, the local gradient surrounding the ravine, the erosional stream power coming from the upland contributing area, or simply the regolith of the ravine. Incision can be analyzed by several different factors, such as volume, slope, ratio of 2-D area to volume. It can also be estimated using manually derived ravine relief represented by elevation change from the uplands surrounding the ravine to the channel elevation nearest the mouth of the ravine. Finally, incision can also be estimated using automatically derived ravine relief (maximum minus minimum elevations from within the ravine digitized boundary).

To analyze the proximity and connectivity of the ravines to the mainstem of the Minnesota River, the distance and the elevation change from the ravine pour points to the Minnesota River were measured. Distance was measured from the ravine mouth to the Minnesota mainstem following the channel length or natural drainage path. Elevation change was measured from the elevation at the ravine mouth to the elevation at the Minnesota mainstem. Another application of this metric is to use it as a proxy for potential energy. Considering the base-level drop the Minnesota River experienced 11,500 yr ago, the tributaries and places where flow accumulated in the uplands experience a sudden base-level drop. This incision set up a potential energy gradient, or knick point, that is now migrating up through the drainage system.

Algorithm to Identify Locations of Ravines

An algorithm was developed using ArcGIS to automatically identify where ravines are located and what proportion of the landscape they cover. This algorithm is a series of steps performed in ArcGIS. While ravines are very distinctive features on a 30-m DEM, the sheer scale of the basin precluded a manual count and/or manual mapping of all ravines in the MRB. In addition, the development of an algorithm reduced the amount of subjectivity in mapping ravine features.

To predict and map the spatial distribution of ravines on the landscape in a sequential series of steps, the Environmental Systems Research Institute's ArcMap and ArcCatalog 9.3 and ArcInfo 9.0 software were used. The efforts of this present study sought to map ravine planform area for all ravines located in the MRB.

The most distinctive features of the ravines are their steep slopes and the symmetrical nature of the opposing hillslopes. The best-fit algorithm developed includes three parameters: a minimum threshold for slope, minimum standard deviation for slope aspect, and minimum and maximum threshold values for flow accumulation. The standard deviation of aspect is included to capture the symmetrical nature of the ravines and to clearly distinguish them from asymmetrical bluffs and terraces. Flow accumulation is used to further distinguish ravines from bluffs and terraces. Thresholds for slope, the standard deviation of aspect and the flow accumulation were determined in an interactive fashion with the nine Blue Earth County ravines described above, which had DEMs based on 3-m LiDAR.

Assessment of Ravine Attributes

A large obstacle to automated ravine identification was the inability to distinguish the curved, high-sloped bluffs and terraces of the sinuous Blue Earth County tributaries from ravine features in the rest of the MRB using only single terrain attributes. Three terrain attributes (aspect, slope, and flow accumulation) were assessed simultaneously to address this issue. The actual thresholds used for these terrain attributes are presented in the discussion of the results and in Table 1.

The first attribute used to discriminate between ravines and bluffs or terraces was the standard deviation of the aspect. This metric was used to determine areas where opposing hillslope faces were in close proximity to one another, as they are in ravines. Aspect measures the direction of the slope face with values representing compass directions in radians. Calculating the standard deviation of the aspect at the center of a nine-celled moving window (approximately 6100

Table 1. Best-fit attributes used in ravine identification algorithm.

Attribute	Threshold
SD aspect	>40 radians
Slope steepness	>7%
Flow accumulation	>200 cells, <7400 cells

m^2) allowed identification of areas with significant heterogeneity in the aspect across a relatively small area. The second attribute used to identify ravine cells was slope, because one of the most readily identifiable features of ravines is their side-wall slopes, which are much steeper than the average slope of the uplands or channel bottoms. The third attribute used to delineate ravines was flow accumulation, which exploits the simple fact that ravines have significantly larger catchment areas than bluffs and terraces. A range of flow-accumulation cells was chosen to represent catchment sizes associated with channel initiation. The lower end of the range eliminated many of the bluffs, terraces, and anomalous, isolated cells in the channels and uplands where small catchment areas might be in close proximity to a cell with sufficient slope. The upper end of the range eliminated larger tributary features on the landscape, which typically have perennial flow and would be subject to different hydrogeomorphic processes than the ravines. The resulting accumulation channel cells were converted to polylines and buffered to 200 m to use as a mask to clip and preserve all cells meeting the requirements for slope and standard deviation of aspect.

Data Resolution: Error Analysis

Many of the primary terrain attributes gathered for the ravines in this study were measured from the mosaic of USGS 30-m DEMs. Ravine boundaries and pour points were manually digitized using a combination of the 30-m DEM and 3-m DOQs available for the State of Minnesota. Before digitizing the ravines in the study, a simple accuracy test was designed and performed to quantify the manual digitizing error associated with coarse resolution of the 30-m DEM and the 3-m DOQ for nine Blue Earth County ravines as described above. A percentage difference was calculated for each of the nine 30-m and 3-m DOQ ravines relative to the 3-m-LiDAR ravine polygons.

Over- and Underestimation of Area

To calculate ravine area for the MRB, it was necessary to determine whether the algorithm (based on 30-m DEMs) was over- or underestimating true ravine area (relative to 3-m DEMs) and to quantify the error in the over- or underestimation. A simple area error analysis was performed on the algorithm by creating buffered sample points for 13 ravines among the six agroecoregions (Hatch et al., 2001) with the highest number of ravines: 3 randomly selected ravines in Alluvium and Outwash, 3 in Steep Stream Banks, 3 in Steep Valley Walls, 2 in Steep Wetter Moraine, 1 on the Coteau, and 1 in Steep Dryer Moraine. Fewer ravines were selected in the last two agroecoregions than in the previous two because the latter agroecoregions have fewer ravines. Agroecoregions were used as an organizing structure because they were created based on the same types of landscape features (geology, soil type, and slope steepness) that would affect ravine development (Hatch et al., 2001). Simply put, agroecoregions are homogeneous landscape units that have similar parent geology, soil types, slopes, precipitation, and land use. They are similar in scale to the USEPA aquatic ecoregions, which place heavy emphasis on presettlement vegetation rather than on soil and landscape features. Agroecoregions are also similar in scale to the USGS eight-digit hydrologic unit code (HUC) major watersheds.

To establish areas within each agroecoregion for analysis, the 13 sample points were each buffered with a 5000-m-diameter circle. All ravines distinguishable by eye on the DEMs within each circle were manually digitized if they were associated with discernible topography, had fluvial networks of considerable flow accumulation, and were not ditches. The digitized ravine areas were then compared to the ravine areas produced by the algorithm polygons within the buffer circles, and a percentage difference was calculated for each circle and then for each agroecoregion. Digitizing was done with the 30-m DEM draped over the DOQ, so the resulting areas were adjusted using the results of the data resolution analysis test.

Small Ravines

One limitation of the algorithm is that many small ravine features (3500–75,000 m^2) are systematically omitted in 30-m DEMs for not having sufficient flow accumulation. Using a lower minimum value for flow accumulation would result in many false-positive indications of small ravines in many agroecoregions. Yet, inclusion of these small features is important to ensure the best representation of ravine distribution throughout the region. A true understanding of that distribution is critical for estimating the potential for sediment production and sediment transport by ravines in the MRB. An analysis was done to quantify the number and area extent of small ravines missed by the algorithm per agroecoregion.

To quantify the error introduced by omission of small ravines, eleven 5000-m-diameter buffer rings were created to define the areas in which little ravines missed by the algorithm were manually digitized: two in Alluvium and Outwash, two in Steep Stream Banks, two in Steep Valley Walls, two in Steep Wetter Moraine, two on the Coteau, and one in Steep Dryer Moraine. From the resultant polygon file, a total count, and total area were calculated for the small ravines missed. The same data resolution area adjustment was made to the final "small ravine" areas as they were digitized from the draped DOQ. These data were added to the algorithm's previous counts and area estimates for the MRB and per agroecoregion to improve the resolution of the gathered data.

Calculations of Area

The algorithm created a raster file of identified ravine cells that was converted to polygon shapes so that area could be easily calculated. The percentage overestimated by digitizing from the digital orthoquad (using the 30-m DEM as a drape to distinguish ravine mouths hidden by vegetation along bluff or terraces) was added to the percentage difference of the algorithm determined by the over-/underestimation error analysis, and a final adjustment made to account for the area of the missed small ravines. The total ravine area was calculated from the algorithm using these adjustments for the ravine area in the MRB as a whole and for ravines in each agroecoregion.

Results and Discussion

Identifying Ravine Locations

Ravine locations have never been mapped in the Minnesota River Basin, nor has their area ever been estimated. An algorithm was developed to automatically identify ravines spatially, estimate the total ravine area in the MRB, and calculate the percentage ravine area in the MRB. Identifying ravine locations and areas is necessary for understanding the potential sediment contribution by ravines to the Minnesota River. The algorithm uses three terrain attributes (slope, aspect, and flow accumulation) calculated from the 30-m DEM. The selection of threshold values for these three terrain attributes was validated using nine ravines in Blue Earth County where DEMS based on 3-m LiDAR were available. The final algorithm for identifying ravines consists of three sequential steps that use ArcGIS commands. First, a threshold of cells with slope greater than 7% was combined with areas having a local standard deviation of the aspect greater than 40 radians. Subsequently, a 200-m buffer was applied to the cells with flow accumulation between 200 and 7400 cells, representing catchment areas between 140,000 and 5,000,000 m² (0.15–5 km², Table 1).

Applying these three attribute thresholds (the algorithm) to the 30-m DEM effectively identifies ravine locations and morphometric features in the MRB (Fig. 4). A total of 65 ravines identified using the algorithm described above were visited in eight counties within the Minnesota River Basin. On the basis of field visits, 59 of these potential ravines were verified as being true ravines, and six were grassed waterways rather than ravines. This is a 90% success rate. Most of the sites that had false-positive ravines were located in the Coteau agroecoregion in the southwestern portion of the Minnesota River Basin. Ravines in this region are typically small and are located on relatively flat topography, making it difficult to distinguish them from grassed waterways.

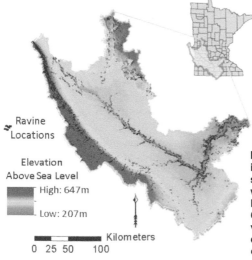

Ravine Locations

Elevation Above Sea Level

High: 647m

Low: 207m

Kilometers
0   25 50   100

Fig. 4. Map showing ravine locations in the Minnesota River Basin. The strong band of ravines to the southwest of the river and running parallel to it (left side of figure) are on the Coteau. Colors indicate relative elevation, with dark greens being lower elevations and orange or brown colors being higher elevations.

Data Resolution: Error Analysis

Nine randomly selected ravines in locations where 3-m LiDAR data were available were analyzed to better understand the errors in the ravine delineation based on the 30-m DEM data available for a majority of the MRB, and to improve the resolution of algorithm area calculations. Results of the initial error analysis show that relative to 3-m LiDAR data, the 3-m DOQ overestimated digitized areas by an average of 50%, with a standard deviation of 68%. The 30-m DEM overestimated area by an average of 134%, with a standard deviation of 115% relative to the 3-m LiDAR (Fig. 5, Table 2).

The slope and relief for the digitized ravine areas were both underestimated with the 30-m DEM (Tables 3 and 4). Slope was underestimated by 60%, with a standard deviation of 8%. Relief was underestimated by 31%, with a standard deviation of 15%.

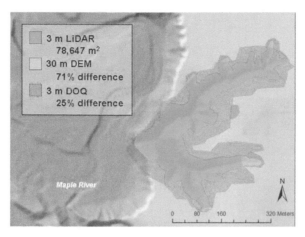

Fig. 5. Percentage-change calculations in digitized ravine areas from 3-m light detection and ranging (LiDAR) digital elevation model (DEM), 30-m DEM, and 3-m digital orthoquad (DOQ).

Table 2. Summary of ravine-area error analysis of nine Blue Earth County ravines for which LiDAR DEMs were available.†

Ravine digitized areas				
3-m LiDAR	3-m DOQ	30-m DEM	Change in DOQ	Change in DEM
m²			%	
17,034	12,929	38,069	−24	123
13,197	17,489	31,562	33	139
5,189	11,457	26,130	121	404
78,647	98,639	134,098	25	71
15,022	28,397	28,271	89	88
4,051	11,125	9,941	175	145
9,880	11,476	27,641	16	180
259,620	381,277	403,665	47	55
38,901	25,797	37,485	−34	−4
		Avg.	50	134
		SD	68	115

† LiDAR, light detection and ranging; DEM, digital elevation model; DOQ, digital orthoquad.

Table 3. Summary of ravine relief error analysis.

Ravine relief measurements†		Difference
3-m LiDAR	30-m DEM	
m		%
39	28	−29
21	11	−48
17	7	−60
42	35	−17
14	10	−27
13	8	−33
12	8	−34
45	40	−13
13	10	−21
	Avg.	−31
	SD	15

† Based on range from digitized areas; LiDAR, light and ranging detection; DEM, digital elevation model.

Table 4. Summary of ravine slope error analysis.

Ravine slope measurements†		Difference
3-m LiDAR	30-m DEM	
%		
44	23	−48
41	12	−72
22	9	−60
43	20	−54
19	6	−68
22	8	−63
22	9	−59
34	17	−48
19	9	−55
	Avg.	−59
	SD	8

† Based on mean from digitized areas; LiDAR, light detection and ranging; DEM, digital elevation model.

Over- and Underestimation by the Algorithm

An additional error analysis was performed on the algorithm to determine the amount that the algorithm over- or underestimated the total ravine area in the MRB. Nine sample points randomly distributed throughout the MRB were buffered with 5000-m-diameter circles (approximately 79 km²), and all visible ravines meeting the requirements (that they are associated with discernible topography, are fluvial networks of considerable flow accumulation, and are not ditched) were manually digitized from the DOQ with a semitransparent DEM overlay. Areas were calculated for the digitized and algorithm-derived polygons and a percentage difference was determined. The algorithm-derived area was found to be 2% larger than DOQ-derived ravine polygon areas (Table 5).

Small Ravines

The final error analysis that had been designed to improve the resolution of the algorithm was intended to quantify the area of small ravines missed by the

Table 5. Summary of algorithm over/underestimation analysis.

Ravine area†		
Algorithm polygons	DOQ digitized polygons	Difference
m²		%
2,736,020	2,675,403	−2

† From nine sample points of approximately 79 km² each.

algorithm due to low flow accumulation. Two 5000-m-diameter buffer circles were placed in each of the six major agroecoregions. Any small ravines missed by the algorithm were manually digitized, and the total area of the small ravines was compared with the total algorithm polygon area (Table 6). An adjustment factor was determined for each agroecoregion sampled. The adjustment percentages range from 3.8% on the Coteau to 12.6% for Steep Dryer Moraine. From the agroecoregion areas, an area-weighted average adjustment of 6.4% was calculated for the MRB. The final ravine area determined by the algorithm was adjusted using this area-weighted factor for small ravines.

MRB Ravine Area

The total area for the ravine algorithm polygons was calculated for each agroecoregion and major watershed in Minnesota's portion of the MRB. The numbers were adjusted down by 50% to account for the DOQ overestimation and then by another 2% to account for the overestimation by the algorithm relative to the DOQ. The small-ravine analysis numbers were applied to the adjusted agroecoregion, and a total ravine area for the MRB was summed. This number represents ravines in the Minnesota agroecoregions only, which are truncated at the Minnesota border with the Dakotas and Iowa (Tables 7 and 8). The algorithm ravine area

Table 6. Summary of small-ravine area error analysis.

Agroecoregion	Total area		Small-ravine adjustment factor
	Small ravine	Algorithm polygon	
	m²		
Alluvium and Outwash	160,870	1712,444	1.094
Steep Stream Banks	119,065	1647,653	1.072
Steep Valley Walls	134,477	2695,919	1.050
Steep Wetter Moraine	73,982	1352,278	1.055
Coteau	50,074	1318,056	1.038
Steep Drier Moraine	41,296	327,603	1.126
Area total and average	579,763	9,053,953	1.064

Table 7. Summary of total ravine area calculation for the Minnesota River Basin (MRB).

MRB total area	Ravine algorithm total area	Orthoquad adjustment (less 50%)†	Algorithm adjustment (less 2%)‡	Small-ravine adjustment (plus 6.3%)§	Area of MRB
		ha			%
4,388,047	37,990	18,995	18,615	19,783	0.45%

† Percentage difference was calculated from comparative study of nine ravines in Blue Earth County.

‡ Percentage difference was calculated from comparative study of manually digitized ravines to algorithm ravines within 13 random buffer circles throughout the MRB.

§ Area weighted from analysis of six agroecoregions with highest ravine density. Percentage difference was calculated from manually digitized ravine areas and algorithm polygons in eleven random buffer circles throughout the MRB.

was also calculated by 12-digit HUC major watersheds within the MRB. The same adjustments were made to the algorithm polygon areas for each major watershed (Table 9). The total ravine area for the MRB determined from the algorithm and three adjustments for accuracy analyses is 197,831,492 m^2, or 0.45% of the area in the MRB.

Ravines are not distributed evenly across watersheds or agroecoregions. Ravines tend to occur more frequently in agroecoregions that are in close proximity to the main channel of the Minnesota River: the Steep Valley Walls and Alluvium and Outwash agroecoregions (Table 9). Ravines also occur frequently in steeper agroecoregions that are farther from the main channel: the Steep Wetter Moraine, Coteau, and Steep Dryer Moraine agroecoregions. Each agroecoregion can cross multiple watersheds. Watersheds that include large proportions of the agroecoregions listed above, not surprisingly, have more frequent occurrences of ravines than other watersheds. These watersheds tend to be located along the main channel of the Minnesota River: the Minnesota River–Headwaters, Minnesota River–Shakopee, and Minnesota River–Mankato (Table 10). Ravines occur in these watersheds near the edge of the Minnesota River Valley where an abrupt fall in elevation occurs toward the Minnesota River channel.

Estimating Geomorphic Attributes of Ravines

In addition to identifying where ravines are located in the MRB, it is also important to develop criteria for determining which ravines are will probably generate the most sediment loading into rivers. Ravines that have larger volumes and greater relief will probably deliver more sediment than smaller ravines with flatter relief.

The small subset of 65 ravines in the MRB described earlier was studied to determine which attributes are good predictors of ravine volume. The study ravines range in area from 7,745 to 1,221,675 m^2, with mean 167,687 and median 87,251 m^2. Ravine 3-D areas range from 7,745 to 1,267,816 m^2, with a mean of 171,453 and a median of 88,558 m^2. Ravine volumes range from 35,240 to 25,629,150 m^3 with a mean of 2,803,300 and a median of 1,057,230 m^3. Ravine relief values, which are calculated automatically in ArcMap by subtracting the minimum from the maximum elevation within the ravine, range from 7 to 68 m, with a mean of 34 m and a median of 35 m.

Scatter plots (Fig. 6 and 7) show there is a very strong relationship between the proximity or elevation change of ravines relative to the Minnesota River base level and ravine area, volume, mean slope, and relief. The proximity or elevation change variable is more consistently correlated with ravine morphometry than with any other attributes. The distance-proximity attribute is based on distance from the ravine mouth to the Minnesota River following the natural channel or drainage path to the Minnesota mainstem. The elevation-change proximity attribute is based on the change in elevation from the ravine mouth to the elevation where the drainage path intersects the mainstem of the Minnesota River.

Figures 6 and 7 illustrate the relationship between either elevation change or distance proximity and ravine volume. The elevation-change attribute is better correlated with ravine size ($R = -0.57$ for area; $R = -0.64$ for volume) than the distance attribute ($R = -0.44$ for area; $R = -0.51$ for volume). This is an important observation, because ravine formation and evolution is directly connected with base-level fall, which originates in the Minnesota River channel. Larger

Table 8. Algorithm ravine polygon area calculations with error-analysis adjustments for the agroecoregions of the Minnesota River Basin (MRB).

Agroecoregion	Algorithm ravine area	Orthoquad area adjustment (less 50%)†	Algorithm overestimation (less 2%)‡	Small-ravine adjustment factor§	Total algorithm area	Agroecoregion area	Area of agroecoregion
	ha	ha			ha	ha	%
Alluvium and Outwash	3834.6	1917.3	1878.9	1.092	2051.9	261,829.5	0.78%
Steep Drier Moraine	2485.3	1242.7	1217.8	1.124	1368.3	328,146.2	0.42%
Poorly Drained Lake Sediment	0.9	0.4	0.4		0.4	1,048.7	0.04%
Dryer Clays and Silts	206.8	103.4	101.3		101.3	234,790.1	0.04%
Stream Banks	134.3	67.1	65.8		65.8	79,708.1	0.08%
Steeper Till	1554.2	777.1	761.6		761.6	552,529.5	0.14%
Rolling Moraine	1315.7	657.8	644.7		644.7	305,288.9	0.21%
Dryer Blue Earth Till	1216.3	608.2	596.0		596.0	674,248.2	0.09%
Steep Valley Walls	5502.4	2751.2	2696.2	1.049	2827.9	84,345.6	3.35%
Steep Wetter Moraine	3351.2	1675.6	1642.1	1.054	1730.1	129,235.1	1.34%
Coteau	3294.1	1647.0	1614.1	1.037	1674.2	343,118.0	0.49%
Wetter Clays and Silts	1320.3	660.1	646.9		646.9	447,145.5	0.14%
Inner Coteau	5.1	2.5	2.5		2.5	994.1	0.25%
Wetter Blue Earth Till	1335.6	667.8	654.4		654.4	374,990.1	0.17%
Steep Stream Banks	1820.4	910.2	892.0	1.071	955.1	45,614.7	2.09%

† Percentage difference was calculated from comparative study of nine ravines in Blue Earth County.
‡ Percentage difference was calculated from comparative study of manually digitized ravines to algorithm ravines within 13 random buffer circles throughout the MRB.
§ Percentage difference was calculated from comparative study of manually digitized ravines to algorithm ravines within 11 random buffer circles throughout the MRB.

Table 9. Algorithm ravine polygon area calculations with error-analysis adjustments for the major watersheds of the Minnesota River Basin (MRB).

Major watershed	Watershed area	Algorithm ravine area	Orthoquad area adjustment (less 50%)†	Algorithm overestimation (less 2%)‡	Inclusion of small-ravine adjustment		
					Small-ravine adjustment factor§	Total algorithm ravine area	Watershed area
		ha	ha			ha	%
Pomme de Terre River	226,611.2	1,242.3	621.1	608.7	1.063	647.1	0.29%
Chippewa River	537,073.4	1,861.4	930.7	912.1	1.063	969.5	0.18%
Minnesota River Headwaters	543,056.5	10,506.9	5253.4	5148.4	1.063	5472.7	1.01%
Minnesota River–Granite Falls	537,301.7	2,752.8	1376.4	1348.9	1.063	1433.8	0.27%
Lac Qui Parle River	284,138.5	2,023.4	1011.7	991.4	1.063	1053.9	0.37%
Minnesota River–Shakopee	471,413.4	8,408.2	4204.1	4120.0	1.063	4379.6	0.93%
Minnesota River–Mankato	349,004.8	5,348.1	2674.0	2620.6	1.063	2785.7	0.80%
Redwood River	182,618.2	788.1	394.0	386.1	1.063	410.5	0.22%
Cottonwood River	340,014.0	2,161.8	1080.9	1059.3	1.063	1126.0	0.33%
Le Sueur River	288,071.9	1,087.3	543.6	532.8	1.063	566.3	0.20%
Watonwan River	227,280.4	471.1	235.6	230.9	1.063	245.4	0.11%
Blue Earth River	401,463.4	1,338.6	669.3	655.9	1.063	697.2	0.17%

† Percentage difference was calculated from comparative study of nine ravines in Blue Earth County.

‡ Percentage difference was calculated from comparative study of manually digitized ravines to algorithm ravines within 13 random buffer circles throughout the MRB.

§ Area weighted from analysis of six agroecoregions with highest ravine density. Percentage difference was calculated from manually digitized ravine areas and algorithm polygons in 11 random buffer circles throughout the MRB.

Table 10. Summary of Pearson correlation coefficients for the proximity attributes.

Ravine attribute	Proximity attribute	
	Elevation change	Distance
3-D Area	−0.57	−0.44
Volume	−0.64	−0.51
Relief	−0.80	−0.65
% ravine area with slope < 15%	0.60	0.59
Slope (mean)	−0.69	−0.61
Longest channel length	−0.51	−0.39
Total channel length	−0.48	−0.33

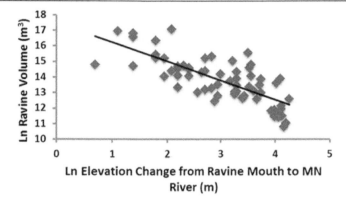

Fig. 6. Plot of elevation change from the mouth of ravine to the Minnesota River against ravine volume (R = 0.76).

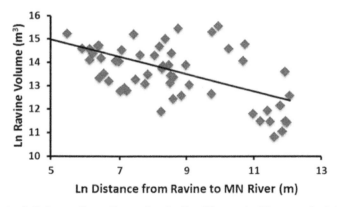

Fig. 7. Plot of distance from the ravine to the Minnesota River against the ravine volume (R = 0.47).

ravines occur in closer proximity to the Minnesota River channel. Table 10 summarizes Pearson correlation coefficients for the primary ravine morphometrics. The strength of elevation change across distance is most probably due to the variability in migration rates of the knick point up the various tributaries away from the location of the original base-level fall. The location of the tributary along the mainstem of the Minnesota River gives each tributary a different potential energy gradient to resolve (incision of the Minnesota River Valley was 45 m at near the headwaters and 70 m near Mankato). Additionally, each tributary has unique flow rates and subsurface materials to erode through, which results in varying migration rates.

Figure 8 illustrates the relationship between elevation change from the ravine mouth to the Minnesota River and the distance to the river. These variables might be expected to be far better correlated than they are ($R = 0.85$). One possible explanation for the scatter is that the knick point is propagating up tributaries at different rates. Ravines at the same distances from the Minnesota River on the Blue Earth and Le Sueur Rivers have different elevation changes. Also, the migration of the base-level fall does not immediately instigate formation of ravines because the flow accumulation on the uplands varies spatially. In some places ravine formation might be delayed, resulting in disproportionately smaller ravines than would be expected, and in other cases ravine formation might be accelerated, fitting the observed trend. A third confounding factor is drainage from the uplands. Considered in conjunction with a migrating knick point, scatter in the distance- or elevation-change plots is reasonable. Overall, the elevation-change proximity attribute is considered the most consistent and reliable metric by which ravines are morphometrically organized on the landscape.

Application of Results

Regional Sediment Production Estimates

The MRB delivers a high proportion of sediment to the Mississippi River and Lake Pepin. A sediment budget for the basin requires basic understanding of the production, routing, and transport of sediment from source areas in the basin as well as of the changes in storage that are happening in channels and elsewhere on the landscape. In addition to identifying and quantifying sediment sources, the rates of sediment movement between storage areas, the quantities and residence times of sediment in storage, and the spatial connection of sediment transport processes on the landscape must be taken into consideration (Walling, 1983; Brooks et al., 2003). Temporal changes in channel flow will also affect the generation, transport, and storage of sediment on the landscape. This study represents

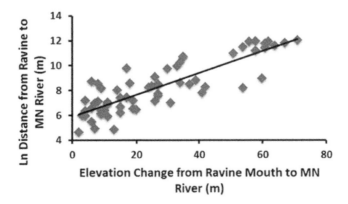

Fig. 8. Plot of elevation change from the mouth of the ravine to the Minnesota River against the ravine distance to the Minnesota River ($R = 0.85$).

one of several steps needed to constrain sediment contributions from ravines to help create a sediment budget for the entire MRB.

The spatial distribution of ravines on the landscape and their morphometry indicates that some ravines produce more sediment than others. Ravines in close proximity to the Minnesota River are experiencing the greatest influence of base-level fall and are incising most aggressively. These ravines are likely generating the highest proportions of sediment (relative to other ravines). These ravines are also most likely to have high sediment-delivery ratios to the Minnesota River, because they are in close proximity, which lowers the likelihood that sediment will be stored within the tributary system before reaching the tributary outlet. In contrast, ravines far from the river exhibit a lower potential for sediment contribution to it because they are undergoing the least amount of incision and are subject to a higher likelihood for sediment storage within the river's tributary systems. For this reason it should be expected that ravines farther from the Minnesota River are contributing proportionally smaller quantities of sediment. Considering the spatial distribution of ravines in the MRB created by the algorithm (Fig. 4), it is important to realize that the same sediment delivery ratio is not applicable to all ravines in the MRB. A proximity threshold—perhaps using the elevation change to Minnesota River in combination with actual sediment loads measured for ravines across the spectrum of ravine types indicated in this work—could effectively be used to estimate the total net contribution of sediment from ravines in the MRB.

Conclusions

Within the context of the potential sediment contribution from ravines to the Minnesota River, this study had two goals: (i) to automatically map the basinwide area extent of the ravines to estimate ravine area and (ii) to identify ravine morphometric features that indicate which ravines will probably have the highest sediment loading to surface waters.

Ravines of the MRB make up approximately 0.45% of the land surface (197,830,000 m^2). This number was derived from the sum of polygon areas identified in the MRB with a terrain-based algorithm. The estimated area was adjusted and refined using the results from a series of error analyses designed to constrain the uncertainty associated with known limitations of the algorithm and resolution differences between 30-m and 3-m DEMs. Ravine boundaries digitized manually from the 30-m DEM overestimated the ravine area digitized from the 3-m LiDAR DEM by an average of 134%, with a standard deviation of 115%. Ravines digitized from the 3-m DOQ overestimated ravine area by 50%, with standard deviation of 68% relative to the 3-m LiDAR DEM. The algorithm itself overestimated ravine area by only 2% relative to the 3-m DOQ. The area of small ravines was underestimated by 6.3% because the ravine algorithm did not consider locations having flow accumulation values below a threshold value of 200 cells. Overall, the adjusted ravine algorithm was accurate in identifying all ravines important for the delivery of sediment to the MRB.

The ravines of the MRB are organized morphometrically on the landscape according to their relative proximity to the base-level fall associated with the downcut channel left by the Glacial River Warren 11,500 yr ago. Statistical analysis of physical ravine characteristics shows that ravines in close proximity to the Minnesota River base level, which have been exposed to the most extreme slope gradients for the longest time, tend to be bigger (mean area, 276,000 m²) and deeper (mean volume, 4,894,500 m³) and have greater relief (mean relief, 46 m) and higher slope means (slope mean, 20%) than ravines further from the mainstem Minnesota River. Ravines in the headwaters of tributaries feeding the mainstem of the Minnesota River have very little gradient-driving incision and are therefore smaller (mean area, 36,800 m²) and shallower (mean volume, 254,000 m³) with lower relief and slope values (13 m and 9% mean values, respectively). Measuring elevation change from a ravine's mouth to the main channel of the Minnesota River makes it possible to differentiate ravines on the basis of their potential for sediment loading to surface waters.

References

American Rivers. 1997. North America's most endangered and threatened rivers of 1997. Annual report of American Rivers. Washington, DC.

Balco, G. 2001. Cosmogenic isotope production rates over large areas. Watershed delineation tool script available from the University of Washington Cosmogenic Isotope Lab: http://depts.washington.edu/cosmolab/P_by_GIS.html (accessed 21 Oct. 2016).

Barnes, R., C. Lehman, and D. Mulla. 2013. An efficient assignment of drainage direction over flat surfaces in raster digital elevation models. Comput. Geosci. 62:128–135. doi:10.1016/j.cageo.2013.01.009

Barnes, R., C. Lehman, and D. Mulla. 2014. Priority-flood: An optimal depression-filling and watershed-labeling algorithm for digital elevation models. Comput. Geosci. 62:117–127. doi:10.1016/j.cageo.2013.04.024

Bates, R.L., and J.A. Jackson, editors. 1984. Dictionary of geological terms. 3rd ed. Anchor/Doubleday, Garden City, NY.

Belmont, S., D.J. Mulla, and J. Nelson. 2015. Identification and analysis of ravines in the Minnesota River Basin with GIS. In: T.G. Mueller and G.F. Sassenrath, editors, GIS applications in agriculture. Vol. 4 of Conservation planning. Taylor and Francis, Boca Raton, FL. p. 37–52

Brooks, K.B., P.F. Folliott, H.M. Gregersen, and L.F. DeBano. 2003. Hydrology and the management of watersheds. 3rd ed. Iowa State Press, Ames.

Coalition for a Clean Minnesota River (CCMR). 2004. Minnesota River watershed drainage policy reform report. CCMR, New Ulm, MN.

Engstrom, D.R., J.E. Almendinger, and J.A. Wolin. 2009. Historical changes in sediment and phosphorus loading to the upper Mississippi River: Mass-balance reconstructions from the sediments of Lake Pepin. J. Paleolimnol. 41:563–588. doi:10.1007/s10933-008-9292-5

Environmental Systems Research Institute (ESRI). 1994. GRID users guide. ESRI, Redlands, CA.

Faulkner, H., R. Alexander, and P. Zukowskyj. 2008. Slope-channel coupling between pipes, gullies and tributary channels in the Mocatan catchment badlands, southeast Spain. Earth Surf. Process. Landf. 33:1242–1260. doi:10.1002/esp.1610

Galzki, J., A.S. Birr, and D.J. Mulla. 2011. Identifying critical agricultural areas with 3-meter LiDAR elevation data for precision conservation. J. Soil Water Conserv. 66:423–430. doi:10.2489/jswc.66.6.423

Gran, K.B., P. Belmont, S.S. Day, C. Jennings, A. Johnson, L. Perg, and P.R. Wilcock. 2009. Geomorphic evolution of the Le Sueur River, Minnesota, USA, and implications for current sediment loading. In: L.A. James, S.L. Rathburn, and G.R. Whittecar, editors, Management and restoration of fluvial systems with broad historical changes and human impacts. Geological Society of America Spec. Paper 451. GSA, Boulder, CO. p. 119–130, doi:10.1130/2009.2451(08).

Hatch, L.K., A.P. Mallawatantri, D. Wheeler, A. Gleason, D.J. Mulla, J.A. Perry, K.W. Easter, P. Brezonik, R. Smith, and L. Gerlach. 2001. Land management at the major watershed-agroecoregion intersection. J. Soil Water Conserv. 56:44–51.

Kelley, D.W., and E.A. Nater. 2000. Historical sediment flux from three watersheds into Lake Pepin, Minnesota, USA. J. Environ. Qual. 29(2):561–568. doi:10.2134/jeq2000.00472425002900020025x

Knox, J.C. 2006. Floodplain sedimentation in the Upper Mississippi Valley: Natural versus human accelerated. Geomorphology 79:286–310. doi:10.1016/j.geomorph.2006.06.031

Minnesota Pollution Control Agency (MPCA). 2005. Minnesota River Basin TMDL Project for Turbidity. Water quality/basins fact sheet 3.33. June.

Montgomery, D.R., and W.E. Dietrich. 1988. Where do channels begin? Nature 336:232–234. doi:10.1038/336232a0

Mulla, D.J., and A.C. Sekely. 2009. Historical trends affecting accumulation of sediment and phosphorus in Lake Pepin, upper Mississippi River, USA. J. Paleolimnol. 41:589–602. doi:10.1007/s10933-008-9293-4

Poesen, J., J. Nachtergaele, G. Verstraeten, and C. Valentin. 2003. Gully erosion and environmental change: Importance and research needs. Catena 50:91–133. doi:10.1016/S0341-8162(02)00143-1

Rada, R.G., J.G. Wiener, P.A. Bailey, and D.E. Powell. 1990. Recent influxes of metals into Lake Pepin, a natural lake on the Upper Mississippi River. Arch. Environ. Contam. Toxicol. 19:712–716. doi:10.1007/BF01183989

Schumm, S.A. 1999. Causes and controls of channel incision. In: S.E. Darby and A. Simon, editors, Incised river channels. John Wiley & Sons., Hoboken, NJ. p. 19–33.

Sekely, A.C., D.J. Mulla, and D.W. Bauer. 2002. Streambank slumping and its contribution to the phosphorus and suspended sediment loads of the Blue Earth River, Minnesota. J. Soil Water Conserv. 57(5):243–250.

Sheeder, S.A., and B.M. Evans. 2004. Estimating nutrient and sediment threshold criteria for biological impairment in Pennsylvania watersheds. J. Am. Water Resour. Assoc. 40(4):881–888. doi:10.1111/j.1752-1688.2004.tb01052.x

Sidorchuk, A. 2006. Stages in gully evolution and self-organized criticality. Earth Surf. Process. Landf. 31:1329–1344. doi:10.1002/esp.1334

Simon, A., and S.E. Darby. 1999. The nature and significance of incised river channels. In: S.E. Darby and A. Simon, editors, Incised river channels. John Wiley & Sons, Hoboken, NJ. p. 3–18.

Simon, A., and M. Rinaldi. 2000. Channel instability in the loess area of the Midwestern United States. J. Am. Water Resour. Assoc. 36(1):133–150. doi:10.1111/j.1752-1688.2000.tb04255.x

Thorleifson, L.H. 1996. Review of Lake Agassiz history. In: J.T. Teller, L.H. Thorleifson, G. Matile, and W.C. Brisbin, editors, Sedimentology, geomorphology, and history of the central Lake Agassiz Basin (field trip B2). Geological Association of Canada, Winnipeg. p. 55–84.

USEPA. 2009. Water quality standards: Regulations and resources. https://www.epa.gov/sites/production/files/2014-12/documents/mnwqs-chapter-7050.pdf (accessed 6 Nov. 2016).

Vandaele, K., J. Poesen, G. Govers, and B. van Wesemael. 1996. Geomorphic threshold conditions for ephemeral gully incision. Geomorphology 16(2):161–173. doi:10.1016/0169-555X(95)00141-Q

Walling, D.E. 1983. The sediment delivery problem. J. Hydrol. 65:209–237. doi:10.1016/0022-1694(83)90217-2

Grassed Waterways

Peter Fiener* and Karl Auerswald

Abstract

Grassed waterways (GWWs) are a widely used best management practice (BMP) in countries with large-area farming. They are broad, shallow, grass-lined channels often located within large fields and have the primary function of draining surface runoff from farmland and preventing gullying along the natural drainage ways. They are rarely found in areas with relatively small fields, for example, in many European countries, where a single GWW would drain a number of fields. Presently, GWWs seem to attract much less attention as BMPs than grassed or vegetated filter strips at the downslope end of fields or along the surface of bodies of water, even though the benefits of GWWs are obvious. Studies clearly show that well-established GWWs effectively prevent gully erosion, reduce sediment and agrochemical delivery, and dampen peak discharge rates. The effects of GWWs on plant and faunal diversity are insufficiently studied, and our knowledge of them is based mostly on analogies from grass strips or set-aside areas. Use of such analogies might be especially misleading with regard to the effects of GWWs on ecological connectivity within arable landscapes. From an agricultural viewpoint, the benefits of GWWs clearly seem to overcompensate for the main disadvantage: loss of land for cultivation. But arriving at general conclusions is difficult due to the huge variation in land prices and establishment costs. In economic terms, GWWs are thought to have a life span of ten years, a misjudgment that may be responsible for less relevance of GWWs since the 1970s. From political and societal viewpoints, outreach programs and financial incentives appear to be good investments. Apart from effectively preventing erosion damage (e.g., muddy floods), society would benefit from other ecological effects that are not included in the economic considerations of farmers. But scientific knowledge regarding the ecological and economic benefits and economic costs of GWWs is incomplete, impeding a holistic assessment of them. This is the case regarding the obviously intended effects on sediment delivery and discharge from agricultural land, because watershed-scale studies are rare and available modeling approaches therefore lack rigid testing.

Abbreviations: BMP, best management practice; GWW, grassed waterway.

P. Fiener, Institut für Geographie, Universität Augsburg, Alter Postweg 118, 86159 Augsburg, Germany; K. Auerswald, Lehrstuhl für Grünlandlehre, Technische Universität München, Alte Akademie 12, 85354 Freising, Germany (auerswald@wzw.tum.de). *Corresponding author (fiener@geo.uni-augsburg.de).

doi:10.2134/agronmonogr59.2013.0021

Precision Conservation: Geospatial Techniques for Agricultural and Natural Resources Conservation
J.A. Delgado, G.F. Sassenrath, and T. Mueller, editors. Agronomy Monograph 59.

Grassed waterways were widely implemented as a BMP to control (ephemeral) gully erosion in the 1970s and 1980s in North America (Atkins and Coyle, 1977; Berg and Gray, 1984; Napier et al., 1984; Bracmort et al., 2004). They are broad, shallow, grass-lined channels often located within large fields and have the primary functions of draining surface runoff from farmland and preventing gullying along the natural drainage ways (Fig. 1). Berg and Gray (1984) reported that in some US counties, up to 90% of all farmers interviewed had GWWs. Grassed waterways were also established in other countries that typically have large-area farming (e.g., Australia; Thomas et al., 2007). Compared with those countries, GWWs are rarely found in Central Europe, even though some authors (e.g., Fiener and Auerswald, 2003b; Evrard et al., 2008b; Boardman and Vandaele, 2016) have shown their potential as a BMP for small-area farming in Europe. But in many areas of the world, increasing globalization has resulted in trends in agriculture—such as larger field sizes and use of a greater proportion of arable land—that increase the demand for GWWs and create the preconditions for their establishment. In general, the emphasis is less on GWWs than on grass or vegetative filter strips at the downslope end of agricultural fields and along open bodies of water. This is especially indicated by a SCOPUS search (October

Fig. 1. (A) Narrow, subsurface grassed waterway (GWW) in Belgium with triangle-shaped cross-section for fast drainage applicable for small drainage areas but with risk of overtopping or incision at high runoff rates (photo by C. Bielders). (B) Wide, flat-bottomed GWW for large runoff rates and rock chute outlet to prevent incision (photo by K. Schneider, USDA Natural Resources Conservation Service). (C) Successional GWW in Germany with large hydraulic roughness and high biotic value due to the provision of habitat, feed, and connectivity for wild species. The margins have been mulched to prevent weed encroachment in the fields and to facilitate runoff inflow into the waterway (photo by G. Gerl). (D) Massive gullying (see person in background for scale) has been initiated by a single runoff event due to the lack of a GWW along the drainage line (photo by R. Brandhuber).

2014) that resulted in approximately seven times more hits for vegetative/grass filter strips than for vegetative/grassed waterways. There are two possible reasons for the difference in interest and scientific knowledge about these two BMPs: (i) the implementation of a GWW along drainage areas is more complex, and hence less common, especially under European conditions, where several fields might drain into one GWW; and (ii) landscape-scale experiments are more challenging (and costly) for experimentally evaluating the effects of GWWs compared with measurements in small grass strips, which are often used in plot experiments.

Assessment of GWWs as a BMP calls for a holistic evaluation of their agronomic (or economic) and ecological effects. The objective of this review is to provide such a view while synthesizing results from studies performed within the last four decades on the economic and ecological effects of GWWs. Political options and constraints on supporting the establishment of GWWs are identified, and gaps in the scientific knowledge are discussed.

The Agricultural Viewpoint

Use of GWWs is in direct proportion to the environmental concern of farmers, their agricultural education, and the acreage farmed (Napier et al., 1984); conversely, the adoption of double cropping while holding share leases reduces the adoption rate of GWWs (Nyaupane and Gillespie, 2011). Grassed waterways are more likely to be amortized on owned land than on land leased for a short or unpredictable period of time. Given that there are different types of GWWs (Fig. 2) and that landscape properties and socioeconomic conditions vary, large differences in costs can be expected. Grassed waterways may be engineered (Fig. 1A, 1B, and 2B–2D), as in the case of perched GWWs, which are not placed at the location where the topography would concentrate the water, or in the cases in which GWWs are used to drain constructed terraces. Alternatively, they may just be set-asides at positions within sloping land where water concentrates naturally and thus do not entail much in the line of construction costs (Fig. 1C and 2A). We will discuss the associated costs, which comprise (i) installation costs that have to be depreciated over the life time of the GWW (including interest rates), (ii) maintenance costs, and (iii) the opportunity cost for the land occupied by the GWW (Kling et al., 2007; Qiu et al., 2009; Chen and Barkdoll, 2013).

Installation Costs

Few or no costs accrue if the area of the GWW is simply taken out of production (Fig. 1C and 2A) or if only a few or small improvements are necessary. Evrard et al. (2008a) calculated the costs of installing earthen detention ponds within GWWs to be €126 ha^{-1}. Seeding of sod-forming species may be necessary when such species would not establish naturally. In contrast, the construction of engineered GWWs usually will cost in the range of $4000 (or €4000) ha^{-1} (Kling et al., 2007; Qiu et al., 2009; Chen and Barkdoll, 2013). Engineered GWWs can be among the most expensive measures of erosion control. Such GWWs usually become necessary when other installations, such as terraces, require the GWW to be in a certain position, at a certain level within the land, and have a certain capacity: all variables that are defined by the other installations. Thus, economically, the costs are not caused by the GWW itself but are subsequent costs resulting from other

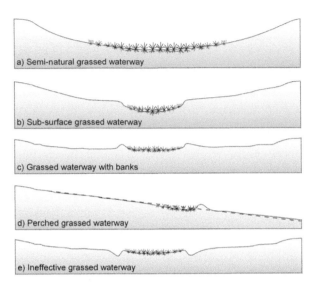

Fig. 2. Typical cross-sections of grassed waterways.

installations; superficially, however, those costs are associated with the construction of the GWW.

Nevertheless, installation costs exist and need to be covered. Whether they
are high or not largely depends on the design life of a GWW. A design life of 10 yr
is commonly assumed, although it is not clear on which basis this estimate was
made. In comparison, the design life of terraces is usually set at 20 yr (Kling et
al., 2007; Chen and Barkdoll, 2013). We are aware of only one study quantifying
the endurance of GWWs. Bracmort et al. (2004) examined GWWs, the majority of
which were built in the 1970s and 1980s, and found that among different structural BMPs, GWWs were the ones most likely to be used after more than 20 yr;
they were functional beyond twice their design life. The GWWs studied by Chow
et al. (1999) and by Fiener and Auerswald (2003a) were still functional 25 yr after
their establishment. A major reason that GWWs persist may be that farmers tend
to discontinue farming practices (e.g., nutrient management) more frequently
than structural practices such as GWWs (Osmond et al., 2012). A design life of 10
yr entails annual costs of about \$400 ha^{-1} yr^{-1} for an engineered GWW, and these
costs would be halved for a design life of 20 yr. Given that the study by Bracmort
et al. (2004) is the only one that empirically determined the lifetime of GWWs,
their design life should be changed to 20 yr, but further studies are clearly needed.
A more sophisticated approach would recognize that their design life is variable
and depends on a number of factors, the most important one perhaps being the
management of the adjacent land.

Many farmers tend to consider the land occupied by GWWs to be a loss of
valuable acreage (Carey et al., 2015a); as a consequence, there is a risk that not
enough land is allotted for the GWW. Usually there is a trade-off between the
amount of land allotted on the one hand and the installation costs, secondary
benefits, and efficiency of a GWW to reduce runoff and sediment export on the
other. Set-asides of natural flow paths usually lead to wide GWWs owing to the
flat bottom of the drainage path (Fig. 1B, 1C, and 2A). Conversely, engineered

GWWs such as subsurface GWWs or GWWs with banks (Fig. 1A, 2B, and 2C) concentrate the flow and thus reduce the need for area. Hence, secondary benefits are reduced and installation costs increase while less land is needed. It will depend on the cost of land (e.g., land rental costs) as to which expense is higher. Installation costs increase with decreasing GWW width because concentrating the flow will increase flow velocity (Fiener and Auerswald, 2005), which calls for a reinforcement of the GWW, and also decrease the retention capacity of the GWW and increase the likelihood of failure (Fig. 3). For a safe GWW, it is recommended that the Froude number (the ratio between the average flow velocity and the wave velocity) should not exceed 1 but should be in the range of 0.8 (Carey et al., 2015), which better describes the hydrodynamic properties of the runoff than simply using flow velocity as in Fig. 3. For a near-rectangular cross-section, wave velocity may be approximated by the square root of gravitational acceleration times flow depth. A Froude number of more than 1 indicates that hydraulic jumps will occur that release much of the flow energy locally and thus may initiate gully formation as typically shown in Fig. 1D. When Froude numbers less than 1 cannot be guaranteed, additional measures such as the reinforcement of the GWW bottom by UV-resistant nets or geotextile mats become necessary (Carey et al., 2015a, 2015b). Although non-engineered GWWs may be adopted by the farmer without technical assistance, the trade-off between GWW width and functionality benefits from the input of an expert who is aware of the hydromechanical properties of a GWW. In some countries, detailed handbooks for the design of GWWs are available that describe how to professionally install GWWs (e.g., in the United States, USDA-SCS, 1954; Temple et al., 1987; in Australia, Carey et al., 2015a), while in other regions of the world such technical support is widely missing.

Maintenance Costs

Maintenance costs are usually small and may range from $10 to $40 ha^{-1} yr^{-1} (Qiu et al., 2009; Chen and Barkdoll, 2013). One of the primary maintenance functions is occasional grass cutting to prevent encroachment of woody species and

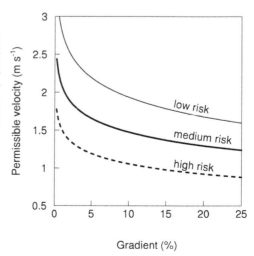

Fig. 3. Permissible runoff velocities in grassed waterways at low risk of erosion (low-erodible soil and 100% cover), medium risk (either low-erodible soil and 70% cover or high-erodible soil and 100% cover), and high risk (high-erodible soil and 70% cover). After Titmarsh and Stone, 1997.

to favor sod-forming species and a dense rooting structure close to the soil surface. Higher maintenance costs occur if the GWW is damaged by sedimentation
or if it loses the capacity to accommodate runoff because of furrows along its
sides (Fig. 2E). In both cases, the problem is not the GWW itself but the improper
management of the surrounding land; hence, the costs would better be assigned
to the land rather than to the GWW. Furrows along the side usually result from
improper tillage that either creates the furrows or that forms a dam that prevents
the runoff from entering the GWW. The runoff will then flow along the unpro-
tected sides of the GWW and produce a rill. Sediment will fill the GWW only if
high erosion rates occur within the delivering watershed. Even then, an obstruc-
tion will probably not form in a GWW because the concentrated flow will always
have a higher transport capacity than the shallow flow on the side slopes, where
sediments are first trapped. The more the hydrodynamic resistance of the grass
is reduced due to sedimentation, the less additional sedimentation will occur in
the GWW. Complete burial of the grass is thus impossible except when the runoff
is slowed down due to other conditions—for example, by detention ponds within
the GWWs (Evrard et al., 2008a), by filter socks (Shipitalo et al., 2012), or by a
meandering flow path (Titmarsh and Stone, 1997). In most of these cases, control
of runoff is not due to the grass in the GWW, and damage to the grass would not
be critical.

Land Costs

Grassed waterways occupy farmland but, despite their size, their share of overall
land is small. Fiener and Auerswald (2006a) analyzed 19 watersheds, ranging in
size from 50 to 1500 ha, and determined the amount of land that would be lost if
a GWW were established wherever the topographical situation would allow for
a seminatural GWW like the one in Fig. 1C and 2A. Assuming that all of it was
used for arable purposes, only 2.3% of land would have been lost to GWWs. Thus,
a farm's net return cannot be reduced by more than 2.3% even if no economic
benefits exist. For the studied watersheds, which comprised mixed-use land, only
0.8% of the land qualified for GWWs (Fig. 4). Even so, construction of GWWs on
this small portion of land would have reduced peak discharge by about 15%.

The cost of land will vary significantly among countries. Chen and Barkdoll (2013) assumed approximately $150 ha^{-1} yr^{-1} (for the USA) for land rental for
GWWs, while Spaan et al. (2010) assumed €60,000 ha^{-1} (for the Netherlands) was
necessary to purchase the land for GWWs. But the question arises whether average costs for renting or purchasing arable land are applicable to GWWs. Usually
a GWW is placed along the natural drainage line, where typically the most fer-
tile, colluvial soils occur that were created by tillage erosion due to the decrease
in slope inclination. The colluvial soils in hollows resulting from tillage may lead
to an erroneous expectation of a low potential hazard from water erosion, while
in fact it often is maximum in this position (Van Oost et al., 2000; Fig. 1D). Due to
the confluence of runoff from both side slopes, the flow width decreases and flow
depth increases, leading to high flow velocities, which cause ephemeral gullies.
Tillage closes the ephemeral gullies but also maintains the field's susceptibility
to gullying. In some cases, no-till may reduce the risk of gullying (Gordon et al.,
2008), but it will not always be sufficient, and it restricts land use across the entire
field. In the case of the watersheds analyzed by Fiener and Auerswald (2003a,
2003b), historical maps showed that a narrow strip of grassland had previously

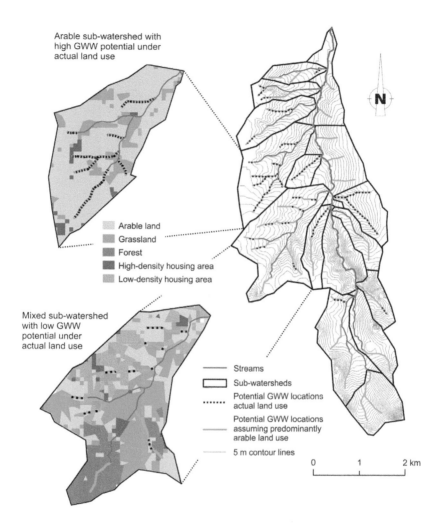

Fig. 4. Lauterbach watershed (16.7 km²) near Bonn in Western Germany. Analyzed for its potential to allocate grassed waterway in differently used sub-watersheds (examples left). Effects on runoff modeled for actual land use and assumed predominantly arable land use in all sub-watersheds (Fiener and Auerswald 2006a).

existed where they established their GWW but that this strip had been plowed and lost with the advent of mechanized agriculture. Mechanization makes it easier to close ephemeral gullies, but it would be misleading to apply the average costs of arable land that does not bear the risk of ephemeral gullying to all areas occupied by GWWs. Some farmers view GWWs as inconveniences that add to production costs and have uncertain economic returns (Chow et al., 1999). Zhou et al. (2009), however, showed that in many cases there was an economic benefit from GWWs even though they assumed costs of up to $11,000 ha⁻¹ and a life span of only 10 yr.

Benefits

While costs are certainly of concern, it is interesting to note that a survey by Berg and Gray (1984) found that up to 90% of all interviewed farmers in some counties in the USA had GWWs. The majority of those farmers said that the main reason for using a particular practice was the expectation that it would reduce operating costs. A focus on the costs of GWWs is thus one-sided if their economic benefits are not considered. The most striking benefit is that GWWs prevent ephemeral gullies. Such gullies cost money to be removed and pose an obstacle during field operations because they often cannot be crossed with farm machinery (Fig. 1D). Conversely, GWWs allow equipment to travel to both sides of a field or across two fields. Additionally, the GWW can be used instead of the field margin for farming operations and thus improves field management at the margin. For example, turning operations can be moved from within the fields onto the GWWs to avoid soil compaction of the headland. However, it needs to be noted that moving traffic into the GWWs might be more appropriate for wide, semi-natural GWWs but more problematic in narrow, more technical ones, because trafficking may decrease infiltration capacity within the GWWs and may also damage the grass sward in areas of concentrated flow. Regarding the damage to the sward, handbooks recommend not using GWWs as turnrows during tillage operations (USDA-SCS, 1954). Generally, GWWs should not be crossed when they are very wet, but except for those GWWs located on very shallow groundwater tables, using them as turnrows should not present a problem when fields are dry enough for field management. The large width in some places of one of the GWWs studied by Fiener and Auerswald (2003b) was unnecessary for the functioning of the GWW. It resulted from the design of the neighboring fields. The design aimed to create field widths that were multiples of farm-machinery widths (15 m). This reduced labor costs and efforts, reduced unintentional application of fertilizers and pesticides outside the field margins, and permitted more precise application of agrochemicals within the field margins and more precise tillage. Problems may arise and redesign may become necessary when machinery widths change, which is not improbable given the long life spans of GWWs. They are often considered an obstacle within the fields that create overlapping or non-target application of agrochemicals because the field boundary becomes more complex (Luck et al., 2011b). However, GWWs can also be designed to have the opposite effect by optimizing field boundaries and thus not creating additional complexity. The complexity is already inherent in the variability of site conditions, particularly when there is uncontrolled erosion. Ignoring this complexity does not remove it but may create additional problems such as compaction of the wetter soils in the depressions, uneven ripening of the crops, or gullying. Engineered GWWs, however, such as those presented in Fig. 2B–2D, may become impassable and thus increase the difficulties for field management (Luck et al., 2011a). Seminatural GWWs (Fig. 2A) do not have this problem and can even be used occasionally as farm roads during dry conditions (Fig. 1C). The best type of GWW to choose thus depends primarily on the relative importance of different cost components. While seminatural GWWs need more land, they have lower installation costs and offer more potential benefits.

The Ecological Viewpoint

From an ecological view, GWWs exhibit a multitude of beneficial effects on different ecosystem functions besides the most obvious and intended one of protecting against gully erosion and the associated reduction in sediment delivery into adjacent ecosystem compartments. Grassed waterways trap incoming sediments and associated particle-bound agrochemicals and nutrients; they reduce runoff volume, especially peak discharge and, as a consequence, also affect the delivery of dissolved nutrients; and they affect a number of ecosystem functions, such as biodiversity and soil carbon stocks.

Reduction in Runoff

Grassed waterways reduce runoff originating from adjacent fields mainly by the following three processes: they create (i) a higher rate of water infiltration due to reduced sealing of continuous grass cover compared to more exposed farmed soils and there is less soil compaction caused by wheeling; (ii) a prolonged time for infiltration because the hydraulic roughness provided by dense grass reduces runoff velocity (Ree, 1949; Ogunlela and Makanjuola, 2000); and (iii) a higher surface storage capacity than drainage ways without GWWs, which also results from the greater hydraulic roughness and greater interception of runoff during periods when drainage ways without GWWs exhibit no vegetation cover.

Compared with the large number of studies that deal with runoff reduction by vegetated filter strips located at the downslope end of fields or along small streams (Dillaha et al., 1989; Muñoz-Carpena et al., 1993; Deletic, 2001; Dass et al., 2011), there are relatively few studies evaluating reduced runoff by GWWs. More such studies are essential because GWWs have more-complex effects than filter strips. Grassed waterways affect the laterally incoming sheet flow in a manner similar to filter strips, but they also affect the concentrated runoff along the drainage way, an effect that does not occur in vegetated filter strips. Paired watershed studies (e.g., Chow et al., 1999; Fiener and Auerswald, 2003b) indicate that the runoff reduction potential of GWWs is large, but it also varies substantially depending on the watershed and the characteristics of the GWW. For example, in two pairs of watersheds with more or less identical land use and cropping patterns but different GWW characteristics, Fiener and Auerswald (2003b) found an average reduction in runoff volume across 8 yr of 90% for a flat-bottomed GWW but a reduction of only 10% for a GWW with a small incision (0.5 m wide and 0.2 m deep) along its drainage way that accelerated the runoff velocity. Chow et al. (1999) reported a reduction in runoff volume of 14% in four vegetation periods when comparing two watersheds: one with up-and-down slope cultivation and another with a terrace/GWW system (note: Chow et al. used only those vegetation periods in which identical crops were cultivated in both watersheds). The seasonal variability of GWWs in reducing runoff volume contributes to the variability of long-term means. Generally, high efficiencies can be expected during phases of small inflow volumes (Fig. 5A). Seasons with high hydraulic roughness due to stiff grasses (mainly during the growing season) or low antecedent soil-water contents will also experience higher efficiencies than other seasons (Fiener and Auerswald, 2006b).

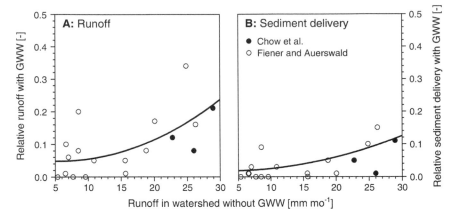

Fig. 5. Relative runoff (A) and sediment delivery (B) from watersheds with grassed waterways (GWWs) versus data from paired watersheds without GWW. Monthly data of paired watersheds with the same soil conservation measures and crop rotations (winter wheat, potatoes, winter wheat, and maize) and a flat-bottomed GWW (from Fiener and Auerswald 2003b), and monthly mean of seasonal data (1 May to 30 November) of paired watersheds with the same crop rotation (potatoes, potatoes, barley, and potatoes) but one with a terrace/GWW system, and the other with up-and-down slope cultivation (Chow et al., 1999) are plotted.

Model-based evaluations (e.g., Hjelmfelt and Wang, 1997; Fiener and Auer-swald, 2006a; Kalantari et al., 2014) predicted smaller reductions in runoff volume than paired watershed studies did (reductions of 5 and 30%, respectively). These predictions resulted in part from the focus on large events rather than on long-term means, but the discrepancy may also be the result of a lack of appropriate parameterizations in the models, for example, high infiltration capacities in GWWs. However, all modeling studies consistently reported pronounced effects on peak discharge because of the larger hydraulic roughness and lower hydraulic radius than those of drainage ways without GWWs (reduction in peak discharges to values between 46 and 85%). Irrespective of potential prediction errors in the runoff-volume reduction for GWWs, models are needed to upscale the efficiency of GWWs for catchment sizes for which experimental evidence is no longer available. Fiener and Auerswald (2006a), for example, modeled a 16.7-km² watershed in which GWWs were placed along the most suitable drainage ways in case of actual land use and assumed predominantly arable land use (Fig. 4). Depending on assumed storm reoccurrence (2–50 yr), season, and land-use scenario, the predicted runoff volume was reduced to values between 70 and 95%, and predicted peak discharge was reduced to values between 60 and 85%. These results indicate the potential for GWWs to reduce peak discharge even from large catchments. Moreover, such modeling approaches also provide tools for deciding on the length, width, and cross-section design of GWWs (Fiener and Auerswald, 2005; Dermisis et al., 2010).

Reduction in Sediment Delivery

Reduction in sediment delivery is mainly the result of (i) a decrease in transport capacity due to reduced runoff velocities and volumes; (ii) potential sieving of large particles (e.g., aggregates) by dense vegetation and litter; and (iii) infiltration of sediment-laden runoff. All three mechanisms are potentially most effective in flat-bottomed, seminatural GWWs (Fig. 2A).

Generally, sediment trapping has been consistently reported to be more effective than runoff reduction (cf. Fig. 5B and Fig. 5A). Mean sediment delivery measured in paired watersheds with and without GWWs (e.g., Chow et al., 1999; Fiener and Auerswald, 2003b) was reduced to 23% or to 3% in the watersheds with GWWs. Also, several modeling studies report that sediment delivery is reduced to 28–35% with 600-m-long GWWs (Hjelmfelt and Wang, 1997; Hjelmfelt and Wang, 1999; Dermisis et al., 2010). The trapping efficiency decreases as more runoff comes from the fields because all three mechanisms of sediment retention become less effective as more runoff concentrates in the GWW (Fig. 5B). Thus, Dermisis et al. (2010) reported much smaller efficiencies for 300-m-long GWWs at high discharge rates (>0.3 m³ s⁻¹) than what Fiener and Auerswald (2003b) reported for GWWs with similar lengths but much smaller inflow rates (estimated mean < 0.01 m³ s⁻¹). Furthermore, the effects will become larger the more room that is allowed for the runoff to pass the GWW. Wide, seminatural GWWs can thus be expected to have larger effects than GWWs that were engineered to reduce the demand for land.

Reduction in Nutrient and Agrochemical Delivery

Obviously, retention of runoff and sediment within the GWWs also affects the delivery of dissolved and particle-bound nutrients and agrochemicals. During a 5-yr paired watershed study, GWWs reduced particulate phosphorus losses to 7 and 23% (Fiener and Auerswald, 2009). The efficiency of particulate-P trapping is directly linked to sediment trapping; however, somewhat smaller effects might be expected because of the typical enrichment of fines and the associated enrichment of phosphorus in delivered sediments (Sharpley, 1980).

Two factors impact the reduction in dissolved substances. The first is the reduction in runoff volume resulting from increased infiltration—and hence the infiltration of dissolved substances. The second effect is more variable. On one hand, the concentration of solubles that originate from the arable land may be reduced by adsorption. This is usually the case for pesticides (Asmussen et al., 1977; Hjelmfelt and Wang, 1997; Briggs et al., 1999). The extent to which the concentration will decrease depends on the extent to which a certain substance interacts with (organic) surfaces. On the other hand, a GWW may also be a source of solubles, such as nutrients, that linger from the former soil use or that accumulate over time in the GWW due to the infiltrating water. Correspondingly, the effects of GWWs on dissolved P may vary. The P concentration did not decrease in one of the GWWs that was analyzed by Fiener and Auerswald (2009), while in the other GWW, the P concentration decreased to 62%. In case of sensitive biotopes downstream, the load and concentration of solubles may be reduced by installing filter socks within the GWW. Shipitalo et al. (2012) showed that such filter socks can decrease the P loads, but other substances by the filter socks. Thus, a GWW can also be a source of solubles. Dissolved organic matter in particular may be released from the debris accumulated within the GWW and thereby raise the concentration compared with field runoff in cases where there is little surface cover in the field to serve as the main source of dissolved organic matter (Glenn and Williams, 1985; Auerswald and Weigand, 1996; Singh et al., 2014). It is important to note, however, that the concentration of any soluble in the throughfall or surface runoff decreases during a rain event due to the depletion of the source. The larger the infiltration before runoff starts, the lower will be the concentrations of solubles in the runoff. A GWW with a high infiltration capacity is also advantageous in this respect.

Potential Groundwater Effects

Additional infiltration along the GWW could potentially turn the GWW into a hotspot of groundwater recharge, which may also affect groundwater quality. The additional recharge is about equal to the reduction in total runoff by the GWW because evapotranspiration of the GWW will increase only to a small degree. Fiener and Auerswald (2003a) estimated an increase in infiltration of about 160 mm yr^{-1} within the area of the monitored GWW, which almost doubled the groundwater recharge within the area of the GWW. The effect on the entire watershed area will be much smaller because a GWW constitutes only a few percent of the total watershed land area.

Usually this additional recharge will be of high water quality for two reasons. First, the loading of the GWW with agrochemicals is small because neither fertilizers nor pesticides are applied on a GWW. Second, the runoff from the surrounding fields that infiltrates within the GWW usually does not contain high concentrations of easily soluble agrochemicals, such as nitrate, that pose a risk to groundwater quality. The reason is that runoff from fields occurs only after the initial abstraction has already leached the easily soluble substances from the very top of the arable soil, which interacts with the runoff (usually not more than a few centimeters; Ahuja et al., 1981).

Effects on Plant and Faunal Diversity

Linear grassland structures within arable fields might play a large role in increasing biodiversity in arable landscapes because they connect different habitats and they provide a different habitat where, in contrast to the surrounding arable land, the application of agrochemicals is low, tillage is missing, and a soil cover is maintained throughout the year (Fig. 1C).

Grassed waterways provide corridors for wildlife and connect patches of fragmented habitats, thus increasing overall habitat quality. In contrast to other linear structural elements—such as buffer strips, riparian areas, terraces, or unplowed strips along field borders—GWWs connect dry upslopes with wet valley bottoms, and they link arable lands that are managed differently. Grassed waterways can thus provide wildlife access to a diversity of habitat resources that are dispersed across the landscape. These effects increase the corridor value of GWWs (Henry et al., 1999). There are, however, few studies of GWWs that focus on these connectivity effects and other ecosystem functions rather than on the potential for soil and water conservation.

The effect on vegetation is usually limited even when the GWW is not sown but is instead developed by succession. Even after 8 yr of succession, the vegetation of such a GWW (Fig. 1C) was dominated by a few fast-growing species commonly found in agricultural landscapes [e.g., *Agropyron repens* (L.) P. Beauv., *Dactylis glomerata* L., *Urtica dioica* L.], some tall herbs (e.g., *Epilobium angustifolium* L., *Galeopsis tetrahit* L., *Galium aparine* L.), and woody plants (e.g., *Salix* spp., *Rubus* spp., *Sorbus* spp.), which contributed between 1 and 15% to the total cover (Fiener and Auerswald, 2003a). The GWW was thus dominated by plants that can commonly be found in intensively used agricultural landscapes. This is not surprising because the colluvial soils along the drainage way promote species that respond to a high concentration of nutrients. Furthermore, an intensively farmed landscape surrounding the GWW does not provide seed sources from other species. The slow invasion of other plants, especial shrubs and trees, offers the advantage of a low-maintenance effort, in which mowing every 10 yr seems to be sufficient to suppress woody species (Fiener and Auerswald, 2003a).

The microbial and faunal effects will be much more pronounced when there is no tillage: the soil microbial biomass will increase and the species composition of the soil fauna will change (Filser et al., 1996). The GWW acts as refuge for epigaeic organisms such as spider and grasshopper species (Agricola et al., 1996) and snakes (Knoot and Best, 2011). As the width of the GWW increases, so does the quality of the habitat quality for all three groups of epigaeic animals. There are also indications that GWWs are a good habitat for birds (Bryan and Best, 1991, 1994; Hultquist and Best, 2001; Fiener and Auerswald, 2003a), but for wide-ranging animals, it becomes more difficult to differentiate between the effects of the various changes in a landscape, such as the introduction of terraces or the reparcellation of the landscape, that usually occur when GWWs are established. To increase the numbers of bird species, especially of those nesting in the GWWs, some studies recommend that these areas should not be mowed until end of August or early September (Bryan and Best, 1991, 1994) and that vegetation should be clipped high (15–30 cm; Bryan and Best, 1994).

The Political Viewpoint

Clearly, to establish a GWW, considerable financial, organizational, and intellectual effort is required on the part of the farmer, but outreach programs (Shelton et al., 2009; Lemke et al., 2011) and financial incentives can help (Berg and Gray, 1984; Lawler et al., 1992; Kling et al., 2007; Lemke et al., 2010, 2011; Chen and Barkdoll, 2013). Although expensive, such aid appears to be a good investment by society, given that the effective life span of GWWs is long and that experts are involved. The proper design of a GWW will improve its functioning and reduce the risk of failure that could discredit this conservation measure. While there are pros and cons of GWWs for the farmer, for society the benefits clearly outweigh the disadvantages. Evrard et al. (2008b) have shown that the damages from muddy floods are ten times higher than the subsidies Belgian farmers receive for the maintenance of grass strips. Also, the positive effects on biodiversity or aquatic environments clearly should be of societal concern even though they do not figure in the economic considerations of a farmer.

Especially in Europe, public interest in GWWs may grow in future because the sizes of fields have increased enormously in the past and are still doing so (Ankenbrand and Schwertmann, 1989; Chartin et al., 2011; Zádorová et al., 2013). Field borders that cross thalwegs and that are able to trap sediment continuously disappear (by a factor of three during the last 50 yr; Boardman and Vandaele, 2016) and will continue to do so. It is not surprising that this loss has resulted in more catchment runoff and peak discharge (Evrard et al., 2007).

The Scientific Viewpoint

Major difficulties arise when evaluating GWWs. They require large-scale (watershed) and long-term (>5 yr) research. Scientific standards (replicates, controls) usually cannot be met because there are no two identical watersheds that can be compared with and without GWWs. Even socioeconomic conditions may differ between watersheds because they are so large (Prokopy et al., 2011). The large measuring effort usually means that no replicates are available, and finally, the installation of GWWs is usually accompanied by other measures, such as a change in field layout, that make it difficult to isolate the effect of the GWW. Furthermore, only studies on functional GWWs exist, and studies describing the conditions, frequency, and remediation of failure ones are missing. An interesting although unique approach that solves at least some of these problems was followed by Shipitalo et al. (2010, 2012), who used only one watershed but diverted the outflow into two GWWs that could then be compared under near-identical conditions without the need to control the agricultural practices within the entire watershed, which would have been necessary if entire watersheds had been compared.

Because of these limitations, many evaluations have been based on modeling, mainly using SWAT (Soil and Water Assessment Tool; Secchi et al., 2007; Lerch et al., 2008; Luo and Zhang, 2009; Makarewicz et al., 2015), WEPP (Water Erosion Prediction Project; Renschler and Lee, 2005; Zhou et al., 2009; Dermisis et al., 2010; Gassman et al., 2010), or a number of other models (Yaramanoglu, 1986; Gassman et al., 2010; Kalantari et al., 2014; Dabney et al., 2015).

Modeling, however, suffers from the same problems as landscape experiments because models cannot be set up, parameterized, and validated without experimental data. Furthermore, most (erosion) models are optimized mainly for distributed runoff on large areas. Such models cannot reflect in enough detail the conditions governing concentrated flow along permanently grassed narrow paths to allow for realistic hydrodynamic modeling. Even those model approaches technically addressing the GWWs as areas of concentrated flow with specific flow-depth-dependent cross-sectional areas (Temple et al., 1987; Kalantari et al., 2014) face major deficits in three areas of data availability and process representation: (i) Data regarding (potentially changing) cross-sectional areas and inflow pathways affected by agricultural management in the surrounding field are often not available in enough detail. For example, Fiener and Auerswald (2005) showed that a small incision (0.5 m wide and 0.2 m deep) resulting from linear erosion during the establishment of a GWW substantially reduced the measured and modeled infiltration capacity within a 370-m-long GWW. This lack of detailed data could be addressed in case of individual GWWs via field surveys and might partly be resolved as a result of the improving quality and resolution of digital elevation models. (ii) Our knowledge is limited in parameterizing the hydrodynamic behavior of different grass and herb communities that have varying stem flexibility and inter-stem porosity (Montakhab et al., 2012). Moreover, little is known regarding the changes in the hydrodynamic properties of vegetation throughout the year, knowledge that calls for season-specific parameter sets. The variability of the hydrodynamic parameters of grasses can be illustrated for hydraulic roughness (expressed as Manning's n). In general, Manning's n in grass-lined channels depends on the density of the grass and is highest and most stable for sod-forming vegetation that has a dense, relatively deep root system (e.g., Bermuda grass [*Cynodon dactylon* (L.) Pers.]; USDA-SCS, 1954). Depending on flow depths and velocity, plants may change their position, start oscillating (USDA-SCS, 1954), or even bend or break if there are high runoff velocities and large runoff depths (Kouwen and Unny, 1973). For dense, sod-forming grasses, Manning's n decreases with increasing flow velocity and depths (often expressed as the product of flow velocity and hydraulic radius), from values between 0.3 and 0.4 s m$^{-1/3}$ (Ree, 1949; USDA-SCS, 1954; Kouwen, 1992) to values between 0.03 and 0.1 s m$^{-1/3}$ (USDA-SCS, 1954; Kouwen and Unny, 1973). It is difficult to predict whether bending or breaking will occur. Both depend not only on the flow velocity and depths, but they are also governed by the seasonally changing hydraulic properties of the grass canopy. Moreover, there is little knowledge regarding the recovery of the hydraulic properties of grasses and herbs after being bent (Fiener and Auerswald, 2006b). (iii) The parameterization of soil properties is limited by the fact that at least during the first few years after establishment, the soil passes through a transition phase, with parameters changing between the former arable use and a typical grassland soil. Compared with common grassland, after the transition phase there may be major differences due to the belt shape of a GWW, which may result in mammal burrows preferentially following the long axis and thereby affecting runoff behavior. This suggestion is, however, speculative.

Surprisingly little experimental evidence exists about the ecological benefits and drawbacks of GWWs (Henry et al., 1999), although this absence should be of major concern to society because it would justify incentives for the installation of GWWs. Experimental evidence would supply knowledge about optimizing design and maintenance in this respect. This deficit is partly caused by the same difficulty that studies on erosion control have; namely, that no two identical watersheds exist that differ only in the existence of a GWW. However, such studies may exist but cannot be found because terms like "set-aside" or "grassland" are used in place of the proper technical term, "grassed waterway." . This may occur because the technical term is unknown to specialists in other fields, such as arachnology, herpetology, or aquatic ecology, or that the term is avoided to attract a wider audience and to make the results more generally applicable. For example, most studies cited above regarding the influence of GWWs on species dynamics (Agricola et al., 1996; Filser et al., 1996; Mebes and Filser, 1997) avoid using "grassed waterway." As a result, the knowledge that these studies concerned GWWs was restricted to the few persons who became aware of the studies during the short period of fieldwork.

The lack of proper ecological studies on GWWs is even more regrettable because GWWs behave differently than set-asides or grasslands in many respects, such as in the corridor shape, the position within the landscape connecting valley bottoms with upslope positions, and the frequent input of water carrying soil, nutrients, and other substances. Hence, GWWs would be mischaracterized if experiences from set-asides or grasslands were applied to them. Alternatively, grasslands and set-asides are equally mischaracterized if, in fact, it is GWWs that have been studied. Also, GWWs and filter strips are frequently treated as similar because both are narrow strips of land covered with permanent vegetation. This is, however, misleading in several hydrological, ecological, and agronomic respects. For example, it is the length of the flow path that governs infiltration, and there is no larger difference in the length of the flow path than that between those of a GWW and a filter strip. Another difference is whether the narrow strip connects different habitats or runs within one riparian habitat. Another difference is whether a narrow strip can be crossed by farm machinery or whether this is impossible due to an adjacent watercourse. Hence, given the diversity of beneficial on-site and off-site effects that seem to exist, there is a clear and pressing need for more direct studies on GWWs and less deduction by analogy.

References

Agricola, U., J. Barthel, H. Laussmann, and H. Plachter. 1996. Struktur und Dynamik der Fauna einer süddeutschen Agrarlandschaft nach Nutzungsumstellung auf ökologischen und integriereten Landbau. Verh. Ges. Ökol. 26:681–692.

Ahuja, L.R., A.N. Sharpley, M. Yamamoto, and R.G. Menzel. 1981. The depth of rainfall-runoff-soil interaction as determined by 32P. Water Resour. Res. 17:969–974. doi:10.1029/WR017i004p00969

Ankenbrand, E., and U. Schwertmann. 1989. The land consolidation project of Freinhausen, Bavaria. In: U. Schwertmann; R.J. Rickson; K. Auerswald, editors, Soil erosion measures in Europe. Proceedings of the European Community Workshop on Soil Erosion Protection, Freising, Germany. 24–26 May 1988. Soil Technology Series 1. Catena Verlag, Reiskirchen, Germany. p. 167–173.

Asmussen, L.E., A.W. White, Jr., E.W. Hauser, and J.M. Sheridan. 1977. Reduction of 2,4D load in surface runoff down a grassed waterway. J. Environ. Qual. 6:159–162. doi:10.2134/jeq1977.00472425000600020011x

Atkins, D.M., and J.J. Coyle. 1977, Grass waterways in soil conservation. USDA Leaflet 477. U.S. Gov. Print. Office, Washington, D.C

Auerswald, K., and S. Weigand. 1996. Ecological impact of dead-wood hedges: Release of dissolved phosphorus and organic matter into runoff. Ecol. Eng. 7:183–189. doi:10.1016/0925-8574(96)00007-9

Bracmort, K.S., B.A. Engel, and J.R. Frankenberger. 2004. Evaluation of structural best management practice 20 years after installation: Black Creek watershed, Indiana. J. Soil Water Conserv. 59:191–196.

Berg, N.A., and R.J. Gray. 1984. Soil conservation: "The search for solutions." J. Soil Water Conserv. 39:18–22.

Boardman, J., and K. Vandaele. 2016. Effect of the spatial organization of land use on muddy flooding from cultivated catchments and recommendations for the adoption of control measures. Earth Surf. Process. Landf. 41:336–343. doi:10.1002/esp.3793

Briggs, J.A., T. Whitwell, and M.B. Riley. 1999. Remediation of herbicides in runoff water from container plant nurseries utilizing grassed waterways. Weed Technol. 12:157–164.

Bryan, G.G., and L.B. Best. 1991. Bird abundance and species richness in grassed waterways in Iowa rowcrop fields. Am. Midl. Nat. 126:90–102. doi:10.2307/2426153

Bryan, G.G., and L.B. Best. 1994. Avian nest density and success in grassed waterways in Iowa rowcrop fields. Wildl. Soc. Bull. 22:583–592.

Carey, B.W., B. Stone, P.L. Norman, and P. Shilton. 2015a. Waterways In: Soil conservation guidelines for Queensland. Chap. 9. Dep. of Science, Information Technology and Innovation, Brisbane. p. 1–19. https://publications.qld.gov.au/dataset/e9116bae-b843-4b61-8050-433898222bd6/resource/125aadac-0c22-4564-98a6-98e61eee1f6e/download/soilconguide3echapter9.pdf (accessed 31 Dec. 2016).

Carey, B.W., B. Stone, P.L. Norman, and P. Shilton. 2015b. Soil conservation in horticulture In: Soil conservation guidelines for Queensland. chap 12. Dep. of Science, Information Technology and Innovation, Brisbane. p. 1–45. https://publications.qld.gov.au/dataset/e9116bae-b843-4b61-8050-433898222bd6/resource/e416f81e-d0e6-4245-a1d1-961070a566a2/download/chapter12horticulture.pdf (accessed 31 Dec. 2016).

Chartin, C., H. Bourennane, S. Salvador-Blanes, F. Hinschberger, and J.-J. Macaire. 2011. Classification and mapping of anthropogenic landforms on cultivated hillslopes using DEMs and soil thickness data: Example from the SW Parisian Basin, France. Geomorphology 135:8–20. doi:10.1016/j.geomorph.2011.07.020

Chen, L., and B.D. Barkdoll. 2013. Optimal location of watershed best management practices for sediment yield, reduction and cost. In: C.L. Patterson, S.S. Struck, and D.J. Murray, editors, Showcasing the future. World Environmental and Water Resources Congress 2013, Cincinnati, Ohio. 19–23 May. American Society of Civil Engineers, Reston, VA. p. 3183–3195. doi:10.1061/9780784412947.315

Chow, T.L., H.W. Rees, and J.L. Daigle. 1999. Effectiveness of terraces/grassed waterway systems for soil and water conservation: A field evaluation. J. Soil Water Conserv. 3:577–583.

Dabney, S.M., D.A.N. Vieira, D.C. Yoder, E.J. Langendoen, R.R. Wells, and M.E. Ursic. 2015. Spatially distributed sheet, rill, and ephemeral gully erosion. J. Hydrol. Eng. 20(6). doi:10.1061/(ASCE)HE.1943-5584.0001120

Dass, A., S. Sudhishri, N.K. Lenka, and U.S. Patnaik. 2011. Runoff capture through vegetative barriers and planting methodologies to reduce erosion, and improve soil moisture, fertility and crop productivity in southern Orissa, India. Nutr. Cycling Agroecosyst. 89:45–57. doi:10.1007/s10705-010-9375-3

Deletic, A. 2001. Modelling of water and sediment transport over grassed areas. J. Hydrol. 248:168–182. doi:10.1016/S0022-1694(01)00403-6

Dermisis, D., O. Abaci, A.N. Papanicolaou, and C.G. Wilson. 2010. Evaluating grassed waterway efficiency in southeastern Iowa using WEPP. Soil Use Manage. 26:183–192. doi:10.1111/j.1475-2743.2010.00257.x

Dillaha, T.A., R.B. Reneau, S. Mostaghimi, and D. Lee. 1989. Vegetative filter strips for agricultural nonpoint source pollution control. Trans. ASAE 32:513–519. doi:10.13031/2013.31033

Evrard, O., E. Persoons, K. Vandaele, and B. van Wesemael. 2007. Effectiveness of erosion mitigation measures to prevent muddy floods: A case study in the Belgian loam belt. Agric. Ecosyst. Environ. 118:149–158. doi:10.1016/j.agee.2006.02.019

Evrard, O., K. Vandaele, C. Bielders, and B. Van Wesemael. 2008a. Seasonal evolution of runoff generation on agricultural land in the Belgian loess belt and implications for muddy flood triggering. Earth Surf. Processes Landforms 33:1285–1301. doi:10.1002/esp.1613

Evrard, O., K. Vandaele, B. Van Wesemael, and C.L. Bielders. 2008b. A grassed waterway and earthen dams to control muddy floods from a cultivated catchment of the Belgian loess belt. Geomorphology 100:419–428. doi:10.1016/j.geomorph.2008.01.010

Fiener, P., and K. Auerswald. 2003a. Concept and effects of a multi-purpose grassed waterway. Soil Use Manage. 19:65–72. doi:10.1111/j.1475-2743.2003.tb00281.x

Fiener, P., and K. Auerswald. 2003b. Effectiveness of grassed waterways in reducing runoff and sediment delivery from agricultural watersheds. J. Environ. Qual. 32:927–936. doi:10.2134/jeq2003.9270

Fiener, P., and K. Auerswald. 2005. Measurement and modeling of concentrated runoff in a grassed waterway. J. Hydrol. 301:198–215. doi:10.1016/j.jhydrol.2004.06.030

Fiener, P., and K. Auerswald. 2006a. Influence of scale and land use pattern on the efficacy of grassed waterways to control runoff. Ecol. Eng. 27:208–218. doi:10.1016/j.ecoleng.2006.02.005

Fiener, P., and K. Auerswald. 2006b. Seasonal variation of grassed waterway effectiveness in reducing runoff and sediment delivery from agricultural watersheds in temperate Europe. Soil Tillage Res. 87:48–58. doi:10.1016/j.still.2005.02.035

Fiener, P., and K. Auerswald. 2009. Effects of hydrodynamically rough grassed waterways on dissolved reactive phosphorus loads coming from agricultural watersheds. J. Environ. Qual. 38:548–559. doi:10.2134/jeq2007.0525

Filser, J., A. Lang, K.-H. Mebes, S. Mommertz, A. Palojärvi, and K. Winter. 1996. The effect of land use change on soil organisms: An experimental approach. Verh. Ges. Ökol. 26:671-679.

Gassman, P.W., J.R. Williams, X. Wang, A. Saleh, E. Osei, L.M. Hauck, R.C. Izaurralde, and J.D. Flowers. 2010. The Agricultural Policy/Environmental eXtender (APEX) model: An emerging tool for landscape and watershed environmental analyses. Trans. ASABE 53:711–740. doi:10.13031/2013.30078

Glenn, S., and G.H. Williams. 1985. Nonpoint source loading of phenolic acids from decomposing crop residue. Soil Tillage Res. 6:45–51. doi:10.1016/0167-1987(85)90005-4

Gordon, L.M., S.J. Bennett, C.V. Alonso, and R.L. Bingner. 2008. Modeling long-term soil losses on agricultural fields due to ephemeral gully erosion. J. Soil Water Conserv. 63:173–181. doi:10.2489/jswc.63.4.173

Henry, A.C., D.A. Hosack, C.W. Johnson, D. Rol, and G. Bentrup. 1999. Conservation corridors in the United States: Benefits and planning guidelines. J. Soil Water Conserv. 54:645–65.

Hjelmfelt, A., and M. Wang. 1997. Using modelling to investigate impacts of grass waterways on water quality. In: F.M. Holly and A. Alsoffar, editors, Water for a changing global community. Proceedings of the 27th International Association of Hydraulic Research (IAHR) World Congress, San Francisco, CA. 10–15 August. American Society of Civil Engineers, Reston, VA. p. 1420–1425.

Hjelmfelt, A., and M. Wang. 1999. Modeling hydrologic and water quality responses to grass waterways. J. Hydrol. Eng. 4:251–256. doi:10.1061/(ASCE)1084-0699(1999)4:3(251)

Hultquist, J.M., and L.B. Best. 2001. Bird use of terraces in Iowa row crop fields. Am. Midl. Nat. 145:275–287. doi:10.1674/0003-0031(2001)145[0275:BUOTII]2.0.CO;2

Kalantari, Z., S.W. Lyon, L. Folkeson, H.K. French, J. Stolte, P.E. Jansson, and M. Sassner. 2014. Quantifying the hydrological impact of simulated changes in land use on peak discharge in a small catchment. Sci. Total Environ. 466-467:741–754. doi:10.1016/j.scitotenv.2013.07.047

Kling, C., S.S. Rabotyagov, M. Jha, H. Feng, J. Parcel, P.W. Gassman, and T. Campbell. 2007. Conservation practices in Iowa: Historical investments, water quality, and gaps. A report to the Iowa farm bureau and partners. http://citeseerx.ist.psu.edu/viewdoc/download?doi=10.1.1.568.7680&rep=rep1&type=pdf (accessed 31 Dec. 2016).

Knoot, T.G., and L.B. Best. 2011. A multiscale approach to understanding snake use of conservation buffer strips in an agricultural landscape. Herpetol. Conserv. Biol. 6:191–201.

Kouwen, N., 1992. Modern approach to design of grassed channels. J. Irrig. Drain. Div., Am. Soc. Civ. Eng. 118:733–743.

Kouwen, N., and T.E. Unny. 1973. Flexible roughness in open channels. J. Hydraul. Div., Am. Soc. Civ. Eng. 99:713–728.

Lawler, D.M., M. Dolan, H. Tomasson, and S. Zophoniasson. 1992. Temporal variability of suspended sediment flux from a subarctic glacial river, southern Iceland. In: J. Bogen, D.E. Walling, and T.J. Day, editors, Erosion and sediment transport monitoring programmes in river basins. Proceedings of the International Symposium, Oslo, Norway. 24–28 August. IAHS Proceedings and Reports, Wallingford, UK. p. 233–243.

Lemke, A.M., K.G. Kirkham, T.T. Lindenbaum, M.E. Herbert, T.H. Tear, W.L. Perry, and J.R. Herkert. 2011. Evaluating agricultural best management practices in tile-drained subwatersheds of the Mackinaw River, Illinois. J. Environ. Qual. 40:1215–1228. doi:10.2134/jeq2010.0119

Lemke, A.M., T.T. Lindenbaum, W.L. Perry, M.E. Herbert, T.H. Tear, and J.R. Herkert. 2010. Effects of outreach on the awareness and adoption of conservation practices by farmers in two agricultural watersheds of the Mackinaw River, Illinois. J. Soil Water Conserv. 65:304–315. doi:10.2489/jswc.65.5.304

Lerch, R.N., E.J. Sadler, N.R. Kitchen, K.A. Sudduth, R.J. Kremer, D.B. Myers, C. Baffaut, S.H. Anderson, and C.H. Lin. 2008. Overview of the Mark Twain Lake/Salt River basin conservation effects assessment project. J. Soil Water Conserv. 63:345–359. doi:10.2489/jswc.63.6.345

Luck, J.D., S.K. Pitla, R.S. Zandonadi, M.P. Sama, and S.A. Shearer. 2011a. Estimating off-rate pesticide application errors resulting from agricultural sprayer turning movements. Precis. Agric. 12:534–545. doi:10.1007/s11119-010-9199-9

Luck, J.D., R.S. Zandonadi, and S.A. Shearer. 2011b. A case study to evaluate field shape factors for estimating overlap errors with manual and automatic section control. Trans. ASABE 54:1237–1243. doi:10.13031/2013.39022

Luo, Y.Z., and M.H. Zhang. 2009. Management-oriented sensitivity analysis for pesticide transport in watershed-scale water quality modeling using SWAT. Environ. Pollut. 157:3370–3378. doi:10.1016/j.envpol.2009.06.024

Makarewicz, J.C., T.W. Lewis, M. Winslow, E. Rea, L. Dressel, D. Pettenski, B.J. Snyder, P. Richards, and J. Zollweg. 2015. Utilizing intensive monitoring and simulations for identifying sources of phosphorus and sediment and for directing, siting, and assessing BMPs: The Genesee River example. J. Great Lakes Res. 41:743–759. doi:10.1016/j.jglr.2015.06.004

Mebes, K.-H., and J. Filser. 1997. A method for estimating the significance of surface dispersal for population fluctuations of Collembola in arable land. Pedobiologia 41:115–122.

Montakhab, A., B. Yusuf, A.H. Ghazali, and T.A. Mohamed. 2012. Flow and sediment transport in vegetated waterways: A review. Rev. Environ. Sci. Biotechnol. 11:275–287. doi:10.1007/s11157-012-9266-y

Muñoz-Carpena, R., J.E. Parson, and J.W. Gilliam. 1993. Numerical approach to the overland flow process in vegetative filter strips. Trans. ASAE 36:761–770. doi:10.13031/2013.28395

Napier, T.L., C.S. Thraen, A. Gore, and W.R. Goe. 1984. Factors affecting adoption of conventional and conservation tillage practices in Ohio. J. Soil Water Conserv. 39:205–209.

Nyaupane, N.P., and J.M. Gillespie. 2011. Louisiana crawfish farmer adoption of best management practices. J. Soil Water Conserv. 66:61–70. doi:10.2489/jswc.66.1.61

Ogunlela, A.O., and M.B. Makanjuola. 2000. Hydraulic roughness of some African grasses. J. Agric. Eng. Res. 75:221–224. doi:10.1006/jaer.1999.0486

Osmond, D., D. Meals, D. Hoag, M. Arabi, A. Luloff, G. Jennings, M. McFarland, J. Spooner, A. Sharpley, and D. Line. 2012. Improving conservation practices programming to protect water quality in agricultural watersheds: Lessons learned from the National Institute of Food and Agriculture-Conservation Effects Assessment Project. J. Soil Water Conserv. 67:122A–127A. doi:10.2489/jswc.67.5.122A

Prokopy, L.S., A.Z. Göçmen, J. Gao, S.B. Allred, J.E. Bonnell, K. Genskow, A. Molloy, and R. Power. 2011. Incorporating social context variables into paired watershed designs to test nonpoint source program effectiveness. J. Am. Water Resour. Assoc. 47:196–202.

Qiu, Z., C. Hall, and K. Hale. 2009. Evaluation of cost-effectiveness of conservation buffer placement strategies in a river basin. J. Soil Water Conserv. 64:293–302. doi:10.2489/jswc.64.5.293

Ree, W.O. 1949. Hydraulic characteristics of vegetation for vegetated waterways. Agric. Eng. 30:184–187.

Renschler, C.S. and T. Lee. 2005. Spatially distributed assessment of of short-and long-term impacts of multiple best management practices in agricultural watersheds. Journal of Soil and Water Conservation 60: 446-456.

Secchi, S., P.W. Gassman, M. Jha, L. Kurkalova, H.H. Feng, T. Campbell, and C.L. Kling. 2007. The cost of cleaner water: Assessing agricultural pollution reduction at the watershed scale. J. Soil Water Conserv. 62:10–21.

Sharpley, A.N. 1980. The enrichment of soil phosphorus in runoff sediments. J. Environ. Qual. 9:521–526. doi:10.2134/jeq1980.00472425000900030039x

Shelton, D.P., R.A. Wilke, T.G. Franti, and S.J. Josiah. 2009. Farmlink: Promoting conservation buffers farmer-to-farmer. Agrofor. Syst. 75:83–89. doi:10.1007/s10457-008-9130-9

Shipitalo, M.J., J.V. Bonta, E.A. Dayton, and L.B. Owens. 2010. Impact of grassed waterways and compost filter socks on the quality of surface runoff from corn fields. J. Environ. Qual. 39:1009–1018. doi:10.2134/jeq2009.0291

Shipitalo, M.J., J.V. Bonta, and L.B. Owens. 2012. Sorbent-amended compost filter socks in grassed waterways reduce nutrient losses in surface runoff from corn fields. J. Soil Water Conserv. 67:433–441. doi:10.2489/jswc.67.5.433

Singh, S., S. Inamdar, M. Mitchell, and P. McHale. 2014. Seasonal pattern of dissolved organic matter (DOM) in watershed sources: Influence of hydrologic flow paths and autumn leaf fall. Biogeochemistry 118:321–337. doi:10.1007/s10533-013-9934-1

Spaan, W., H. Winteraeken, and P. Geelen. 2010. Adoption of SWC measures in South Limburg (The Netherlands): Experiences of a water manager. Land Use Policy 27:78–85. doi:10.1016/j.landusepol.2008.10.015

Temple, D.M., K.M. Robinson, R.M. Ahring, and A.G. Davis. 1987. Stability design of grass-lined open channels. Agric. Handb. 667. USDA-ARS, Washington, DC. http://www.fcd.maricopa.gov/pub/docs/scanfcdlibrary%5C101_302_StabilityDesignofGrass_linedOpenChannels.pdf (accessed 31 Dec. 2016).

Thomas, G.A., G.W. Titmarsh, D.M. Freebairn, and B.J. Radford. 2007. No-tillage and conservation farming practices in grain growing areas of Queensland: A review of 40 years of development. Aust. J. Exp. Agric. 47:887–898. doi:10.1071/EA06204

Titmarsh, G.W., and B.J. Stone. 1997. Runoff management: Techniques and structures. In: A.L. Clarke and P.B. Wylie, editors, Sustainable crop production in the sub-tropics: An Australian perspective. Department of Primary Industries, Queensland, Australia. p. 181–194.

USDA Soil Conservation Service (USDA-SCS). 1954. Handbook of channel design for soil and water conservation. USDA-SCS, Washington, DC. http://www.wcc.nrcs.usda.gov/ftpref/wntsc/H&H/TRsTPs/TP61.pdf (accessed 31 Dec. 2016).

Van Oost, K., G. Govers, and P. Desmet. 2000. Evaluating the effects of changes in landscape structure on soil erosion by water and tillage. Landsc. Ecol. 15:577–589. doi:10.1023/A:1008198215674

Yaramanoglu, M. 1986. Microcomputer-based interactive design of grassed waterways. Comput. Electron. Agric. 1:173–184. doi:10.1016/0168-1699(86)90005-0

Zádorová, T., V. Penižek, L. Šefrna, O. Drábek, M. Mihaljevič, S. Volf, and T. Chuman. 2013. Identification of Neolithic to modern erosion-sedimentation phases using geochemical approach in a loess covered sub-catchment of South Moravia, Czech Republic. Geoderma 195-196:56–69. doi:10.1016/j.geoderma.2012.11.012

Zhou, X., M. Helmers, M. Al-Kaisi, and M. Hanna. 2009. Cost-effectiveness and cost-benefit analysis of conservation management practices for sediment reduction in an Iowa agricultural watershed. J. Soil Water Conserv. 64:314–323. doi:10.2489/jswc.64.5.314

Terracing and Contour Farming

Allen Thompson* and Ken Sudduth

Abstract

One of the major challenges in proper land management is limiting surface runoff and sediment loss. Crop production by its very nature involves disturbing the soil. Therefore practical approaches that support crop production while effectively protecting the soil are needed. Two such practices that help to accomplish this include terracing and contouring. Today's advances in precision conservation offer improvements in properly locating terraces and in maintaining acceptable contour grade in the field.

Terraces are earthen embankments constructed across the prevailing field land slope. They have been used in differing forms for thousands of years in an attempt to protect steep land slopes from runoff-induced erosion. Various cross-sections are possible, and in general consist of a cut section and downslope earthen berm designed to intercept and remove runoff in a controlled fashion (Fig. 1). In arid regions, they are often used for harvesting rainfall to increase soil moisture retention in the field. As such, they may be nearly level, with end sections blocked to facilitate infiltration slowly over time. In humid regions, terraces serve as a conservation practice by limiting concentrated flow length down the slope. Runoff water is directed out of the field along the cut channel to a vegetative waterway or through riser pipes connected to underground outlets (UGO). On flatter field slopes, terraces may be cropped on both the up and downslope sections of the embankment. As field slope increases, cut sections constructed to provide material for the embankment may become excessive, such that one or both side slopes must be maintained in permanent grass.

Contouring involves formation of small soil ridges and furrows placed at right angles to the field slope. They are formed during tillage and planting, and are used to intercept surface runoff by directing flow along the hill slope to facilitate increased infiltration. They can be effective when maintained close to the contour, but lose their effectiveness quickly if contour grade cannot be maintained. They are often limited to flatter slopes and shorter slope lengths due to the increased likelihood of overtopping on steeper grades and longer field lengths. Field implementation is straightforward in uniformly-sloped fields, but can be

Abbreviations: DEM, digital elevation model, GIS, geographic information systems; GPS, global positioning systems; LiDAR, light detection and radar; RTK, real-time kinematic; TDT, terrace design tool; UGO, underground outlets.

A. Thompson, Bioengineering Department, University of Missouri, Columbia, MO 65211; K. Sudduth, USDA-Agricultural Research Service, Cropping Systems and Water Quality Research, Columbia, MO 65201. *Corresponding author (ThompsonA@missouri.edu).

doi:10.2134/agronmonogr59.2016.0010

Precision Conservation: Geospatial Techniques for Agricultural and Natural Resources Conservation
J.A. Delgado, G.F. Sassenrath, and T. Mueller, editors. Agronomy Monograph 59.

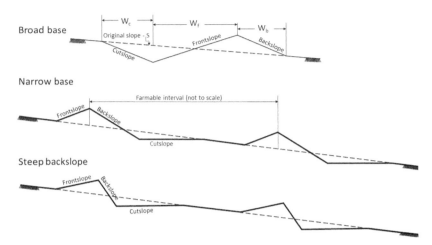

Fig. 1. Example cross-sections for broadbase, narrow base, and steep backslope terrace cross-sections; W_c = cutslope width, W_f = frontslope width, and W_b = backslope width.

difficult to fully accomplish for fields that are dissected or where the slope varies widely along a given lateral position across the field width. In such cases, use of grassed waterways may improve the overall usefulness of the practice.

Agronomic and Environmental Considerations

Terracing can be a costly investment, and is often only used when other con-servation measures are inadequate to control soil loss from overland flow, or concentrated flow results in ephemeral or gully formation. Horizontal spacing intervals between terraces should be designed to accommodate cropping equip-ment while maintaining proper slope length limits to provide effective erosion control. The horizontal intervals between adjoining terraces should be limited to 210 m on 0 to 4% slopes, 180 m for 4 to 6% slopes, 150 m for 6 to 12% slopes, 120 m for 12 to 18% slopes, and 90 m for slopes greater than 18% (ASABE, 2012). Regard-less of the final spacing, maximum flow rate in the terrace channel should not exceed 0.8 m s^{-1}, and for erodible soils may be limited to 0.5 m s^{-1} (ASABE, 2012). Estimated flow velocities can be determined using overland flow relationships based on contributing runoff area, slope, and terrace channel length. Addition-ally, terrace spacing can be estimated through use of the vertical interval equation or evaluated based on sediment loss estimates to limit soil loss to T-values using programs such as RUSLE2 (ASABE, 2012; USDA-NRCS, 2010a).

Because maintenance and proper management are vital to long-term success of terraces, terrace layout should provide a field system that facilitates typical farming practices. This includes adequate radius of curvature using curves that are smooth and practical, particularly in view of today's wider field equipment. Terraces are suited for land capability classes II, III, and IV, which represent good quality farmable land with slopes over 2% where moderate to very severe erosion may occur (Das, 2009). Installations on slopes up to 18% are possible, but cut sec-tions may become excessive and other forms of soil erosion protection should be considered. On flatter field slopes, terraces are useful in reducing sheet and rill erosion. For steeper slopes, terraces are most beneficial in reducing ephemeral gully erosion or gully formation. Terraces are typically not recommended for soils with limited topsoil since removal of material needed to construct embankments further reduces the productive potential by exposing subsoil. Productivity of most terraces, particularly those in shallow soils, can be enhanced by replacing removed topsoil after construction is completed. The added cost of replacing the topsoil can often be justified by considering overall soil health and improvement in crop performance as well as reduced surface runoff.

Terraces are classified by alignment, cross-section, grade, or outlet. For alignment, terraces can either be parallel or nonparallel. Parallel alignment is preferred since it eliminates point rows and results in a more efficiently farmed system. However, parallel layouts often require more soil to be moved, resulting in increased construction costs. Nonparallel systems are aligned with the contour, and may be located with either constant or variable grade depending on the terrace length and given topography. They are simpler to lay out, but for fields with varying slope, they will be more difficult to farm since adjacent terraces will not be spaced an equal distance apart along their entire length. It has been shown that non-parallel systems require about 20% more time to farm than parallel systems (Steichen and Powell, 1985).

Terrace cross-section is often based on field slope, with broadbase terraces used on flatter slopes, and steep backslope or narrow based terraces used on steeper slopes (Fig. 1). The cutslope section for broadbase terraces (i.e., the source location for soil to construct the berm) is on the upslope side of the embankment which results in an increase in the effective slope as measured from the ridge of the upslope terrace to the upslope toe of the adjoining downslope terrace. This increase in effective slope limits these cross-sections to field slopes less than 6 to 8% (ASABE, 2012). The cutslope for steep backslope and narrow base terraces is normally taken from the area directly below the terrace embankment, effectively reducing the average slope over the length of the farmable terrace interval. Although steep backslope and narrow based cross-sections remove land area from crop production, the benefits are reduced soil erosion while farming steeper field slopes, and a potential increase in wildlife habitat from the premanently grassed backslope and/or frontslope sections. It is preferred that the channel bottom width be at least 0.9 m wide to facilitate water flow, and the ridge should be at least 0.9 m wide to facilitate the required design capacity (ASABE, 2012). For illustrative purposes, the terrace cross-sectional schematic in Fig. 1 is drawn with a sharp apex for the ridge and channel, but over time, sedimentation in the channel and erosion on the embankment may give a wider, more parabolic effective bottom width and a rounded ridge. Therefore terrace height should be based on the elevation difference between the top of the embankment and bottom of the channel where each is measured to be at least 0.9 m wide. This helps avoid overestimating the effective height of the terrace. Where rainfall is low, level terraces with blocked outlets may be used to conserve rainwater, whereas graded terraces are used where erosion control is the primary goal. Terrace design capacity should be sufficient to handle the 10-yr frequency 24-hr duration storm (ASABE, 2012).

Vegetative waterways and UGO systems can be used with nearly all graded terrace systems, the selection being based on cost and ease of farming. Underground outlets are often chosen for parallel terrace systems to take greater advantage of the improved farmability of the layout. If UGO systems are used, all water should be removed from the terrace channel within 24 to 48 h to avoid drowning the vegetation (ASABE, 2012). Since UGO systems directly remove water from the channel, opportunity for filtering is minimized and left for possible removal at the lower outlet point of the conduits where the water may be discharged onto grassed areas at the base of the field. One of the benefits of vegetative waterways is that they filter sediments and nutrients from the runoff. A disadvantage to grassed waterways is the loss of crop production from the field area removed from production and the time delay to get vegetation established. Specifically, waterways must be constructed prior to terrace construction; therefore a delay of 1 to 2 yr may be needed before terraces can actually be installed.

Contour farming, where tillage and planting create ridges and furrows at nearly right angles to the field slope, is encouraged to facilitate reduction in sheet and rill erosion with or without terraces. Historically, contouring has been most universally applied in the eastern wheat belt region of the United States across Nebraska, Kansas, Oklahoma, and Texas. In arid areas, contouring may improve rainfall infiltration. The practice of contour farming is limited in slope length for steeper slopes due to the greater potential to overtop, and is most effective for slope lengths between 30 and 120 m (USDA-NRCS, 2010b). The overall benefits of contouring are directly related to field slope, with the greatest benefit found on field slopes between 3 and 8% (Wischmeier and Smith, 1958). For slopes greater than 8%, the tendency of runoff water to overtop the contour ridges increases, whereas for slopes less than 3%, the relative benefit of contouring compared to other land surface practices is diminished since the flatter slopes do not result in as great a runoff rate. The risk of overtopping should be carefully considered since overtopping can increase gullying. Maximum row grade along the contour should be as uniform as possible, limited to one-half of the hill slope percent (not to exceed 10%) or not greater than 0.2% for increased water infiltration (USDA-NRCS, 2010b). One of the challenges with contouring is the ability of the equipment operator to stay within the effective contour grade across the entire field width. As with terraces, point rows caused by nonuniform field slopes may impede farmability and reduce the desirability of this practice to producers. The benefits of contouring typically decrease during the year due to deposition of sediment by wind and rain, and reduction in infiltration due to surface sealing (Harold, 1947). This may be partially offset by soil surface protection from increased crop foliage or residue during the growing season.

Planning and Design with Geospatial Analysis and Data

Precision farming concepts and supporting technologies, including global positioning systems (GPS) and geographic information systems (GIS), have modified agricultural practices by introducing new opportunities to apply spatial precision to field management (Stafford, 2000). In addition to precise control of cropping operations, these opportunities extend to planning, design, and implementation of conservation practices, beginning with visualization and evaluation of the topography of the field or watershed in question.

An example of applying precision conservation tools to evaluate a watershed area is shown in Fig. 2, where elevation data, based on a light detection and radar (LiDAR) digital elevation model (DEM) surface, were shade-highlighted and superimposed on an ortho photographic image. Existing terraces are clearly identified in the image. By selecting the desired outlet point, the watershed boundary (red) and maximum flow length (dark blue) were automatically traced, and the resulting watershed area, maximum flow length, and land slope determined. Data layers of soil survey feature class and land use feature class are also available.

Additional watershed hydrologic characteristics, such as overland flow paths, can be further analyzed by selecting the minimum areal contribution to overland flow as shown in Fig. 3 (light blue lines). Collectively, these provide powerful tools to support conservation decisions.

Traditionally, terrace layout has been done manually using field engineering surveys. The procedure typically requires a field technician to stake out points along each terrace line based on topographic maps, which are then verified with a laser level or similar survey equipment. Time may be needed for additional field measurements to finalize an acceptable design. The design and layout must consider the cost effectiveness and manageability of the terrace systems in addition to the soil, topography, and field shape (Schottman and White, 1993; Tollner, 2002). These types of determinations are best made when a landowner can see various design possibilities; however, conservation professionals today typically do not have the tools to easily provide alternative design visualizations. This often results in the establishment of terrace systems that are difficult for the landowner to maintain. Unfortunately once installed, modifications are generally limited to minor adjustments along the ends of terraces. Ultimately a terrace system that a landowner has difficulty supporting will have reduced conservation effectiveness.

Fig. 2. Drainage area delineation and watershed hydrology characteristics using LiDAR DEM surface, orthophotography, soil survey feature class and land use feature class; completed using ESRI ArcMap 10.0 GIS software.

Georeferenced terrace designs allow for use of position guidance and machine control during construction utilizing either laser or GPS based systems. The development of high-resolution DEMs can assist in making the initial layout plans (Fangmeier et al., 2006). The digital elevation data used to create high-resolution DEMs may be obtained efficiently through LiDAR or Real Time Kinematic (RTK) GPS equipment used in precision farming. Although specific results may vary with equipment and operational characteristics, studies have shown the two data sources to be of similar accuracy

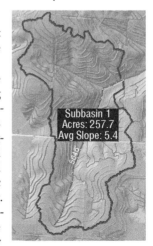

Fig. 3. Overland flow paths added to example drainage area delineation and watershed hydrologic characteristics.

(Kumhálová and Moudrý, 2014). LiDAR surface DEMs can aid in soil loss estimates by providing an accurate field slope model without requiring an in-field survey, broadening the time frame for terrace design, and eliminating possible delays due to weather, workload, or interference of field crops. Improved ground surface data increases the accuracy of earthwork as well as UGO storage computations vs. previous methods of survey and design that did not account for terrace curvature or ground surface estimates from widely spaced cross-sectional data.

Precision conservation methods, aided by georeferenced ground topography, soil information, land use, field imagery, and ground survey points, are now available to assist in terrace layout. For example, Guo and Maas (2012) developed a procedure for parallel terrace layout for relatively flat field slopes by incorporating GIS with an automated RTK guidance system. Bay et al. (2014) developed a web-based conservation planning tool, WebTERLOC, to provide multiple terrace layout options using DEM and GIS. The program enables automation of the terrace layout procedure, providing visual images of numerous layout options for the user or landowner to consider based on equipment width and user selected locations of waterways or UGOs. Tools can assist in maximizing the area treated per foot of terrace and support a layout selection based on conservation, overall terrace length, and ease in farming.

Once terraces are constructed, precision conservation tools are available to support terrace management. Georeferenced positions of the terrace system can facilitate variable rate nutrient applications in the altered soils disturbed during construction of the terrace channel and ridge. Evaluation of soil texture and productivity disturbances due to terrace construction can be monitored with mobile soil apparent electrical conductivity sensors such as those described by Sudduth et al. (2003). Terrace ridge and channel elevations can also be monitored over time using RTK GPS to determine when and where maintenance is needed. This is important since long-term effectiveness of terraces is dependent on continual maintenance. Precision planter technology with automatic row or section control can be used to eliminate double planting of point rows in terraced fields, which may provide a significant economic advantage (Velandia et al., 2013). Precision sprayer technology with automatic section control can be used to minimize overlaps and to eliminate overspray of grassed terrace sections and grassed waterways. This can provide a significant cost savings (Luck et al., 2010) and can also increase the operational life of the terrace system.

Precision conservation methods also hold promise for improvement in contour farming practices. In addition to the use of a straight reference line, many commercial auto-guidance systems allow users to specify a reference curve to which all other operation passes will be parallel, an option that can be used to establish a contour farming pattern. While such control is straightforward and easily implemented on relatively flat fields, additional complexities are introduced by machine dynamics and soil-machine interactions on sloping fields. For more precise contour farming on steep slopes it may be necessary to control the motion of the implement relative to the tractor (Heege, 2013). This can be done using any one of a number of commercially available systems.

Ortiz et al. (2012) evaluated contour planting and digging of a peanut crop using auto-guidance compared to manual steering. In five of six field tests, peanut yields were greater in the autoguidance treatment. Precision technologies also enable a "path planning" approach, where an optimum set of operation curves could be developed before field operations. For example, Jin and Tang (2011) developed a three-dimensional path planning system for use in rolling fields. Their system allowed assigning costs to erosion, which were higher for non-contoured paths, and to machine operating costs, which were higher for contoured paths. Relative weights could be assigned to arrive at an optimal design for farmability and erosion control. On average their three-dimensional algorithm compared to a two-dimensional algorithm saved 22% on the overall weighted costs of headland turning, soil erosion, and skipped area due to radius of curvature being shorter than equipment width could access.

Contour seeding approaches using DEMs and GPS-based guidance systems are also currently being developed. Some of the challenges that must be overcome include the tendency of the planter to slide downslope independent of the tractor as it attempts to remain on contour. Williams et al. (2011) compared elevation transects on Pacific Northwest fields having undulating topography and complex slopes, and found that using 9 m spaced DEM transects was adequate for applying precision contouring for autonomous travel. Although collecting elevation values to develop the DEMs can be time consuming, they suggested that such data could be determined by collecting it with a GPS receiver during other farming operations such as tillage and harvesting. Use of LiDAR to develop DEMs would also be possible. They further demonstrated that a precisely located contour strip of deep-furrow seeding (approximately 20 cm) on the upper shoulder above a steep slope section could hold sufficient runoff to protect downslope areas from a 100-yr, 24-h storm if the strip was approximately 2% of the runoff collection area.

Because slopes are generally not constant across fields, contour farming layouts will often result in point rows similar to those found in many terraced fields. The precision planting and spraying control technologies described above will therefore be important for maximizing farmability and minimizing excess inputs in contour farming.

Case Study for Terrace Layout

As part of the decision process for improved land management, a resource inventory should be conducted when deciding whether or not to construct terraces. This should consider field topography, existing drainage features, soil and soil condition, habitat, and existing infrastructure. Estimates of soil loss based on RUSLE2, or concerns of potential ephemeral gully erosion should be determined. Alternatives such as changes in cropping systems, mechanical practices, or land use conversion should also be considered. The cost of installation, operation, and maintenance will be part of this decision process. Zhou et al. (2009) provide an example of such an evaluation of the effectiveness of various conservation measures using DEM data and erosion modeling.

The following case study illustrates use of precision conservation tools for terrace layout. The case assumes that the previously mentioned decision process has been followed, and terraces were selected as the preferred alternative. The example field area with topographic map is shown in Fig. 4. The contour map with 0.3 m contour intervals was prepared from remotely sensed LiDAR DEM data using ArcGIS 10.0 (ESRI, Inc, Redlands, CA). The color shading is based on elevation gradation from LiDAR DEM data, with the field boundary (14 ha) delineated in red and overland flow lines (minimum drainage area of 0.20 ha) in blue. Adding such coloring helps to highlight potential waterway or UGO placement in the field making it easier to design the layout. In this example, the field has an average slope between 2.0 and 2.5%, and has been maintained in row crop production.

A broadbase terrace cross-section was selected for this field since slopes are relatively low and this cross-section supports crop production on all portions of the terrace. A conventional graded terrace layout system was designed for this field by USDA-NRCS technicians using standard layout methods (NRCS, 2010). This required a field site visit to set up and read survey equipment as needed to help in choosing the waterway and/or underground outlet (UGO) placements. Terrace locations were then flagged along their respective lengths in the field. Terraces were subsequently constructed in the field per the NRCS specifications. Following construction, terrace coordinates were recorded using a Sokkia Set 2 total station (accuracy: angle = 2 s, tilt = 1 arc s, distance EDM = ± (2 mm + 2 ppm × Distance)) and handheld GPS (Hemisphere XF101, 1 m accuracy) equipment (Sokkia, Olathe, KS) to ensure the terrace layout met the intended design. The horizontal and vertical data were then uploaded into ArcGIS for comparison with computer designs.

A conventional graded terrace system layout using broadbase terraces was subsequently designed using the *Web*TERLOC program (Bay et al., 2014). The program results were then compared to the layout design developed using the standard NRCS procedures. A flow chart of the terrace layout process using the *Web*TERLOC program is shown in Fig. 5.

The first step in the terrace layout process is to upload a DEM of the field contour map superimposed on an aerial image of the field, and conduct a visual overview of the topography. The program prompts the user for input on equipment width, terrace cost per unit length, and other supporting information. Then using the computer cursor and mouse, the user can quickly trace in the ridgeline and terrace field boundaries. Next, the desired number and location of divides and waterways and/or underground outlet locations can be added to the graphically displayed contour map. Once all information is submitted, the program locates conventional constant grade terraces and displays them on the screen. If the layout does not meet the needs of the landowner, the location of waterways and divides can be quickly modified using the mouse and cursor. This process can be repeated until a satisfactory conventional terrace layout is achieved. Using this terrace layout, the user then selects which conventional terraces to use as key terraces to complete a parallel layout. Each option of parallel terraces is subsequently displayed based on the selection of key terraces. This process is continued until a final acceptable parallel layout is reached. A hard copy of all terrace layout information is provided as well as creation of a LandXML format file output useful in supporting the field layout installation. What would typically require several hours of field work to select final waterway and/or underground (UGO) locations and key terraces for parallel layouts can be done in about an hour or two in the convenience of the office by using this or similar precision conservation tools.

Unlike traditional terrace layout, using precision conservation tools such as the *Web*TERLOC program permits the landowner to evaluate several layout options before narrowing the terrace design to the preferred system. Not only does this save time, but it also permits the landowner to be more directly involved

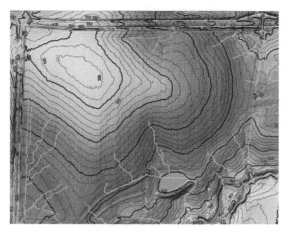

Fig. 4. Example field area with contour map constructed using ESRI ArcMap 10.0 GIS software from light detection and ranging (LiDAR) digital elevation model (DEM) data.

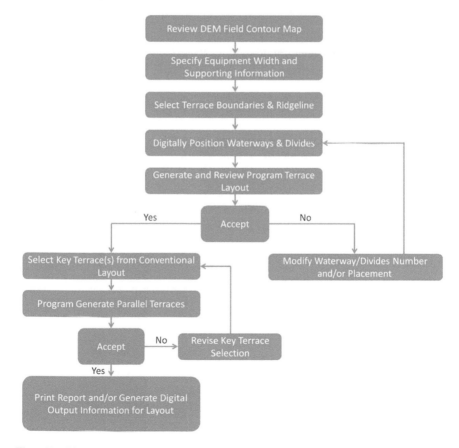

Fig. 5. Flow chart showing modeling sequence for WebTERLOC terrace layout procedure.

in creating the field design, and should improve acceptance of the layout to facilitate long-term support in maintaining the terrace system.

To complete the case study, a comparison of the two layouts is shown in Fig. 6. The NRCS manual layout, illustrated in red, shows the alignment of the terraces as constructed in the field. The *Web*TERLOC layout is shown in yellow. The manual layout contains six conventional grade terraces spaced between 27.5 and 54.9 m apart. The *Web*TERLOC layout contains four conventional grade terraces, having a slightly wider spacing and not including the two short terraces placed at the top and bottom of the slope in the manual layout. Average terrace spacing for the *Web*TERLOC layout was within 4% of the standard layout method. Terrace channel grade was approximately 0.4% for both systems.

Based on the field layout coordinates shown in Fig. 6, the NRCS Engineering Field Tool (EFT) Terrace Design Tool (TDT) was used to design the *Web*TERLOC terrace cross-section (plan view shown in Fig. 7) and the standard terrace layout as actually constructed in the field (plan view shown in Fig. 8). Note that a fifth, optional, short terrace was added at the lower field section of the *Web*TERLOC design (similar to the one for the standard layout method) at the discretion of NRCS field staff. Additionally, three of the terraces were slightly extended to match the field boundary of terraces from the standard design method. These types of modifications can be added by an experienced individual, and help to illustrate the benefit of precision conservation tools in aiding consideration of several different options with the flexibility to be modified for additional improvements. The overall shape and terrace placement are very similar for the two systems. However, total terrace length and construction cost per ha are nearly 15% less, and total length of UGO conduit approximately 3% less for the *Web*TERLOC design compared to the standard layout design method. Although it is not possible to do a direct comparison between the standard NRCS layout and the *Web*TERLOC program based on sediment loss, the similarity in spacing and terrace length would result in similar levels of erosion protection.

Terrace data output from the *Web*TERLOC program is in LandXML, compatible with many programs. As in the standard layout design method, these data

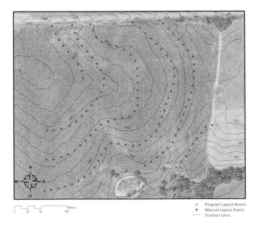

Fig. 6. Comparison of standard manual (red) and WebTERLOC program (yellow) terrace layout for field in Saline County, MO. (Bay et al., 2014).

can be entered into the NRCS Terrace Design Tool (TDT) to design the required terrace cross-sections along each respective terrace.

Summary

The tools and techniques of precision agriculture can provide important advantages, both in designing and in utilizing terracing and contour farming approaches to conservation management. Computer-enabled design methods,

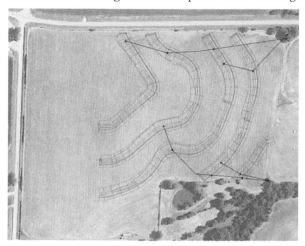

Fig. 7. Terrace layout with UGOs based on WebTERLOC; the cyan color represents pool area of water behind the terrace ridges draining into the underground conduits (black lines). Image based on NRCS Engineering Field Tool (EFT) Terrace design.

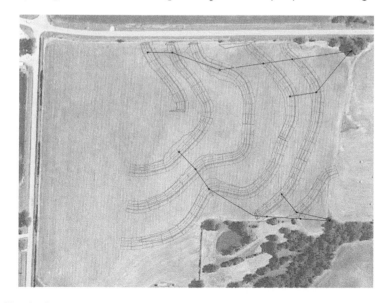

Fig. 8. Standard terrace layout method with UGOs; the cyan color represents pool area of water behind the terrace ridges draining into the underground conduits (black lines). Image based on NRCS Engineering Field Tool (EFT) Terrace design.

such as *Web*TERLOC, have been estimated to reduce terrace layout time up to 50%, providing savings in time and money. Additional cost savings can be realized by the grower through implementation of a more efficient design using *Web*TERLOC. In addition, such design methods provide flexibility in considering numerous potential layout options to best meet conservation and landowner needs.

Use of design tools that rely on spatial data will assist in implementing management practices and designs that contribute to soil and water conservation in agricultural and natural ecosystems while supporting crop production goals. Importantly, precision field machines that utilize GPS guidance and automatic row or section control will contribute to the farmability and economic sustainability of these conservation practices.

References

ASABE. 2012. Standard: ASABE Standard S268: Design, layout, construction and maintenance of terrace systems. American Society of Agricultural and Biological Engineers, St. Joseph, MI.

Bay, M.K., A.L. Thompson, K.A. Sudduth, and C.J. Gantzer. 2014. An online planning tool for designing terrace layouts. Appl. Eng. Agric. 30:1–15.

Das, G. 2009. Hydrology and soil conservation engineering including watershed management 2nd ed. PHI Learning Private Limited, New Delhi, India.

Fangmeier, D.D., W.J. Elliot, S.R. Workman, R.L. Huffman, and G. Schwab. 2006. Soil and water conservation engineering. 5th ed. Thomson Delmar Learning, Clifton Park, NY.

Guo, W., and S.J. Maas. 2012. Terrace layout design utilizing geographic information system and automated guidance system. Appl. Eng. Agric. 28:31–38. doi:10.13031/2013.41283

Harold, L.L. 1947. Land-use practiceson runoff and erosion from agricultural watersheds. Agric. Eng. 28:536–566.

Heege, H.J. 2013. Precision in guidance of farm machinery. In: H.J. Heege, editor, Precision in crop farming: Site specific concepts and sensing methods: Applications and results. Springer, Dordrecht, Netherlands. p. 35–50. doi:10.1007/978-94-007-6760-7_4

Jin, J., and L. Tang. 2011. Coverage path planning on three-dimenstional terrain for arable farming. J. Field Robot. 28:424–440. doi:10.1002/rob.20388

Kumhálová, J., and V. Moudrý. 2014. Topographical characteristics for precision agriculture in conditions of the Czech Republic. Appl. Geogr. 50:90–98. doi:10.1016/j.apgeog.2014.02.012

Luck, J.D., S.K. Pitla, S.A. Shearer, T.G. Mueller, C.R. Dillon, J.P. Fulton, and S.F. Higgins. 2010. Potential for pesticide and nutrient savings via map-based automatic boom section control of spray nozzles. Comput. Electron. Agric. 70:19–26. doi:10.1016/j.compag.2009.08.003

NRCS. 2010. Terrace Conservation Practice Standard, Code 600. USDA- NRCS, Washington, D.C. https://www.nrcs.usda.gov/Internet/FSE_DOCUMENTS/stelprdb1046935.pdf (verified 8 June 2017).

Ortiz, B.V., G. Vellidis, K.B. Balkcom, H. Stone, J.P. Fulton, and E. van Santen. 2012. Evaluation of the advantages of using GPS-based auto-guidance on rolling terrain peanut fields. In: Proceedings of the 11th International Conference on Precision Agriculture. Monticello, IL. 15-18 July 2012. International Society of Precision Agriculture, Monticello, IL.

Schottman, R.W., and J. White. 1993. Choosing terrace systems. Agricultural Publication G1500. Univ. of Missouri Extension. http://extension.missouri.edu/publications/DisplayPub.aspx?P=G1500 (verified 8 June 2017).

Steichen, J.M., and G.M. Powell. 1985. Measuring farmability of terrace systems. Trans. ASAE 28:1130–1134. doi:10.13031/2013.32400

Stafford, J.V. 2000. Implementing precision agriculture in the 21st century. J. Agric. Eng. Res. 76:267–275. doi:10.1006/jaer.2000.0577

Sudduth, K.A., N.R. Kitchen, G.A. Bollero, D.G. Bullock, and W.J. Wiebold. 2003. Comparison of electromagnetic induction and direct sensing of soil electrical conductivity. Agron. J. 95:472–482. doi:10.2134/agronj2003.0472

Tollner, E.W. 2002. Natural resources engineering 1st ed. Iowa State Press, Ames, IA.

USDA-NRCS 2010a. Natural Resources Conservation Service Conservation Practice Standard, Terrace, NRCS MOFOTG 600. USDA-NRCS, Washington, DC.

USDA-NRCS 2010b. Natural Resources Conservation Service Conservation Practice Standard, Contour Farming, NRCS MOFOTG 330.USDA-NRCS, Washington, DC.

Velandia, M., M. Buschermohle, J.A. Larson, N.M. Thompson, and B.M. Jernigan. 2013. The economics of automatic section control technology for planters: A case study of middle and west Tennessee farms. Comput. Electron. Agric. 95:1–10. doi:10.1016/j.compag.2013.03.006

Williams, J.D., D.S. Long, and S.B. Wuest. 2011. Capture of plateau runoff by global positioning system-guided seed drill operation. J. Soil Water Conserv. 66:355–361. doi:10.2489/jswc.66.6.355

Wischmeier, W.H., and D.D. Smith. 1958. Rainfall energy and its relation to soil loss. Trans., Am. Geophys. Union 39:285–291. doi:10.1029/TR039i002p00285

Zhou, X., M.J. Helmers, M. Al-Kaisi, and H.M. Hanna. 2009. Cost-effectiveness and cost-benefit analysis of conservation management practices for sediment reduction in an Iowa agricultural watershed. J. Soil Water Conserv. 64:314–323. doi:10.2489/jswc.64.5.314

Elements of Precision Manure Management

Peter J.A. Kleinman,* Anthony R. Buda, Andrew N. Sharpley and Raj Khosla

Abstract

Manure management, in its most comprehensive form, involves practices that span most of a livestock production system, from feed management to barnyard management to manure storage and handling to land application. We propose that precision manure management involves careful manipulation of these components, with an eye to interactions between these components as well as to factors beyond the farm gate. Great strides have been made in many areas of manure management, from more efficient use of feed inputs (and therefore lower manure nutrients), to innovations in storage and handling, to tools improving the use of manure nutrients in crop production. Despite these gains, barriers exist to adoption of many practices. This chapter reviews advances made in manure management and reflects on the challenges and opportunities involved in developing holistic strategies for precision manure management that balance production and conservation priorities.

Manure management has long been of local concern from nuisance and health perspectives (Fig. 1), and in recent decades, has been at the heart of various environmental regulations, including the historic USEPA-USDA agreement to regulate Concentrated Animal Feeding Operations (CAFOs) under the Non-Point Discharge Elimination System (US Environmental Protection Agency, 2012). More recently, manure management has been the focus of watershed mitigation programs (Kleinman et al., 2011) and is at the foundation of the USDA Natural Resource Conservation Service's revisions of their national nutrient management standard (USDA Natural Resource Conservation Service, 2011a, 2011b). Advances in manure management offer a broad range of opportunities to better control manure sources and end uses, balancing the objectives of efficient livestock production with environmental and human health concerns and helping to overcome some of the historical barriers to practice adoption.

Given the acceptance of precision management concepts in other areas of agriculture and conservation, it is reasonable to apply the chief tenets of precision management to the major components of manure management systems: livestock production, farmstead operation, agronomy, and environmental conservation. We

P.J.A. Kleinman and A. Buda, USDA-ARS, Pasture Systems and Watershed Management Research Unit, University Park, PA; A.N. Sharpley, Dep. Crop, Soil and Environmental Sciences, Division of Agriculture, Univ. Arkansas, Fayetteville, AR; R. Khosla, Dep. Soil and Crop Sciences, Colorado State University, Fort Collins, CO.*Corresponding author (peter.kleinman@ars.usda.gov)
Mention of trade names does not constitute endorsement by the USDA. USDA is an equal opportunity employer.

doi:10.2134/agronmonogr59.2013.0023

Precision Conservation: Geospatial Techniques for Agricultural and Natural Resources Conservation
J.A. Delgado, G.F. Sassenrath, and T. Mueller, editors. Agronomy Monograph 59.

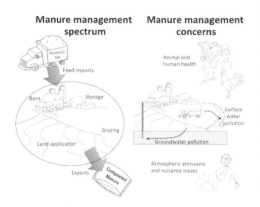

Manure management spectrum

Manure management concerns

Fig. 1. Major components and concerns of manure management systems. Drawings courtesy of A. Mowery.

posit that precision manure management targets key components of this continuum, matching cost-effective management options to specific production needs, optimizing system level efficiency while focusing on individual processes (often within the farm), and considering the unintended consequences of alternative management decisions. It is important to reiterate that because manure generation, handling, storage and end use are influenced by decisions extending well beyond the farm gate, precision manure management must consider the broadest array of options, from feeding strategies that affect manure generation to barnyard practices that affect manure quality to manure storage practices that influence options for land application, export or import. Indeed, if applied holistically, or comprehensively, precision manure management should transcend multiple scales: field, farm, region, and industry.

It goes without saying that livestock feed utilization, barnyard management, manure storage and handling, and manure application and/or export vary widely across and within livestock industries. In the United States, the vertically integrated nature of most poultry and swine production creates opportunities for uniform coordination of manure management within these production systems. However, this requires that precision manure management is compatible with production objectives that are often beyond the concerns, motivations, and feasibilities of the individual producer. Swine and poultry production systems have shown great strides in feed nutrient use efficiency, as highlighted by the widespread use of phytase in their feed, reducing on-farm inputs of feed phosphorus (P) and therefore P loss in excreta (Havenstein et al. (1994) and 2003). Manure storage and handling systems of swine and poultry industries vary dramatically, requiring different strategies for land application and export; swine manures are predominantly liquid (i.e., pit and lagoon storage) whereas broiler poultry manures tend to be dry (< 30% moisture to minimize foot lesions) and the semi-solid manures of battery cage layer houses are rapidly transitioning to drying technologies. Greater internal diversity exists in beef and dairy industries, which do not include the same degree of vertical integration, elevating the consideration of on-farm variables and individual producer goals, as reflected by a range of semisolid pack, slurry and liquid cattle manures.

This chapter seeks to apply precision management concepts to the components of manure management systems. Our intent is to highlight general achievements and opportunities for precision management, while acknowledging the great variety of factors facing producers and others charged with managing manures. This is not a comprehensive review of manure management options, and undoubtedly, important areas of concern and achievement will be missed. In addition to highlighting key opportunities and barriers to adopting important management practices, we focus broadly across the spectrum of live-stock management practices that contribute to manure management, divided into three general areas:

(1) manure generation and storage (feed management, farmstead management, storage and handling);

(2) land application of manure (source, rate, timing, method) in confinement and grazing operations;

(3) export and import of manures.

Manure Generation and Storage

Precision Feeding

The concept of precision feeding has a strong foundation in livestock production, with a steady historical emphasis on improving feed use efficiency for economic profitability. It is therefore not surprising that precision feeding can be used to regulate livestock excreta, making precision feeding an important component of precision manure management. Indeed, as illustrated for the broiler chicken industry (Fig. 2), dramatic reductions in both nitrogen (N) and P excretion have been achieved over the past five decades, with average nutrient excretion in 2001 less than 20% of the 1959 average. These historical declines in nutrient excretion are the combined product of better feed formulation and improved breeding, reducing the nutrient content of feed and the efficiency with which consumed nutrients are metabolized by livestock.

Over the past several decades, the addition of the enzyme phytase to swine and poultry feed has emerged as an excellent example of the potential for innova-tions in feed formulation to contribute to precision manure management. Phytate, or phytic acid, is the predominant form of P in conventional corn (~70%) and soy-beans (~60%) and is not readily metabolized by monogastric livestock (Ravindran et al., 1995). While bacteria in the rumen of cattle produce the phytase (i.e., endog-enous phytase), which breaks down phytate into component inorganic phosphate molecules, endogenous phytase is not naturally produced by swine and poul-try (and humans for that matter), due to the absence of symbiotic populations of

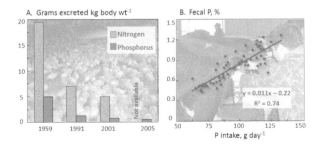

Fig. 2. A. Industry average N and P excretion by broiler chickens from 1959–2001, in addition to average P excretion by broiler chickens on the Delmarva Peninsula in 2005, and B. relationship between P intake (in feed) and extraction (in manure) for lactating dairy cows. Data for poultry adapted from Havenstein et al. (1994; 2003) for 1959, 1991, and 2001, and Angel (2005) for 2005 (Delmarva Poultry only) and for dairy from Wu et al. (2001).

phytase-generating bacteria. In the mid-1990s, phytase was successfully licensed by the BASF Corporation as a feed additive. Today, there is widespread use of phytase in poultry and swine feed. While addition of phytase alone does not significantly reduce P loss in feces, the enzyme does enable lesser quantities of supplemental P to be added to poultry and swine diets due to greater bioavailability of the phytate in the feed (Applegate et al., 2003; Baxter et al., 2003; Cromwell et al., 1993). As a result, cost savings through lower supplemental P costs have enabled feed mills to add phytase with little change in feed prices. Additional potential for reduced P excretion has been documented with the development of low phytate varieties of corn and soybean (Raboy, 2001; Spencer et al., 2000) as well as with phase feeding.

Keeping with the advances in feed P management, compelling ties between feed N content and livestock N excretion serve as the foundation for improved feed N management for both monogastric and ruminant livestock. In general, strategies to minimize livestock N excretion have focused on regulating crude protein content of feed. For beef and dairy cattle, adjusting the protein content of feed rations to match their requirements for essential amino acids has been shown to substantially reduce N excretion (e.g., 15% compared with traditional diets). Indeed, strategies even exist for grazing cattle by adjusting supplementation of N (Mulligan et al., 2004), pasture fertilization regimes (Dufrasne et al., 2010) or the timing of grazing rotations to periods when forage grasses are at more mature stages and have a higher fiber content compared with protein content. For swine, decreasing the crude protein of rations and adding supplemental essential amino acids can reduce N excretion by 20 to 25% (as summarized by Sutton and Beede, 2003). Similar strategies of combining lowered crude protein and adding essential amino acids have been effective for poultry, too (Aletor et al., 2000).

In the face of dramatic improvements in precision feeding and continued opportunities for greater feed use efficiency, a remaining priority of precision manure management is to address recalcitrant pockets within the livestock industry where adoption of precision feeding practices is poor. In particular, there remains a need to expand the awareness of producers, particularly those independent farmers who own the animal and formulated feed rations (i.e., independent dairies), nutritionists and other advisors (e.g., veterinarians) that historical "safety margins" built into feed formulations at the farm or local level result in wasted nutritional resources and unintended environmental consequences. For instance, for decades there has been a concerted effort to counter the perception that P supplementation promotes dairy cattle fecundity. Continued education is needed to confront feed supplementation recommendations from advisors that are based on faulty perceptions of risk, rather than real risk.

Farmstead Infrastructure

The farmstead, comprised of the infrastructure housing animals, their feed, and their waste, presents one of the most fundamental problem areas in manure management. With older facilities, farmstead infrastructure has been situated, sized and designed for a myriad of reasons other than prudent manure management. Indeed, it is not uncommon to find barn and/or barnyards in flood plains or along waterways, with limited options for excluding runoff waters, establishing functional buffers and developing modern manure storage structures. Similarly, neighborhoods and ordinances change, such that long-established livestock operations may need to contend with nuisance odor or environmental health concerns that were not present when the farms were originally established. When confronted with complaints or regulatory mandates, the expense of tackling farmstead infrastructure is often prohibitive, especially for smaller individual operations, requiring support (expertise and financial) from action agencies and other sources. Therefore, precision management must place cost effectiveness at the top of targeting farmstead management options.

Managing odor from livestock facilities can be difficult, with seasonal variations in temperature and humidity creating different potentials for odor transfer. Even so, standard odor management principles exist for most livestock facilities (Table 1). Using these principles, the development of an odor management strategy, or plan, can provide other benefits, from improved livestock health to reduced emissions of nutrients. To minimize nuisance odor complaints with new facilities, a premium should be placed on finding sites where the downwind fetch avoids residential development and associated nuisance odor complaints. In addition, establishing tree plantings around facilities can help to diffuse nuisance odors as well as to serve as a wind break (Powers, 1999).

Nutrient losses from manure in livestock facilities occur via a number of loss pathways and can be substantial due to the concentration of manure. As illustrated for N (Table 2), a pollutant that includes atmospheric and hydrologic (surface and ground) pathways, losses from livestock facilities can range widely within and between housing and manure collection categories, with losses as high as 90% from open air facilities (e.g., cattle feed lots) and as low as 4% when manures are intensively managed (e.g., belt dried layer chicken manure).

Strategies for conserving N in manure involve some of the same principles as those for preventing odor emissions (Table 1), from drying to minimize volatile losses of NH_3 and improve manure transport properties, to minimizing manure exposure to the atmosphere, rain and runoff waters. Clearly, fewer options exist for open facilities than for closed facilities, but even in closed facilities control of N emissions is difficult, as exemplified by advances in poultry production. Modern broiler poultry barns have evolved considerably from the open walled barns of the past, with highly regulated ventilation systems that are so efficient that loss of power to these systems can result in rapid asphyxiation of the birds in the barn. Even so, Moore et al. (2011) determined that ammonia (NH_3) emissions from broiler poultry barns were equivalent to 0.03 kg NH_3 bird^{-1} over the approximately seven-week period of their growth, compared with 0.02 kg NH_3 bird^{-1} emitted from stored litter and 0.01 kg NH_3 bird^{-1} from land applied litter.

Table 1. Odor management principles for livestock facilities. Adapted from Pennsylvania Conservation Commission (2013).

Minimize dust and feed accumulation in pens, aisles and on animals.

Ventilation to provide fresh air flow so that animals and facility surface are clean and dry.

Minimize damp, exposed manure.

Remove mortalities daily and dispose appropriately.

Match feed nutrients to animal nutrient requirements to avoid excess nutrient excretion.

Manure storage facility should minimize exposed surface areas and off-site odor transfer.

Adapted from Pennsylvania Conservation Commission (2013)

Table 2. Typical N loss from animal housing facilities as a percentage of N excreted. Adapted from the literature summary of Rotz (2004).

Facility and/or manure type	Typical loss of total N	Range of observations	Forms of N
	% of N excreted		
Poultry, high rise	50	40–70	NH_3
Poultry, deep litter	40	20–70	NH_3, N_2O, N_2
Poultry, cage and belt	10	4–25	NH_3
Poultry, aviary	30	15–35	NH_3, N_2O
Swine, slatted floor	25	15–30	NH_3
Swine, deep litter	50	50–60	NH_3, N_2O, N_2
Swine, free range	35	25–40	NH_3, NO_3, N_2O, N_2
Cattle, tie stall	8	2–35	NH_3, N_2O, N_2
Cattle, free stall	16	10–20	NH_3
Cattle, bedded pack	35	25–40	NH_3, N_2O, N_2
Cattle, feed lot	50	40–90	NH_3, NO_3, N_2O, N_2

Direct capture of dust and NH_3 from ventilation fans remains a priority concern to air and water quality, with few options currently available. For instance, a U.S. Environmental Protection Agency study conducted in Kentucky estimated that a single broiler house emits greater than 1700 kg of dust per year (Burns et al., 2008). The United States produces approximately 8.4 billion broiler chickens annually for consumer consumption (USDA National Agricultural Statistics Service, 2012a). Considering that the average broiler house is 40' by 400', housing 25,000 birds per flock, cycling 5.5 flocks per year, and there are approximately 61,301 broiler houses in the United States, this equates to an estimated 104 million kilograms of dust emitted annually from broiler poultry houses. While the U.S. Environmental Protection Agency has conducted studies to quantify dust emissions, little data is available regarding the nutrient content of the dust as it relates to water quality impairment. Dust from poultry farms is the fine particulate portion of poultry litter and is a combination of feathers, fecal material, skin, spilled feed, mold spores, bacteria, fungus, and bedding fragments (Wicklen and Czarick, 1997). The nutrient content of the dust is expected to vary depending on bedding type, bedding management between flocks, and feed management as it relates to the manure excreted. Significant dust and NH_3 can be trapped by shelter belts (i.e., tree plantings) established around the barn (Adrizal et al., 2008). However, even though these plantings do capture some N, P, and dust, they largely serve to disrupt air flows, which can easily disrupt ventilation fan efficiency and diffuse concentrated emissions (Tyndall and Colletti, 2007).

Farmsteads are a key point source of manure nutrients and pathogens to ground and surface waters, largely because of the prevalence of manure on exposed, impervious surfaces. Effective control of farmstead runoff requires concerted, regular cleaning of manure sources, maintenance of gutters, downspouts and other infrastructure required to route clean rain water away from manure sources, and installation of practices that minimize diffuse runoff, trap sediment and even take up select nutrients (e.g., vegetated filter strips, constructed wetlands, and detention basins). As a result of expense, farmstead manure sources can pose an overwhelming barrier to achieving offsite remediation goals. However, when these infrastructure needs are left unaddressed, broader management goals can be unobtainable. In a long-term study of nutrient pollution within the LaPlatte River watershed, Vermont, United States, Meals (1990 and 1993), it was determined that developing barnyard infrastructure was one of the most essential, and cost-effective means of improving downstream water quality. Similarly, the success of New York City's Watershed Agriculture Program to mitigate diffuse P pollution loadings to a the Cannonsville drinking water reservoir from roughly 200 small (< 100 cow) dairy farms required, in no small part, the conversion of traditional, earthen barnyards and cattle housing facilities to modern structures that met USDA Natural Resource Conservation Service design criteria, excluding clean water from animal heavy use areas, improving removal of barnyard waste, and installing filtration and diffusion practices for farmstead runoff.

The performance of filters around farmstead areas can be quite mixed, reflecting the concentrated nature of the manure source, the intense hydraulic loading rates that can occur from poorly designed or maintained hardscapes, and the high maintenance demands of most treatment options. While good nutrient, pathogen and sediment removal efficiencies have been reported for some runoff treatment practices in the short term (e.g., Schwer and Clausen, 1989; Young et al., 1980, House et al., 1994), these practices can be readily overwhelmed over the long term. For instance, diminished P sorption capacities of wetland sediments and seasonal fluctuations in oxidation and/or reduction can make wetlands ineffective in controlling P runoff (Mitsch and Gosselink, 1993; Novotny and Olem, 1994), turning practices that were designed to control pollution into sources of pollution. Pries and McGarry (2002) found that a constructed wetland placed below a feedlot and monitored for 4 yr varied in its performance to such a degree that, over time, it actually increased total P concentrations by 30% above influent concentrations.

In some cases, the P sorption capacity of treatment areas can be enhanced by added by-product material to precise areas of the landscape where manure runoff may be occurring. As with experimental manure amendments, a large number of materials produced as byproducts from a range of manufacturing, mining, energy production, and drinking water treatment processes are produced, many with the potential to bind P because of their residual concentrations of active iron and aluminum compounds (Buda et al., 2012; Callahan et al., 2002; Penn and McGrath, 2011; Vohla et al., 2011). Use of these amendments on critical areas in a watershed has the potential to reduce P loss significantly (Stout et al., 1999; Penn et al., 2007; Delgado and Berry, 2008). There are several industrial byproducts that have the potential to bind large amounts of P by different chemical reactions, dependent of the material itself. These materials include fly ash, steel slag, acid mine drainage residuals, drinking water treatment residuals and flue gas desulfurization gypsum. Some acid mine drainage residuals have been shown to sorb appreciable amounts of P (Dobbie et al., 2009; Fenton et al., 2009; Heal et al., 2005; Sibrell et al., 2009; Penn et al., 2011).

Handling and Storage

Manure handling and storage systems are central to modern livestock facility design, from cleaning systems that efficiently remove manure from barns to treatment systems that change the properties or value of the manure, to storage systems that enable producers to accumulate manure so that it does not need to be land-applied or hauled off site on a frequent basis. Retrofitting old handling and storage systems, however, can be very tricky. At each step of the design process (new or renovation), options for manure management must be weighed against trade-offs that extend well beyond cost and labor, and include such considerations as recycling of bedding material, animal welfare and health, availability of spreading equipment and future facility plans. Fortunately, an increasing number of practices exist within the barn to improve the handling of manure as well as to provide ancillary benefits and alternative end uses.

Manure storage is a critical component in the management of manure supply, as well as in more precisely timing the land application of manures (described below). Although many small livestock farms, particularly dairy farms, continue to operate without significant manure storage (e.g., Dou et al., 2001), there has been a long, consistent expansion of manure storage infrastructure within all animal production industries. A primary concern with the absence of manure storage is that farms with minimal or no storage capacity must rely on daily hauling and spreading of manure as it accumulates on the farm, resulting in land application during periods when there is no crop demand for manure nutrients and when environmental losses may be greatest (e.g., runoff from frozen ground). This concern has driven considerable subsidy for manure storage construction as part of conservation initiatives (e.g., slurry storages on Vermont dairy farms or litter storage sheds in Maryland broiler operations), but has repeatedly raised the question as to whether the profound expense of constructing a storage is cost-effective and whether funds spent on manure storage could be better applied to other facets of manure management.

Undoubtedly, as manure storage is increasingly installed on dairy, swine, and poultry operations, farmers are better able to manage the supply of their manures. In addition, there are also measurable benefits to manure quality. For example, covering lagoons with liquid dairy manure and swine slurries greatly reduces N losses to the atmosphere, thus preserving manure N for crop production when it is needed (Rotz, 2004). For dry manures such as poultry litters, storage under roofed structures allows time for air-drying, chemical treatment, and composting, all of which enhance the spreading characteristics of the litter and improve nutrient availability for crops (Moore et al., 1995). Implementing manure storage on farms gives farmers greater control over manure supply and quality and therefore should be considered a key component of precision manure management.

Manure moisture content is one of the most important properties affecting manure end use, and therefore precision management. More options exist for dry manures than for liquid manures and slurries. Therefore, manipulating the moisture content of manure can present great advantages to handling, storage and transport. One of the most significant shifts in managing the moisture of manure has taken place with layer chickens. Traditionally, broiler poultry generate dry litter (< 30% moisture), whereas egg layers (i.e., battery cage operations) produce semi-solid manure (40– 60% moisture). A growing number of modern laying facilities now employ drying belts that yield powder-dry manure. While the cost of developing or retrofitting poultry facilities with drying belts is considerable (> 50% more than conventional battery cage systems, when normalized on a per hen basis), as are the postconstruction costs of maintenance and energy, advantages to belt drying layer manure include fly control, less exposure to dust, NH_3 and other gases in the barn, greater productivity by the chickens, and greater export and/or transport options (Xin et al., 2011). Recently, a litterless "poultry house of the future" has been under development (Harter-Dennis, 2010), eliminating the need for bedding through a slatted floor that quickly dries feces. This technology, like many other pilot manure management technologies, has not come to commercial fruition because of cost. Comparable approaches have been explored in other livestock industries, including the potential to separate feces and urine in dairy barns, but benefits have not been as stark. For instance, initial research with dung and/or urine separation has not shown significant reductions in NH_3 emissions (e.g., Vaddella et al., 2010).

Treating manure with amendments has received considerable attention, and, in some industries, is well established. For instance, aluminum sulfate $[Al_2(SO4)_3]$ has been used to effectively curb NH_3 emissions in poultry production facilities, including as an amendment to broiler chicken litter and as an aerosol in battery layer houses where improvements in the barn atmosphere with liquid alum spray systems have been shown to reduce NH_3 emissions by 30% (Wilson, 2004). Poultry Litter Treatment sodium bisulfate $(NaHSO_4)$ is also well established as an amendment, providing NH_3 conservation from its weak acidity. Still, other amendments are intended to adsorb NH_3 (e.g., zeolites) or inhibit N transformation to volatile or highly mobile forms (e.g., urease and nitrification inhibitors). In addition, various salts (acidic and alkaline) have been evaluated for the purpose of decreasing P solubility in manures, thereby lowering the potential for dissolved P losses in runoff following land application of the manure (Moore et al., 2000). In addition to the amendments mentioned above, other commonly used or tested manure amendments include: ferric chloride, sodium bisulfate, calcium hydroxide, gypsum, and various industrial byproducts (Xin et al., 2011). In many U.S. states, this can have a direct benefit to precision manure management in terms of increased N content when used as a fertilizer and several states use water extractable P of manure as a determinant of allowed land application rates (Sharpley et al., 2003; 2012b). In the latter case, greater amounts of litter can be land-applied with the same risk of P runoff.

Notably, many manure amendments present trade-offs beyond the economic costs that must be considered as part of precision manure management strategies. For instance, amendments containing sulfur (S) may pose a hazard when used with stored liquid manures that become anaerobic and undergo chemical reduction. Under reducing conditions, the conversion of sulfates to sulfides will generate highly poisonous hydrogen sulfide (H_2S), which is regularly implicated in accidents (livestock and human) around liquid manure storage structures (e.g., Hooser et al., 2000). A less acute example of a trade-off is the effect of alkaline amendments on the N/P ratio of manures. Alkaline amendments (e.g., lime) will exacerbate NH_3 volatilization, decreasing the N content of the manure, further skewing the N/P ratio which is already poorly balanced for crop demand in most manures, and, most importantly, devaluing the manure as an N fertilizer resource. As a result, weakly acidic to neutral amendments are preferable as they conserve NH_3.

As with other areas of farm management, a range of commercial amendments can be found that purport to improve manure properties and manure management. In many cases, specific properties of these products are shrouded by claims of proprietary formulation. It is impossible for objective entities such as government agencies, land grant institutions and conservation districts to evaluate all of these products, and the internet is replete with examples of supposed testimonials stating the benefits of amendments and other technologies in eliminating manure odors, curtailing environmental losses of manure nutrients, improving the handling qualities of manure, etc. These claims are directed at the manure and nutrient management advisors as much as they are at farmers themselves. Therefore, a skeptical, critical perspective is particularly important in assessing novel amendments and other technologies used in manure management, with prudent judgment grounded in the adage: "buyer beware", and if it "sounds too good to be true," it probably is.

Value-added Processes

Techniques to add to value to manure can be traced back to the advent of modern fertilizer manufacturing in the early 1800s, when sulfuric acid was added to manure, fossilized dung, and bone to produce superphosphate. Today, the greatest fertilizer value of manure remains as a soil amendment near its source. As a result, there has been great interest and innovation, in value-added steps that transform manure into products that can be sold or exported off farm. Examples range from decomposable planters derived from pressed dairy solids, to organic fertilizer products such as composts, vermicomposts, and pelletized litters. These value-added options do work at a local scale, with many anecdotal successes, but larger initiatives, that is, initiatives that would include substantial swaths of a regional livestock industry, have almost universally required outside subsidy, support, or mandate. Ever since the boom days of early fertilizer manufacturing, demand for manure products has simply not been able to match the ever growing supply of manure.

Similar cautionary tales exist for manure energy projects, especially those designed to collect manures from across a region. One company, Fibrowatt LLC has pursued various manure combustion projects in the United Kingdom and United States, including the 55 MW Fibrominn plant in the heart of Minnesota's turkey production region, with roughly 50% of the region's turkey litter mandated to serve as a biofuel. Fibrominn's business model includes sale of ash to be processed into commercial fertilizer. Given the novelty of projects such as these, potential risks are manifold and projects proposed for other regions have sometimes not come to fruition (e.g., Pennsylvania's Cove Digester and other projects by Fribrowatt in the Chesapeake Bay region). For example, in northwest Arkansas, litter combustion for energy has not occurred despite the area being second only to Georgia for annual broiler production (at about 2 billion birds yr^{-1}). This is due to the plant requirement of generated, long-term supply of litter and farmers' reluctance to sign long-term contracts to deliver their litter often at costs below that which they can get from export sales to row crop production.

At a finer scale, on-farm and community manure energy projects have shown considerable potential, with pioneering projects forced to contend with severe investment hurdles, regulations, resistance from the electrical industry and political vicissitudes. Anaerobic digesters have become a regular feature in large dairy and swine operations, offsetting local power costs, generating various cap-and-trade credits (nutrient and carbon), and offering opportunities for further treatment of manure. Start-up and maintenance costs for these projects must be tied to the complete manure management system. Furthermore, sustainable end uses for residual liquids and solids must be identified, otherwise these projects, which require substantial societal investment, will not address the environmental concerns associated with local manure accumulation.

Manure nutrient recovery is often envisioned as a fundamental requirement of sustainable food production systems that recycle nutrients from areas of livestock production to areas of feed production. Here, the concept is to extract manure nutrients and convert them into marketable forms. Nutrient recovery options are largely in trial stages. Recovery of struvite (NH_4MgPO_4) has potential in liquid manure storage systems. However, precipitation of struvite requires careful control of solution stoichiometry, and scaling struvite reactors from benchtop to farm manure storage has proven very difficult. In South Carolina, a three-stage nutrient recovery system was developed for swine lagoons (Pons, 2005). This system employs bulk solid separation, biological N removal, and Ca–P precipitation. Approximately 99% of manure N and 95% of manure P are removed by this process. Manure solids and precipitated Ca–P are exported off-farm for end uses in compost or low-solubility fertilizer. Liquid manure with low N and P content is left on farm where it can be used in fertigation (Vanotti et al., 2007; 2009).

Land Application

Given the emphasis of this book on precision conservation techniques, it is natural to associate precision manure management with land application of manures. Great opportunity exists to transform land application decision making processes, technologies and resource use efficiency with precision techniques. To this day, land application of manure is typically concentrated around livestock production areas (Fig. 3a). Indeed, the majority of animal manure generated in the US is land-applied to cropland as fertilizer; however, only 5% of these lands (15.8 out of 315.8 million acres) actually receive manure due to the high costs of shipping manure off-site and frequent environmental regulations governing its use (MacDonald et al., 2009). At a smaller scale, fields near production areas often contain substantially more P than fields more distant from the production site. Concentrating manure application around intensive livestock operations leads to a range of unintended environmental and social consequences, including well-documented impacts to soil, water and air resources (Ribaudo et al., 2003). Some of the most notable examples of these impacts include enduring pollution legacies such as the accumulation of P in soils (Sharpley et al., 2003; 2013) and the build-up of NO_3–N levels in ground waters (Tesoriero et al., 2013; Sanford and Pope, 2013) (Fig. 3b and 3c). Striking a balance between farm productivity and

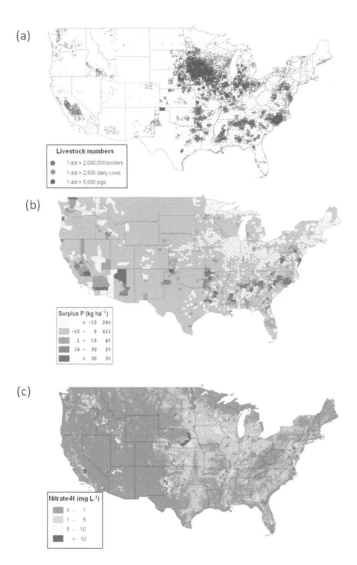

Fig. 3. County level estimates of the distribution of major livestock concentrations in the United States (a) agricultural P balances (b) and groundwater NO₃–N concentrations, adapted from USDA National Agricultural Statistics Service (2012b), Kellogg et al. (2000) and Nolan and Hitt (2006), respectively.

unwanted impacts to the environment and to society is central to a farmer's decision making process when applying animal manure to agricultural lands.

Over the past several decades, a growing number of decision support tools have been developed to guide the application of manure resources to crop and pasture lands (Karmakar et al., 2007). The vast majority of these tools are strategic in nature, providing generalized recommendations. The best established and most widely adopted tools are those focused on nutrients and pathogen losses to surface or ground waters. For instance, the P Index, first proposed by

Lemunyon and Gilbert (1993), is now used by 48 states to identify agricultural fields with the greatest risk of P loss based on the chief source and transport factors governing P movement (Sharpley et al., 2012b). Similar tools for N, such as the N Index (Delgado et al., 2006; 2008) and Adapt-N (Melkonian et al., 2008), are intended to mitigate N leaching losses to ground water. While nutrients have been the mainstay of decision support tools for land application of manure, new tools have emerged to deal with evolving issues such as greenhouse gas emissions (Karmakar et al., 2010) and odor problems (Pennsylvania Conservation Commission, 2013).

At the heart of all recommendations for land applying manure is the 4R nutrient stewardship concept (right source, right rate, right timing, and right placement), which considers manure properties (right source), rate of application relative to agronomic and environmental availability (right rate), timing of application relative to crop need and environmental loss potential (right time), and method of manure placement (right placement). The 4Rs were originally promoted by the Fertilizer Institute, the Canadian Fertilizer Institute, the International Plant Nutrition Institute, and the International Fertilizer Industry Association as a means to achieving more responsible fertilizer usage by farmers and certified crop advisers (Ehmke, 2012). Today, the 4R concept is a common thread running through major nutrient management legislation such as USDA's national nutrient management standard (USDA Natural Resources Conservation Service, 2011a; 2011b), and its precepts readily extend to the land application component of precision manure management.

Source

For farmers, choosing the right manure source is not as simple as with mineral fertilizers, which can be purchased with exact formulations of nutrients to match the requirements of a specific crop. Livestock manures are bulky, have variable nutrient composition, and contain substantial amounts of water (Jensen, 2013). In addition, animal production systems are often specialized on a local or regional basis, limiting the range of manures that are accessible to crop farmers seeking to use them as fertilizers. Even so, many of the manure feeding, barn management and handling and/or storage practices described above can be used to adjust the quality of manure sources available for land application: feeding strategies that affect manure nutrient content, amendments (acids, enzyme inhibitors) that conserve manure N and in some cases, reduce manure P solubility, soil separation and drying that reduces manure water content and improves transportability.

Manure testing is undoubtedly at the foundation of precision management of manure sources. Due to variability over time in diet, bedding, storage and climate, a single livestock facility can generate manures with a wide range of properties. In fact, manure properties can change significantly within a storage facility, requiring either excellent mixing to promote homogenous conditions or detailed subsampling to assess the range of properties within a stored batch. Timely sampling of manure and immediate testing are needed to ensure that results are representative of conditions at time of application. To date, few viable options exist for real time monitoring of manure properties during the land application processes; however, successful quick tests have been developed to enable on-site assessment of a manure source (Van Kessel and Reeves, 2000; Reeves, 2007). In particular, spectroscopic quick tests for N and carbon (C) have proven to be the most accurate and repeatable (Chen et al., 2013). Therefore, regular sampling of manures and testing via analytical laboratories remains a fundamental requirement of precision land application of manure.

Rate

Too often, manure application rate is determined by the need to evacuate or clean out a manure storage container or barn rather than to prescribe manure nutrients to crop requirement. Precision manure management requires testing of both soil fertility and manure quality to ensure that the right rate of manure application is achieved. Notably, there are interactions with other aspects of manure management, especially placement, which can dramatically impact nutrient delivery to crops (see below). Nutrient management recommendations call for routine testing of field soils based on appropriate sampling schemes to ensure that inferences from soil tests match field recommendations. Crop requirements for manure nutrients can be quite variable; confounded by dynamic temporal demands that may not match manure nutrient transformations that affect nutrient availability (Pierzynski and Logan, 1993). In fact, it is now recommended that farmers know the N and P content of manures prior to application so that rates can be matched as closely as possible to plant requirements. The main issue here is the disconnect between N/P ratio of manures (~2:1) to plant needs (~8:1). Thus, farmers are left with the precision management dilemma of either applying enough N for the crop and too much P or enough P and insufficient N that has to be made up by costly mineral N fertilizer. Precise manure application at rates that match crop needs are often supplanted by the risk of P runoff or N leaching. Risk assessment for precise manure application is dictated in most US states by the USDA Natural Resources Conservation Service's Nutrient Management (590) Standard (USDA Natural Resources Conservation Service, 2011a, b; Sharpley et al., 2013). For instance, in northwest AR where 2004 litigation required farmers to base land application on the P Index assessment and follow a threshold soil P level of 300 mg Mehlich-3 P kg^{-1} (which was lowered to 150 mg kg^{-1} in 2014), rates have declined from prelitigation levels of 6 to 8 tons ha^{-1} yr^{-1} to a current average of 2.6 tons ha^{-1} yr^{-1} (Sharpley et al., 2012a). However, this does leave a shortfall in N for optimal forage production, which is either made up by mineral fertilizer or legumes, or in some cases a reduction in beef–cow carrying capacity of pastures.

Great strides have been made in precision nutrient management with fertilizers, particularly on the variable rate application front, but these advances have yet to be fully realized with manures, with liquid manures showing the greatest promise. In large part, this lag can be attributed to the paucity of reliable technologies allowing for variable-rate application of most manures, including real-time monitors that would account for heterogeneity in manure characteristics and flow control technologies that would adjust manure rates on the fly. While slurry meters (Tunney, 1986), conductivity sensors (Provolo and Martínez-Suller, 2007), ion-specific electrodes (Scotford et al., 1998) and infrared spectrometers (Chen et al., 2013) exist, their integration into functioning variable rate spreading technologies has primarily been at the experimental level. Mallarino and Wittry (2010) experimented with varying application rate using a flow controller while spreading liquid swine manure with a tanker without accounting for variations in manure quality within the spreader. They were able to match manure application to within-field soil P variability, applying less manure to areas with high soil test P. While their trials did not affect yield, they did find that variable-rate application helped to lower within-in field soil test P over time when compared with constant rate application. Such glimmers of hope, however, belie the current levels of on-farm testing needed to widely transfer this technology.

Method of Placement

Precise placement of manure includes consideration of how the manure is applied to soil as well as where the manure is applied on the landscape. This section emphasizes the technologies used to place manure into the soil. As described above, the use of decision support tools in manure management is essential to controlling off-site problems associated with land application of manure: N leaching; P runoff; ammonia and odor emission; and greenhouse gases. Decision support tools serve an important role in identifying which fields should receive manure and which fields should not. These tools are often included in comprehensive planning programs such as Purdue's Manure Management Planner, the Wisconsin Interactive Soils Program, and the European Waste Engineering Expert System (Karmakar et al., 2007). In general, prudent agronomic management should guide the application of manure to soils, such that manure nutrients are applied where they are needed. However, this generalization is greatly complicated by the land area and availability of fields to receive manure during periods of spreading. As a result, tools such as the P Index seek to guide manure application to fields where runoff losses of manure P are minimized (rather than to fields where the resource would be used optimally).

Considerable interest exists in variable rate application of manures, now relatively common with commercial fertilizers. Indeed, variable-rate application of

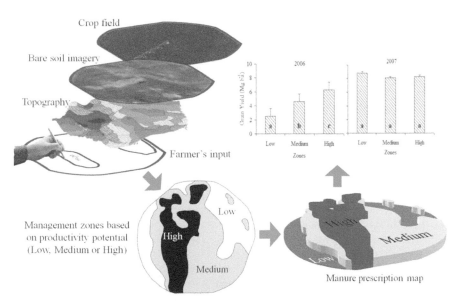

Fig. 4. A schematic of precision manure management study by Moshia et al. (2014), **illustrating (i) data layers used to develop the management zones, (ii) final management zones map, (iii) corresponding manure prescription map and (iv) crop response to variable rate manure application.**

commercial fertilizer nutrients within fields has been shown to improve nutrient use efficiency, maintain or increase crop yield, and reduce off-site pollution (Khosla et al., 2002; Hornung et al., 2003; Delgado et al., 2005). The potential for variable-rate application of manure was demonstrated by Moshia et al. (2014), who evaluated precision manure management strategies across within-field "Management Zones." As illustrated in Fig. 4, Management Zones were delineated using a number of spatially distributed factors: (i) soil properties, (ii) Field microtopography and (iii) any data from past management experience (e.g, Khosla et al., 2002). Moshia et al. (2014) reported that corn grain yield of the low-yielding Management Zone was enhanced with precision manure management. However, they also found that variable application across Management Zones could result in either excess or deficit N and pose a potential threat for nitrate leaching, given the properties of the manure they studied. Their findings highlight the need for precision manure management to fit within broader nutrient management objectives, such as to meet crop nutrient requirements at various stages of growth or to support soil health.

By definition, precision manure management must consider off-site concerns associated with land application and placing a priority on decision support tools that help balance practical concerns with environmental impact. With that said, achieving widespread adoption of such tools by farmers can often be problematic, especially when farmers are not involved in the creation, testing, or implementation of the system (Lynch et al., 2000). Therefore, participatory approaches are needed (e.g., Oliver et al., 2012; Walters and Shrubsole, 2013) to assure that decision support tools in precision manure management are perceived by farmers as useful (i.e., that they enhance on-farm management of manures or help farmers meet environmental quality objectives/regulations) and easy-to-use (i.e., that they save farmers time, money, effort, etc.).

Nowhere are the trade-offs of precision manure management more apparent than in the selection of manure application technologies. Broadcasting manure remains the most efficient means of applying manure quickly, a major factor affecting voluntary adoption of manure application options, but broadcasting leaves manure exposed on the soil surface and vulnerable to major environmental processes contributing to off-site impacts: volatilization of NH_3 and odor; surface runoff of P and pathogens. Broadcasting manure minimizes disturbance to the soil surface, and is therefore readily amenable to the objectives of reduced tillage management and preserving perennial cover; but, broadcasting manure promotes severe vertical stratification of manure nutrients, aggravating long-term environmental losses and placing nutrients above the plant rooting depth. Historically, incorporation of broadcast manure into soil has required some form of tillage, which is again anathema to reduced tillage, perennial cover and soil health management. However, a large variety of low-disturbance manure incorporation technologies have been developed over the past three decades, many of which incorporate manure into the upper 10 cm of soil and therefore avoid the concern of exacerbating NO_3 leaching that was tied to traditional, deep injection technologies. Shallow, low-disturbance manure incorporation technologies offer one of the most visible forms of precision manure management.

The vast majority of innovative, low disturbance manure incorporation technologies are designed for liquid manures and slurries, typically with < 12% moisture. These manures flow well through plumbing under pressure, and therefore can be distributed across a large number of hoses and directed onto soil in bands or into soil furrows. Manure injection remains the most familiar and popular technique for low disturbance manure incorporation, with injection points spaced more closely for perennial forages (e.g., 15 cm) than for row crops (e.g., 75 cm). In most cases, a steel disk or shank is used to cut residue, shatter soil and create a furrow into which manure is injected. In the best cases, impacts on residue and soil are generally comparable to modern no-till planters. For perennial forages, a high pressure system has been developed in Norway that employs pressure alone to force manure into small pockets. Alternatives to injection include the use of aerators to improve infiltration of manure into soil, and the placement of manure below the canopy of grass (e.g., banding or trailing shoe), benefitting from the boundary formed by the grass sward (Fig. 5).

An increasing number of options exist for dry manures, although these technologies are in their early state of development and on-farm testing. Most celebrated has been the poultry litter Subsurfer (BBI Spreaders, Inc.), an applicator that employs common no-till planter components to create a furrow for litter that is conveyed and ground by a system of augers. This technology shows great potential, but performs best with a very narrow range of manure properties (low moisture, low plasticity, low aggregation). The Subsurfer also travels much more slowly than conventional broadcast spreaders, pointing to one of a suite of constraints of low disturbance manure incorporation alternatives that have, to date, served as barriers to voluntary adoption: these precision technologies are generally more expensive to purchase, maintain and operate than conventional broadcast applicators and require additional effort to adapt to existing manure management systems.

The economic and labor liabilities of low disturbance manure incorporation technologies are not universal and must be considered in the face of broader manure management objectives, including nutrient conservation, odor reduction and air and/or water quality protection. Over the short-term, odor intensity and NH_3 volatilization can be greatly curtailed with manure incorporation when compared with broadcast application (Table 3; Brandt et al., 2011; Dell et al., 2011). As a rule of thumb, the greater the degree of incorporation, the greater the reduction in atmospheric emissions. In addition to emission reductions, low-disturbance incorporators can reduce the potential for fly infestation and the unsightly appearance of recently manured fields. Similar benefits have

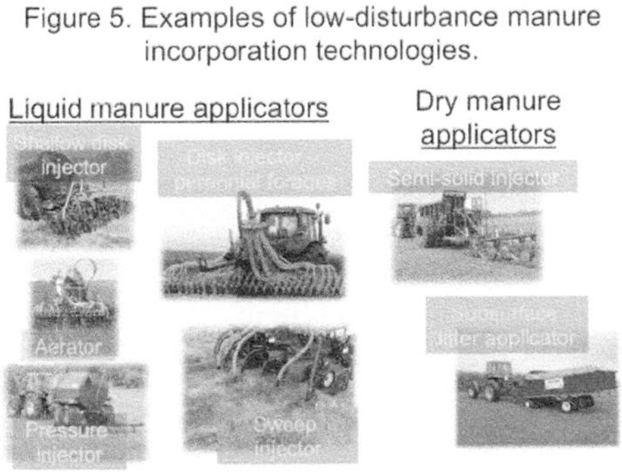

Figure 5. Examples of low-disturbance manure incorporation technologies.

Liquid manure applicators

Dry manure applicators

Fig. 5. Examples of low disturbance manure incorporation technologies.

Table 3. Odor at time of manure application relative to broadcast manure. Adapted from Maguire et al. (2011).

Method		Compared	
		Broadcasting†	Aeration‡
	Chisel with sweep	85% less	67% less
	spike and/or knife	72-79% less	57-94% less
Disk Injection	Shallow disk		11% less
	Tandem disk	74% less	
Aerator		38-75% less	
Tillage	By moldboard		
	chisel by disk	79% less	

†Compiled from Bittman (2003), Lorimor (1998), Hannah et al., (2000), Lau et al., (2003)
‡Compiled from Sexton et al. (2005), Lau et. al (2003)

been observed for losses of P in surface runoff. The substantial improvements in NH_3 conservation with manure incorporation can translate into greater nutrient use efficiency and higher crop yields, but these benefits can be obscured over the short-term in farming systems with a history of overapplication of manure, that is, those where manure has historically been applied at rates well in excess of crop requirement (Ribaudo et al., 2011).

Timing

Precisely timing the application of manure resources to agricultural lands is a matter of nutrient use efficiency, with the goal being to balance the effective use of nutrients by crops with the protection of air and water quality. Losses of manure nutrients to air and water are generally concentrated in the period shortly after manure is applied, with the length of that period determined by factors such as placement method (described above), antecedent and post-application weather conditions, and crop growth. Over the first few days after manure is applied, the main pathway of N loss is by NH_3 volatilization to the atmosphere, which is modified by air temperature, wind speed, and relative humidity (Rotz, 2004). Up to 80% of applied N can be lost during that period (Dell et al., 2012). In contrast, the most critical factor influencing P loss is the timing of manure application relative to the timing of a rainfall event that produces surface runoff (Sharpley et al., 2002). Notably, the risk of "incidental" P losses in runoff tend to diminish with time as the applied P in manure is translocated into soil by rainwater and invertebrates where it is converted to more recalcitrant forms (Edwards and Daniel, 1993), whereas with N, the risk of N transfers to water tends to increase as organic N is converted to NO_3, which can then be leached to ground water (Rotz, 2004). Nitrate leaching from manure application is a seasonal phenomenon, with lower risks during the growing season, and much greater risks in the fall and winter months (van Es et al., 2006).

One area of advancement in precision manure management is the advent of short-term decision support tools that leverage state-of-the-art weather data to forecast the likelihood of runoff events that increase the risk of nutrient losses following manure application. A key water quality concern is to protect against "incidental transfers" of applied nutrients, that is, the punctuated contribution of applied nutrients to runoff immediately following application of a manure or fertilizer (Preedy et al., 2001). These transfers, also referred to as manure or fertilizer "wash-off," produce the highest observed concentrations of nutrients, and pathogens, in runoff, but generally are restricted to a short period following application (Kleinman and Sharpley, 2003). From the standpoint of nutrient use efficiency, application of manure prior to an infiltrating rainfall event is highly desirable, as infiltrating rainfall translocates soluble nutrients, particularly N, into the soil (Penn State Extension, 2007). This process, known as "wash-in", lowers the potential that the nutrients will be volatilized or washed off and improves their delivery to crop root systems. Prudent nutrient application decisions discriminate between events that will result in the liability of rainfall runoff or the benefit of rainfall infiltration.

One of the first weather forecasting tools to guide land application of manure was implemented in the state of Wisconsin in 2010, where NOAA's River Forecast Center model (Sacramento Soil Moisture Accounting Model) was configured to provide daily (24-, 48-, and 72-hr) forecasts of runoff risk on large watersheds (~ 750 km^2) (Wisconsin Manure Management Advisory System, 2014;Goering, 2013). Similar such tools have been trialed in the US (Dahlke et al., 2013) and Europe (Holden et al., 2007; Kerebel et al., 2013) using daily water balance models to forecast soil moisture conditions that are optimal for manure spreading. In the Chesapeake Bay watershed, efforts are underway to incorporate probabilistic modeling approaches that provide information on the likelihood of wash-in versus wash-off events (Buda et al., 2013). While decades of hydrologic and water quality research clearly support the need for short-term, operational decision support tools such as the ones described above, widespread adoption of these new technologies by farmers can admittedly be difficult to achieve (Lynch et al., 2000; Lynch and Gregor, 2004). Recent experience with Ireland's Sustainable Nutrient Management Decision Support System (SNM-DSS) shows that participatory engagement of farmers throughout the life cycle of tool development and implementation is a crucial factor influencing the perception and adoption of such tools by farmers, as well as the potential for these tools to increase the precision of day-to-day manure management decisions on individual farms (Kerebel et al., 2013).

Grazing

The exclusion of grazing livestock from streams is a perennial issue, particularly with smaller dairy and beef operations, with considerable resistance stemming from factors from convenience, to aesthetics, to maintenance, to resentment over environmental regulations (Armstrong et al., 2011). A solid consensus exists that excluding livestock from streams results in local, if not downstream, water quality improvement. For instance, Meals (2000) conducted a paired watershed study in north central Vermont to assess the effect of stream restoration and pasture fencing on the water quality of streams. In one watershed, the combined treatment (stream restoration and fencing) was implemented to exclude 97% of the grazing cattle, resulting in 30 to 50% reductions in watershed export of N, P, and sediment. Elsewhere, James et al. (2007) observed that grazing dairy cattle deposited a disproportionate number of fecal pats in and near streams intersecting pastures. They estimated that the direct deposit of fecal P was equivalent to 12% of the entire watershed load attributed to agriculture. Rotational grazing can provide farmers with precision management options by redistributing manure across pastures. This can also be done by moving hay feeding areas within a field to redistribute manure and associated nutrients. The former practice does require a higher level of knowledge of pasture growth but can reap rewards in avoiding over grazing of pastures.

Curtailing runoff losses from pastures takes work. Precision management of fecal and urine nutrient losses from pastures requires concerted focus on a variety of factors. In a review of practices to decrease N loss from grazing animals, Rotz (2004) identified a suite of key practices including (i) optimizing stocking rates to minimize damage to pasture grasses and prevent nutrient overloading, (ii) regularly moving of water, feed and shade devices to distribute excreta across the pasture, (iii) timing grazing to avoid periods when plant uptake is low, (iv) installing fencing and stream exclusions and (v) synchronizing carbohydrate source with high N pasture in the rumen to recapture greater amounts of NH_3. Many of these practices would lower P and sediment losses from pastures too, as confirmed by an abundance of environmental research from grazing lands in New Zealand (e.g., McDowell and Kleinman, 2011). The accumulation of dung on paths and loafing areas, even those developed as best management practices (e.g., improved crossings of streams), is highly vulnerable to wash-off due to the relatively impermeable nature of the soil surface.

Farm Export and Import

Precision management of manure suggests a focus on narrow areas of manure management, but should by no means be defined as such. In fact, precision management cannot exist without including strategies to tackle the imbalance of nutrients promoted by concentrating animal production. Due to the accumulation of manure nutrients at farm and regional scales (Fig. 3), there is a great need to expand the trade of manure off-farm and onto crop farms or other locations (e.g., reclaimed mine lands). To-date, most trading of manure has involved dry manure, particularly poultry litters. The relatively high nutrient density and lower transport costs of dry manures provides a substantial advantage for export and value added processing. Furthermore, poultry litter is often a preferred organic fertilizer source by crop producers, given its mix of macro- and micronutrients and relatively benign nuisance properties (odor, fly attraction).

Not surprisingly, there has been significant industry and government investment in poultry litter trade. These programs have had mixed success, but have helped to promote innovation in exporting manure, including the baling of poultry litter to facilitate storage and ensure timely availability of the litter, as well as brokering programs that tie sources of manure with end users. In Arkansas' Illinois River watershed, more than 70% of poultry litter is exported via an innovative brokering program that ties producers in the watershed with crop farmers around the region (AR and OK). This program, mandated as part of law suit, simultaneously provides poignant examples of the potential for success, as well as the need for impetus outside the existing market system to drive such solutions (Herron et al., 2012). In fact, the success of this program has now created a situation where demand for litter exceeds supply. To be sure, there are many examples of failure, even with well subsidized manure trading programs.

With liquid manures, there are very few options for manure export short of local trading of raw manure, processing to concentrate solids in exportable form, or value added steps such as the bioenergy and nutrient recovery projects described earlier. In areas with substantial regulations restricting on-farm use of manures, manure export may take the form of expanding a farm's land base through additional land acquisition, such as purchase or rental. This option is not always available, making export critical to the persistence of animal production. Such critical conditions can be found in peri-urban areas with expanding populations and inflated land prices.

In the case of composts and pelletized manure products, marketing and supply chain barriers can be great, requiring attention to product quality, adaptation to shifting market demands and efficient distribution strategies. Vermicomposting in particular has seen a boom and bust of interest, although niches remain for high quality vermicomposts that can be used in compost tea. The Perdue Agri-Cycle plant in Seaford, Delaware (United States) has the capacity to process litter from nearly 85 million broiler chickens (roughly 14% of the local production), but has struggled to run at capacity and maintain viability, particularly during periods of inflated fertilizer prices. The AgriCycle plant has required financial support from federal and state subsidies, as well as from the poultry industry. Even so, the AgriCycle plant has remained open for over a decade and produces a consistent product that is shipped to buyers around the United States.

Conclusions- Barriers and opportunities

Precision manure management should be considered the detailed consideration of components of the manure management spectrum, understanding that with the precise manipulation of specific practices brings consequences across a spectrum of scales and responses, some of which are synergistic and others presenting tradeoffs. Precision manure management must, first and foremost, recognize that the fertilizer value of manure declines with distance from barn, placing a priority on activities that minimize the generation of manure, then minimize the on- and off-site impacts of that manure. It also declines with its water content. Consequently, one of the main challenges to precise manure management is reducing the water content of manures in a cost-beneficial manner for any given farming system. While there are numerous examples of successful research on precision-based manure management and development of added-value byproducts, very few have made it to the real world of private farming.

Under purely voluntary conditions, market forces serve to divide manure export from continued application of manure to soils above crop requirement. Even when prompted by regulation or litigation, manure export requires novel markets and infrastructure to sustain the trade of manure from areas of nutrient surfeit to nutrient deficit. As the economic backdrop to agricultural production, energy costs, increasing mineral fertilizer prices changes to enhance the value of manures as an organic, multi-beneficial fertilizer (e.g., organic matter, water holding capacity, and drought tolerance of soils can all increase with manure additions) will also increase. This will serve to open and expand new markets and opportunities for precision management of manures.

A mounting body of experience now exists with examples of success and failure in developing cap-and-trade market incentives, auction and brokering systems, transport systems and regulatory policy. These experiences highlight the condemning role of uncertainty in timing, availability and quality. Education and outreach are paramount, pre-empting conflicts over the use of a valuable resource and promoting the best practices for land application when manures are imported. These experiences point to previously unknown demands for certain manures such as dry poultry litters. Unfortunately, success stories remain relatively limited in number and scope. Making manure trading work requires the perspective of precision management, with all details considered so that unintended consequences and unforeseen obstacles are avoided. The fact that one is simply not transferring a nutrient surplus and associated surface water quality concerns from one area to another must be addressed in any manure trading or transfer program.

To date, manure management has been viewed through the lens of production and environmental concerns, but it is impossible to ignore the long-term role of manure management in national food security, and to a lesser extent, energy. Considering that the majority of grain in the United States is used to support livestock and that roughly two-thirds of the nutrients consumed by livestock are unmetabolized, more efficient cycling of manure nutrients should be viewed in the context of food–energy–water nexus–security (Jarvie et al., 2014). Until recently, more than half of the N fertilizers used in U.S. agriculture was imported. As U.S. rock phosphate supplies diminish, the nation is expected to shift from net exporter of P fertilizer to a net importer of P. There is no hyperbole in concluding that efficient recycling and reuse of manure nutrients means less dependence on foreign sources. It is in this context that barriers to precision management of manures must be removed and developmental incentives promoted.

Acknowledgments
Special thanks are extended to Allison K. Mowery, USDA Agricultural Research Service, for assistance with artwork.

References

Adrizal, A., P.H. Patterson, R.M. Hulet, R.M. Bates, C.A. Myers, G.P. Martin, R.L. Shockey, M. van der Grinten, D.A. Anderson, and J.R. Thompson. 2008. Vegetative buffers for fan emissions from poultry farms: 2. ammonia, dust and foliar nitrogen. J. Environ. Sci. Health B 43:96–103. doi:10.1080/03601230701735078

Aletor, V.A., I.I. Hamid, E. Nie, and E. Pfeffer. 2000. Low-protein amino acid-supplemented diets in broiler chickens: Effects on performance, carcass characteristics, whole-body composition and efficiencies of nutrient utilization. J. Sci. Food Agric. 80:547–554. doi:10.1002/(SICI)1097-0010(200004)80:5<547::AID-JSFA531>3.0.CO;2-C

Angel, R. 2005. Field study to validate research results on Phosphorus (P) requirements and use of feed additives in broilers. Center for Agro-Ecology, Inc. http://agresearch.umd.edu/sites/default/files/_docs/locations/wye/Angel%20Full%20Report%20MCAE%20Pub-2005-06.pdf (Verified 18 June 2014).

Applegate, T.J., B.C. Joern, D.L. Nussbaum-Wagler, and R. Angel. 2003. Water-soluble phosphorus in fresh broiler litter is dependent upon phosphorus concentration fed but not on fungal phytase supplementation. Poult. Sci. 82:1024–1029. doi:10.1093/ps/82.6.1024

Armstrong, A., E. James, R. Stedman, and P. Kleinman. 2011. Influence of resentment in the New York City Conservation Reserve Enhancement Program. J. Soil Water Conserv. 66:337–344. doi:10.2489/jswc.66.5.337

Baxter, C.A., B.C. Joern, D. Ragland, J.S. Sands, and O. Adeola. 2003. Phytase, high available phosphorus corn and storage impacts on phosphorus levels in pig excreta. J. Environ. Qual. 32:1481–1489. doi:10.2134/jeq2003.1481

Brandt, R.C., H.A. Elliott, M.A.A. Adviento-Borbe, E.F. Wheeler, P.J.A. Kleinman, and D.B. Beegle. 2011. Field olfactometry assessment of dairy manure land application methods. J. Environ. Qual. 40:431–437. doi:10.2134/jeq2010.0094

Buda, A.R., G.F. Koopmans, R.B. Bryant, and W.M. Chardon. 2012. Emerging technologies for removing nonpoint phosphorus from surface water and groundwater: Introduction. J. Environ. Qual. 41(3):621–627. doi:10.2134/jeq2012.0080

Buda, A.R., P.J.A. Kleinman, R.B. Bryant, G.W. Feyereisen, D.A. Miller, P.G. Knight, and P.J. Drohan. 2013. Forecasting runoff from Pennsylvania landscapes. J. Soil Water Conserv. 68:185–198. doi:10.2489/jswc.68.3.185

Burns, R., H. Li, L. Moody, H. Xin, R. Gates, D. Overhults, and J. Earnest. 2008. Quantification of particulate emissions from broiler houses in the southeastern United States. Proceedings of the 2008 Livestock Environment VIII Conference. ASABE Pub 701P0408.

Callahan, M.P., P.J.A. Kleinman, A.N. Sharpley, and W.L. Stout. 2002. Assessing the efficacy of alternative phosphorus sorbing soil amendments. Soil Sci. 167:539–547. doi:10.1097/00010694-200208000-00005

Chen, L., L. Xing, and L. Han. 2013. Review of the application of near-infrared spectroscopy technology to determine the chemical composition of animal manure. J. Environ. Qual. 42:1015–1028. doi:10.2134/jeq2013.01.0014

Cromwell, G.L., T.S. Stahly, R.D. Coffey, H.J. Monegue, and J.H. Randolph. 1993. Efficacy of phytase in improving the bioavailability of phosphorus in soybean meal and corn-soybean meal diets for pigs. J. Anim. Sci. 71:1831–1840. doi:10.2527/1993.7171831x

Dahlke, H.E., Z.M. Easton, D.R. Fuka, M.T. Walter, and T.S. Steenhuis. 2013. Real-time forecast of hydrologically sensitive areas in the Salmon Creek Watershed, New York State, using an online prediction tool. Water 5:917–944. doi:10.3390/w5030917

Delgado, J.A., and J.K. Berry. 2008. Advances in precision conservation. Adv. Agron. 98:1–44. doi:10.1016/S0065-2113(08)00201-0

Delgado, J.A., R. Khosla, D.G. Westfall, W. Bausch, and D. Inman. 2005. Nitrogen fertilizer management based on site-specific management zones reduces potential for NO3-N leaching. J. Soil Water Conserv. 60:402–410.

Delgado, J.A., M. Shaffer, C. Hu, R.S. Lavado, J. Cueto-Wong, P. Joosse, X. Li, H. Rimski-Korsakov, R. Follett, W. Colon, and D. Sotomayor. 2006. A decade of change in nutrient management: A new Nitrogen Index. J. Soil Water Conserv. 61:66A–75A.

Delgado, J.A., M. Shaffer, C. Hu, R.S. Lavado, J. Cueto-Wong, P. Joosse, D. Sotomayor, W. Colon, R. Follett, S. Del Grosso, X. Li, and H. Rimski-Korsakov. 2008. An index approach to assess nitrogen losses to the environment. Ecol. Eng. 32:108–120. doi:10.1016/j.ecoleng.2007.10.006

Dell, C.J., J.J. Meisinger, and D.B. Beegle. 2011. Subsurface application of manure slurries for conservation tillage and pasture soils and their impact on the nitrogen balance. J. Environ. Qual. 40:352–361. doi:10.2134/jeq2010.0069

Dell, C.J., P.J.A. Kleinman, J.P. Schmidt, and D.B. Beegle. 2012. Low disturbance manure incorporation effects on ammonia and nitrate loss. J. Environ. Qual. 41:928–937. doi:10.2134/jeq2011.0327

Dobbie, K.E., K.V. Heal, J. Aumonier, K.A. Smith, A. Johnston, and P.L. Younger. 2009. Evaluation of iron ochre from mine drainage treatment for removal of phosphorus from wastewater. Chemosphere 75:795–800. doi:10.1016/j.chemosphere.2008.12.049

Dou, Z., D.T. Galligan, C.F. Ramberg, Jr., C. Meadows, and J.D. Ferguson. 2001. A survey of dairy farming in Pennsylvania: Nutrient management practices and implications. J. Dairy Sci. 84:966–973. doi:10.3168/jds.S0022-0302(01)74555-9

Dufrasne, I., V. Robaye, L. Istasse, and J.L. Hornick. 2010. Nitrogen excretions in dairy cows on a rotational grazing system: Effect of fertilization type, days in paddock and time period. Grassland Science in Europe 15:1034–1036.

Edwards, D.R., and T.C. Daniel. 1993. Drying interval effects on runoff from fescue plots receiving swine manure. Trans. ASABE 36:1673–1678. doi:10.13031/2013.28510

Ehmke, T. 2012. The 4Rs of nutrient management. Crops & Soils 45(5):4–10.

Fenton, O., M.G. Healy, and M. Rodgers. 2009. Use of ochre from an abandoned metal mine in the south east of Ireland for phosphorus sequestration from dairy dirty water. J. Environ. Qual. 38:1120–1125. doi:10.2134/jeq2008.0227

Goering, D.C. 2013. Decision support for Wisconsin's manure spreaders: Development of a real-time runoff risk advisory forecast. M.S. Thesis. University of Arizona. Tucson, AZ. p. 265.

Harter-Dennis, J. 2010. Poultry house of the future concept. In: Chesapeake Goal Line 2025: Opportunities for Enhancing Agricultural Conservation. October 5-6, 2010. Hunt Valley, MD. The Science and Advisory Committee, Edgewater, MD. Http://www.chesapeake.org/OldStac/agconservationtools/dennis.pdf (Verified 17 June 2014).

Havenstein, G.B., P.R. Ferket, S.E. Scheideler, and B.T. Larson. 1994. Growth, livability and feed conversion of 1991 vs 1957 broilers when fed "typical" 1957 and 1991 broiler diets. Poult. Sci. 73:1785–1794. doi:10.3382/ps.0731785

Havenstein, G.B., P.R. Ferket, and M.A. Qureshi. 2003. Growth, livability and feed conversion of 1957 vs 2001 broilers when fed representative 1957 and 2001 broiler diets. Poult. Sci. 82:1500–1508. doi:10.1093/ps/82.10.1500

Heal, K.V., K.E. Dobbie, E. Boika, H. McHaffie, A.E. Simpson, and K.A. Smith. 2005. Enhancing phosphorus removal in constructed wetlands with ochre from mine drainage treatment. Water Sci. Technol. 51:275–282.

Herron, S., A. Sharpley, S. Watkins, and M. Daniels. 2012. Poultry litter management in the Illinois River Watershed of Arkansas and Oklahoma. Report FSA9535. University of Arkansas Cooperative Extension. Little Rock, AR. http://www.uaex.edu/publications/pdf/FSA-9535.pdf (verified 12 June 2014).

Holden, N.M., R.P.O. Schulte, and S. Walsh. 2007. A slurry spreading decision support system for Ireland. In: N.M. Holden, T. Hochstrasser, R.P.O. Schulte, and S. Walsh, editors, Making science work on the farm, A Workshop of Decision Support Systems for Irish Agriculture. Agmet, Oakwood Village, OH. p. 24–33.

Hooser, S.B., W. Van Alstine, M. Kiupel, and J. Sojka. 2000. Acute pit gas (hydrogen sulfide) poisoning in confinement cattle. J. Vet. Diagn. Invest. 12:272–275. doi:10.1177/104063870001200315

Hornung, A., R. Khosla, R. Reich, and D.G. Westfall. 2003. Evaluation of site-specific management zones: Grain yield and nitrogen use efficiency. In: J. Stafford and A. Werner, editors, Precision agriculture. Wageningen Academic Publ., Wageningen, the Netherlands. p. 297–302.

House, C.H., S.W. Broome, and M.T. Hoover. 1994. Treatment of nitrogen and phosphorus by a constructed upland-wetland wastewater treatment system. Water Sci. Technol. 29:177–184.

James, E., P. Kleinman, T. Veith, R. Stedman, and A. Sharpley. 2007. Phosphorus contributions from pastured dairy cattle to streams of the Cannonsville Watershed, New York. J. Soil Water Conserv. 62:40–47.

Jarvie, H.P., A.N. Sharpley, D. Flaten, and P.J.A. Kleinman. 2014. Phosphorus sustainability and the water-energy-food security nexus. Environ. Sci. Technol.

Jensen, L.S. 2013. Animal manure fertilizer value, crop utilization and soil quality impacts. In: S.G. Sommer, M.L. Christensen, T. Schmidt, and L.S. Jensen, editors, Animal manure recycling: Treatment and management. John Wiley & Sons, Ltd., United Kingdom. p. 295–328. doi:10.1002/9781118676677.ch15

Karmakar, S., C. Laguë, J. Agnew, and H. Landry. 2007. Integrated decision support system (DSS) for manure management: A review and perspective. Comput. Electron. Agric. 57:190–201. doi:10.1016/j.compag.2007.03.006

Karmakar, S., M.N. Ketia, C. Laguë, and J. Agnew. 2010. Development of expert system modeling based decision support system for swine manure management. Comput. Electron. Agric. 71:88–95. doi:10.1016/j.compag.2009.12.009

Kellogg, R.L., C.H. Lander, D.C. Moffitt, and N. Gollehon. 2000. Manure nutrients relative to the capacity of cropland and pastureland to assimilate nutrients: Spatial and temporal trends for the United States. USDA–GSA National Forms and Publication Center, Fort Worth, TX. doi:10.2175/193864700784994812

Kerebel, A., R. Cassidy, P. Jordan, and N.M. Holden. 2013. Farmer perception of suitable conditions for slurry application compared with decision support system recommendations. Agric. Syst. 120:49–60. doi:10.1016/j.agsy.2013.05.007

Khosla, R., K. Fleming, J.A. Delgado, T. Shaver, and D.G. Westfall. 2002. Use of site-specific management zones to improve nitrogen management for precision agriculture. J. Soil Water Conserv. 57:513–518.

Kleinman, P.J.A., and A.N. Sharpley. 2003. Effect of broadcast manure on runoff phosphorus concentrations over successive rainfall events. J. Environ. Qual. 32:1072–1081. doi:10.2134/jeq2003.1072

Kleinman, P.J.A., A.N. Sharpley, R.W. McDowell, D.N. Flaten, A.R. Buda, and L. Tao. 2011. Managing agricultural phosphorus for water quality protection: Principles for progress. Plant Soil 349:169–182. doi:10.1007/s11104-011-0832-9

Lemunyon, J.L., and R.G. Gilbert. 1993. The concept and need for a phosphorus assessment tool. J. Prod. Agric. 6:483–496. doi:10.2134/jpa1993.0483

Lorimor, J. 1998. Iowa odor control demonstration project: Soil injection. Iowa State University Extension Publication PM 1754E, Ames, IA. Available online: https://store.extension.iastate.edu/Product/pm1754e-pdf (verified Nov 30, 2017).

Lynch, T., S. Gregor, and D. Midmore. 2000. Intelligent support systems in agriculture: How can we do better? Aust. J. Exp. Agric. 40:609–620. doi:10.1071/EA99082

Lynch, T., and S. Gregor. 2004. User participation in decision support systems development: Influencing system outcomes. Eur. J. Inf. Syst. 13:286–301. doi:10.1057/palgrave.ejis.3000512

MacDonald, J., M. Ribaudo, M. Livingston, J. Beckman, and W. Huang. 2009. Manure use for fertilizer and for energy: Report to congress. Administrative Publication No. (AP-037). USDA–ERS. p. 53

Maguire, R.O., P.J.A. Kleinman, and D.B. Beegle. 2011. Novel manure management technologies in no-till and forage systems. J. Environ. Qual. 40:287–291. doi:10.2134/jeq2010.0396

Mallarino, A.P., and D.J. Wittry. 2010. Crop yield and soil phosphorus as affected by liquid swine manure phosphorus application using variable rate technology. Soil Sci. Soc. Am. J. 74:2230–2238. doi:10.2136/sssaj2009.0215

McDowell, R.W., and P.J.A. Kleinman. 2011. Efficiency of phosphorus cycling in different grassland systems. In: G. Lemaire, J. Hodgson, and A. Chabbi, editors, Grassland productivity and ecosystem services. CABI Publishing, Oxfordshire, UK. p. 108–119. doi:10.1079/9781845938093.0108

Meals, D.W. 1990. LaPlatte River watershed water quality monitoring and analysis program: Comprehensive final report. Program Report No. 12. Vermont Water Resources and Lake Studies Center, Univ. of Vermont, Burlington, VT.

Meals, D.W. 1993. Assessing nonpoint phosphorus control in the LaPlatte River watershed. Lake Reservoir Manage. 7:197–207. doi:10.1080/07438149309354271

Meals, D.W. 2000. Lake Champlain Basin agricultural watersheds Section 319 National Monitoring Program Project, Year 6 annual report: May, 1998-September, 1999. Vermont Dep. of Environmental Conservation, Waterbury, VT, p. 182.

Melkonian, J.J., H.M. van Es, A.T. DeGaetano, and L. Joseph. 2008. ADAPT-N: Adaptive nitrogen management for maize using high-resolution climate data and model simulations. In: R. Kosla, editor, Proceedings of the 9th International Conference on Precision Agriculture, July 20-23, 2008, Denver, CO (CD-ROM).

Mitsch, W.J., and J.G. Gosselink. 1993. Wetlands, Second ed. Van Nostrand Reinhold, New York. p. 722.

Moore, P.A., Jr., T.C. Daniel, D.R. Edwards, and D.M. Miller. 1995. Effect of chemical amendments on ammonia volatilization from poultry litter. J. Environ. Qual. 24:293–300. doi:10.2134/jeq1995.00472425002400020012x

Moore, P.A., Jr., T.C. Daniel, and D.R. Edwards. 2000. Reducing phosphorus runoff and inhibiting ammonia loss from poultry manure with aluminum sulfate. J. Environ. Qual. 29:37–49. doi:10.2134/jeq2000.00472425002900010006x

Moore, P.A., Jr., D. Miles, R. Burns, D. Pote, K. Berg, and I.H. Choi. 2011. Ammonia emission factors from broiler litter in barns, in storage, and after land application. J. Environ. Qual. 40:1395–1404. doi:10.2134/jeq2009.0383

Moshia, M.E., R. Khosla, L. Longchamps, D.G. Westfall, J.G. Davis, and R. Reich. 2014. Precision manure management across site-specific management zones: Grain yield and economic analysis. Agron. J. doi:10.2134/agronj13.0400

Mulligan, F.J., P. Dillon, J.J. Callan, M. Rath, and F.P. O'Mara. 2004. Supplementary concentrate type affects nitrogen excretion of grazing dairy cows. J. Dairy Sci. 87:3451–3460. doi:10.3168/jds.S0022-0302(04)73480-3

Nolan, B.T., and K.J. Hitt. 2006. Vulnerability of shallow groundwater and drinking-water wells to nitrate in the United States. Environ. Sci. Technol. 40:7834–7840. doi:10.1021/es060911u

Novotny, V., and H. Olem. 1994. Water quality: Prevention, identification, and management of diffuse pollution. Van Nostrand Reinhold, New York.

Oliver, D.M., R.D. Fish, M. Winter, C.J. Hodgson, A.L. Heathwaite, and D.R. Chadwick. 2012. Valuing local knowledge as a source of expert data: Farmer engagement and the design of decision support systems. Environ. Model. Softw. 36:76–85. doi:10.1016/j.envsoft.2011.09.013

Penn, C.J., and J.M. McGrath. 2011. Predicting phosphorus sorption onto steel slag using a flow-through approach with application to a pilot scale system. Journal of Water Resources and Protection 3:235–244. doi:10.4236/jwarp.2011.34030

Penn, C.J., R.B. Bryant, P.J.A. Kleinman, and A.L. Allen. 2007. Removing dissolved phosphorous from drainage ditch water with phosphorous sorbing materials. J. Soil Water Conserv. 6:269–276.

Penn, C.J., R.B. Bryant, M.A. Callahan, and J.M. McGrath. 2011. Use of industrial by-products to sorb and retain phosphorus. Commun. Soil Sci. Plant Anal. 42:633–644. doi:10.1080/00103624.2011.550374

Pennsylvania Conservation Commission. 2013. Pennsylvania Odor BMP Reference List (Version 2.0). Pennsylvania Conservation Commission. http://www.agriculture.state.pa.us/portal/server.pt/gateway/PTARGS_0_2_24476_10297_0_43/agwebsite/Files/Publications/PAOdorBMPReferenceListVersion20.pdf (verified 29 May 2014).

Penn State Extension. Penn State agronomy guide. 2007. Penn State College of Agricultural Sciences. University Park, PA. p. 331.

Pierzynski, G.M., and T.J. Logan. 1993. Crop, soil, and management effects on phosphorus soil test levels. J. Prod. Agric. 6:513–520. doi:10.2134/jpa1993.0513

Preedy, N., K. McTiernan, R. Matthews, L. Heathwaite, and P. Haygarth. 2001. Rapid incidental phosphorus transfers from grassland. J. Environ. Qual. 30:2105–2112. doi:10.2134/jeq2001.2105

Pries, J., and P. McGarry. 2002. Feedlot stormwater runoff treatment using constructed wetlands. Http://agrienvarchive.ca/bioenergy/download/WEAO_2001_Pries.pdf (verified 30 May 2014).

Pons, L. 2005. Blue lagoons on Pig Farms? Agric. Res. 53(3):14–15.

Powers, W.J. 1999. Odor control for livestock systems. J. Anim. Sci. 77:169–176. doi:10.2527/1999.77suppl_2169x

Provolo, G., and L. Martínez-Suller. 2007. In: situ determination of slurry nutrient content by electrical conductivity. Bioresour. Technol. 98(17):3235–3242. doi:10.1016/j.biortech.2006.07.018

Raboy, V. 2001. Seeds for a better future: 'Low phytate' grains help to overcome malnutrition and reduce pollution. Trends Plant Sci. 6(10):458–462. doi:10.1016/S1360-1385(01)02104-5

Ravindran, V., W.L. Bryden, and E.T. Kornegay. 1995. Phytates: Occurance, bioavailability and implications in poultry nutrition. Poult. Avian Biol. Rev. 6:125–143.

Reeves, J.B. 2007. The present status of "quick tests" for on-farm analysis with emphasis on manures and soil: What is available and what is lacking. Livest. Sci. 112:224–231. doi:10.1016/j.livsci.2007.09.009

Ribaudo, M., N. Gollehon, M. Aillery, J. Kaplan, R. Agapoff, L. Christensen, V. Breneman, and M. Peters. 2003. Manure management for water quality: Costs of animal feeding operations of applying nutrients to land. Agricultural Economic Report 824. Washington, DC: USDA Economic Research Service.

Ribaudo, M., J. Delgado, L. Hansen, M. Livingston, R. Mosheim, and J. Williamson. 2011. Nitrogen in agricultural systems: Implications for conservation policy. Economy Research Report, P1-82. USDA-ERS. Washington, DC.

Rotz, C.A. 2004. Management to reduce nitrogen losses in animal production. J. Anim. Sci. 82(E. Suppl.): E119-E137.

Sanford, W.E., and J. Pope. 2013. Quantifying groundwater's role in delaying improvements to Chesapeake Bay water quality. Environ. Sci. Technol. 47:13330–13338. doi:10.1021/es401334k

Schwer, C.B., and J.C. Clausen. 1989. Vegetative filter treatment of dairy milkhouse waste water. J. Environ. Qual. 18:446–451. doi:10.2134/jeq1989.00472425001800040008x

Scotford, I., T. Cumby, L. Han, and P. Richards. 1998. Development of a prototype nutrient sensing system for livestock slurries. J. Agric. Eng. Res. 69(3):217–228. doi:10.1006/jaer.1997.0246

Sharpley, A.N., P. Richards, S. Herron, and D. Baker. 2012a. Comparison between litigated and voluntary nutrient management strategies. J. Soil Water Conserv. 67(5):442–450. doi:10.2489/jswc.67.5.442

Sharpley, A.N., P.J.A. Kleinman, R.W. McDowell, M. Gitau, and R.B. Bryant. 2002. Modeling phosphorus transport in agricultural watersheds: Processes and possibilities. J. Soil Water Conserv. 57(6):425–439.

Sharpley, A.N., D. Beegle, C. Bolster, L. Good, B. Joern, Q. Ketterings, J. Lory, R. Mikkelsen, D. Osmond, and P. Vadas. 2012b. Phosphorus indices: Why we need to take stock of how we are doing. J. Environ. Qual. 41:1711–1719. doi:10.2134/jeq2012.0040

Sharpley, A.N., H.P. Jarvie, A.R. Buda, L. May, B. Spears, and P.J.A. Kleinman. 2013. Phosphorus legacy: Overcoming the effects of past management practices to mitigate future water quality impairment. J. Environ. Qual. 42(5):1308–1326. doi:10.2134/jeq2013.03.0098

Sharpley, A.N., J.L. Weld, D.B. Beegle, P.J.A. Kleinman, W.J. Gburek, P.A. Moore, and G. Mullins. 2003. Development of phosphorus indices for nutrient management planning strategies in the U.S. J. Soil Water Conserv. 58:137–152.

Sibrell, P.L., G.A. Montgomery, K.L. Ritenour, and T.W. Tucker. 2009. Removal of phosphorus from agricultural wastewaters using adsorption media prepared from acid mine drainage sludge. Water Res. 43:2240–2250. doi:10.1016/j.watres.2009.02.010

Spencer, J.D., G.L. Allee, and T.E. Sauber. 2000. Phosphorus bioavailability and digestibility of normal and genetically modified low-phytate corn for pigs. J. Anim. Sci. 78:675–681. doi:10.2527/2000.783675x

Stout, W.L., A.N. Sharpley, W.J. Gburek, and H.B. Pionke. 1999. Reducing phosphorus export from croplands with FBC flyash and FGD gypsum. Fuel 78:175–178. doi:10.1016/S0016-2361(98)00141-0

Sutton, A., and D. Beede. 2003. Feeding strategies to lower nitrogen and phosphorus in manure. Best Environmental Management Practices Circular ID-304. https://www.extension.purdue.edu/extmedia/id/id-304.pdf (verified 18 June 2014).

Tesoriero, A.J., J.H. Duff, D.A. Saad, N.E. Spahr, and D.M. Wolock. 2013. Vulnerability of streams to legacy nitrate sources. Environ. Sci. Technol. 47:3623–3629. doi:10.1021/es305026x

Tunney, H. 1986. Manure nutrient composition: Rapid methods of assessment. In: P. L'Hermite, editor, Processing and use of organic sludge and liquid agricultural wastes. D. Reidel Publishing Company, Dordrecht, Holland. p. 243–257. doi:10.1007/978-94-009-4756-6_17

Tyndall, J., and J. Colletti. 2007. Mitigating swine odor with strategically designed shelterbelt systems: A review. Agrofor. Syst. 69:45–65. doi:10.1007/s10457-006-9017-6

USDA Natural Resource Conservation Service. 2011a. Conservation Practice Standard, Nutrient Management 590. http://www.nrcs.usda.gov/Internet/FSE_DOCUMENTS/stelprdb1046177.pdf (verified 30 June 2014).

USDA Natural Resource Conservation Service. 2011b. Title 190– National Instruction (Title 190-NI, Amend., December 2011) 302-X.1, Part 302– Nutrient Management Policy Implementation. http://www.nrcs.usda.gov/Internet/FSE_DOCUMENTS/stelprdb1046177.pdf (verified 3 June 2014).

USDA National Agricultural Statistics Service. 2012a. Poultry summary available at: http://www.uspoultry.org/economic_data/ (verified 3 June 2014).

USDA National Agricultural Statistics Service. 2012b. 2012 Census of Agriculture. http://www.agcensus.usda.gov/ (verified 17 June 2014).

US Environmental Protection Agency. 2012. NPDES permit writers manual for concentrated animal feeding operations. http://www.epa.gov/npdes/pubs/cafo_permitmanual_frontmatter.pdf (verified 3 June 2014).

Vaddella, V.K., P.M. Ndegwa, H.S. Joo, and J.L. Ullman. 2010. Impact of separating dairy cattle excretions on ammonia emissions. J. Environ. Qual. 39:1807–1812. doi:10.2134/jeq2009.0266

van Es, H.M., J.M. Sogbedji, and R.R. Schindelbeck. 2006. Effect of manure application timing, crop, and soil type on nitrate leaching. J. Environ. Qual. 35:670–679. doi:10.2134/jeq2005.0143

Van Kessel, J.S., and J.B. Reeves, 3rd. 2000. On-farm quick tests for estimating nitrogen in dairy manure. J. Dairy Sci. 83:1837–1844. doi:10.3168/jds.S0022-0302(00)75054-5

Vanotti, M.B., A.A. Szogi, P.G. Hunt, P.D. Millner, and F.J. Humenik. 2007. Development of environmentally superior treatment system to replace anaerobic swine lagoons in the USA. Bioresour. Technol. 98:3184–3194. doi:10.1016/j.biortech.2006.07.009

Vanotti, M.B., A.A. Szogi, P.D. Millner, and J.H. Loughrin. 2009. Development of a second-generation environmentally superior technology for treatment of swine manure in the USA. Bioresour. Technol. 100:5406–5416. doi:10.1016/j.biortech.2009.02.019

Vohla, C., M. Kõiv, H.J. Bavor, F. Chazarenc, and Ü. Mander. 2011. Filter materials for phosphorus removal from wastewater in treatment wetlands—A review. Ecol. Eng. 37:70–89. doi:10.1016/j.ecoleng.2009.08.003

Walters, D.F., and D. Shrubsole. 2013. Assessing the implementation of Ontario's nutrient management decision support system. Can. Geogr. 58:203–216. doi:10.1111/j.1541-0064.2013.12058.x

Wicklen, G.L., and M. Czarick. 1997. Particulate emissions from poultry housing. ASAE Annual International Meeting, Minneapolis Convention Center, Minneapolis, MN, 10-14 August 1997.

Wilson, M.G. 2004. Development and testing of a liquid alum delivery system for high rise hen houses. M.S. Thesis. University of Arkansas. Fayetteville, AR.

Wisconsin Manure Management Advisory System. 2014. Runoff risk advisory forecast. Wisconsin Manure Management Advisory System. http://www.manureadvisorysystem.wi.gov/app/runoffrisk (Verified 17 June 2014).

Wu, Z., L.D. Satter, A.J. Blohowiak, R.H. Stauffacher, and J.H. Wilson. 2001. Milk production, phosphorus excretion, and bone characteristics of dairy cows fed different amounts of phosphorus for two or three years. J. Dairy Sci. 84:1738–1748. doi:10.3168/jds.S0022-0302(01)74609-7

Xin, H., R.S. Gates, A.R. Green, F.M. Mitloehner, P.A. Moore, Jr., and C.M. Wathes. 2011. Environmental impacts and sustainability of egg production systems. Poult. Sci. 90(1):263–277. doi:10.3382/ps.2010-00877

Young, R.A., T. Hutrods, and W. Anderson. 1980. Effectiveness of vegetative buffer strips in controlling pollution from feedlot runoff. J. Environ. Qual. 9:483–487. doi:10.2134/jeq1980.00472425000900030032x

Irrigation Management

David L. Bjorneberg,* Robert G. Evans, and E. John Sadler

Abstract

Competition for limited water supplies continues to restrict water available for irrigation. Irrigated agriculture must continually improve irrigation management to continue producing food, fiber, and fuel for a growing world population. Precision irrigation is the process of applying the right amount of water at the right time and place to obtain the best use of available water. Precision irrigation management is needed on large irrigation projects so water delivery matches irrigation needs and on individual fields to apply the right amount of water at the right time and place. Technology is commercially available to precisely apply water when and where crops need it; however, user-friendly decision tools are still needed to quantify specific irrigation needs and control water application within fields. Integrating information from various sensors and systems into a decision support program will be critical to highly managed, spatially varied irrigation.

Irrigation supplements precipitation in the hydraulic cycle with the primary goal of meeting the transpiration needs of crops. Ideally, all applied irrigation water would be beneficially used by crops. However, irrigation systems cannot be controlled and operated to apply the exact amount of water that each individual plant requires. Applying too much water in an area can cause deep percolation and leaching. Applying water faster than it can infiltrate results in ponding and runoff. Ponded water can evaporate and therefore is not beneficially used by crops. Runoff water either flows from the field, becoming on off-site problem, or infiltrates in other areas of the field causing nonuniform soil water content in the field. Continued nonuniform irrigation with in-field runoff can cause deep percolation in areas where runoff infiltrates. In addition to being a nonbeneficial use of water, deep percolation and off-site runoff often transport nutrients and sediment.

Irrigation research and development has historically attempted to uniformly irrigate fields even though fields are seldom uniform. Even large irrigation projects often deliver water uniformly to all fields within the project regardless

Abbreviations: ET, evapotranspiration; ICID, International Commission on Irrigation and Drainage; SSIM, site-specific irrigation management.

David L. Bjorneberg, USDA-ARS, 3793 N 3600 E, Kimberly, ID 83341-5076. Robert G. Evans, 32606 W. Knox Road, Benton City, WA 99320 (robertevans910@gmail.com). E. John Sadler, Rm. 269 Agric. Eng. Bldg., University of Missouri, Columbia, MO 65211 (john.sadler@ars.usda.gov). *Corresponding author (Dave.Bjorneberg@ars.usda.gov).

doi:10.2134/agronmonogr59.2013.0025

Precision Conservation: Geospatial Techniques for Agricultural and Natural Resources Conservation
J.A. Delgado, G.F. Sassenrath, and T. Mueller, editors. Agronomy Monograph 59.

of crops grown or irrigation systems used. The goal of precision irrigation is to apply the right amount of water at the right time and place at any scale. Precision irrigation often confers images of center-pivot irrigation systems with sprinklers pulsing on and off as the machine rotates through the field. Varying irrigation rates within a field, however, is only one facet of precision irrigation. Precision irrigation can be applied at any scale or with any type of irrigation. At the irrigation project scale, managers need to know crop water requirements so irrigation diversions can be adjusted to meet the irrigation demand. Irrigators at the farm scale need similar information to ensure that their water supply can meet the needs of the crops on their farm. On individual fields, farmers need to know crop water requirements to schedule irrigation and have irrigation systems with the capability and flexibility to apply the right amount of water in the right place and time.

In humid, and even semiarid areas, crop irrigation requirements must be determined throughout the growing season to account for precipitation and variations in crop growth to precisely irrigate. In arid areas, precipitation during the growing season is usually minimal; however, arid areas still benefit from in-season measurements of crop water use to precisely apply irrigation water. Precision irrigation requires knowledge of crop water needs, soil water holding capacity, irrigation system capacity, and available water supply. If any one of these factors is not considered, irrigation water can be over- or underapplied. Insufficient irrigation will reduce crop yield and will cause unacceptable quality for some crops. Irrigating too much wastes water and can cause runoff or deep percolation.

All of the precision seeding, tillage, fertility, and conservation concepts for nonirrigated farming apply to irrigated production with the advantage—and complication—of the ability to vary water application. Precision irrigation is perhaps the most complicated aspect of precision agriculture because irrigation decisions are made multiple times each season, and irrigation interacts with fertilizer and disease management. Crop water use, for example, can be greater at higher N fertilization rates (Lenka et al., 2009) or similar between low and high rates (Holmen et al., 1961). Pandey et al. (2000) showed that maize (*Zea mays* L.) yield reductions to water shortage were greater at high N fertilization rates than low rates. Crop response to applied water cannot be assumed to be the same under uniform irrigation management and site-specific irrigation (Sadler et al., 2002). Intuitively, lower-yielding areas in an irrigated field should require less water and fertilizer. However, low-yielding areas may require more water and nutrients per unit of production. If water supply is limited, the best practice may be to not irrigate low-yielding areas and focus irrigation on higher-yielding areas. Precisely applying irrigation water can also alleviate some soil issues by ensuring that water application rates match soil infiltration rates. Technology to precisely apply irrigation water is wasted if the water does not infiltrate into the soil where it was applied and remain in the soil profile for the crop to use.

The purpose of this chapter is to discuss aspects and technologies affecting precision irrigation. One major aspect affecting precision irrigation is the availability of water to irrigate. Managing an irrigation system connected to a high-capacity well is not restricted by water supply. Off-site water suppliers, however, often restrict irrigation amounts and timing, providing an additional challenge to applying the right amount of water at the right time. Low-capacity wells or regulatory limitations on groundwater withdrawal can also restrict water supply. Another major aspect of precision irrigation is determining crop water needs so adequate water can be supplied to an entire irrigation project or applied on irrigated farms and fields. Finally, technology for managing irrigation application on fields is discussed.

Background

Irrigation is the process of artificially applying water to soil with the intention of improving crop yield and quality. Humans have used irrigation to help provide more stable food production for 8000 yr, initially diverting and channeling flood-waters from the Nile or Tigris and Euphrates Rivers. Irrigation equipment and techniques have evolved from hand-dug ditches and wild flooding to carefully controlled microirrigation and variable-rate, center-pivot machines that can be operated remotely from cell phones or computers. Irrigation enhances the magnitude, quality, and reliability of crop production. According to the Food and Agriculture Organization of the United Nations, irrigation contributes to ~40% of the world's food production on <20% of the world's crop production land. In the United States, only 8% of the total cropland is on farms where all crops are irrigated (USDA-NASS, 2014a). These farms produce 27% of the market value of crops and 12% of the total market value of all livestock. Half of the crop value is produced on farms with some irrigated land, and these farms account for only 28% of the total cropland in the United States (USDA-NASS, 2014b).

Certain crops rely heavily on irrigation (Table 1). All rice (*Oryza sativa* L.) in the United States is grown with irrigation, while only 9% of the soybean [*Glycine*

Table 1. Total and irrigated crop area in the United States.

Crop	Total area†	Total irrigated crop area‡		Pressurized irrigation‡§	Gravity irrigation‡
	ha	ha	%	———— ha ————	
Corn (grain)	35,389,897	5,189,906	15	4,458,346	921,916
Corn (silage)	2,913,615	665,488	23	487,503	212,642
Sorghum (grain)	2,081,821	253,606	12	277,842	49,923
Wheat	19,854,343	1,363,630	7	1,074,673	219,862
Soybean	30,811,652	2,897,738	9	1,636,397	1,364,930
Dry edible bean	665,100	191,703	29	129,713	45,830
Rice	1,090,591	1,090,591	100	13,640	1,254,738
Other small grains (barley, oats, rye, etc.)	21,727,990	489,687	2	359,500	130,187
Alfalfa (hay and silage)	6,731,106	2,340,591	35	1,435,385	796,547
All other hay	16,371,840	1,575,706	10	592,238	744,154
Peanuts	656,531	211,204	32	145,953	4,506
Cotton	3,799,223	1,543,909	41	882,562	313,574
Orchards, vineyards, and nut trees	2,105,153	1,723,495	82	1,274,689	207,955
All vegetables	1,692,668	1,212,479	72	1,074,192	137,791
All berries	117,374	88,798	76	89,635	2,532
All cropland	157,769,398	22,600,094	14	16,106,891	8,706,350

† From USDA-NASS 2012 Census of Agriculture (USDA-NASS, 2014a).

‡ From USDA-NASS 2013 Farm and Ranch Irrigation Survey (USDA-NASS, 2014b).

§ Pressurized irrigation includes sprinkler, micro, and drip irrigation.

max (L.) Merr.] and 7% of the wheat (*Triticum aestivum* L.) in the United States are grown with irrigation. In addition, irrigation is used on 82% of the land in orchards, 72% of land growing vegetables, and 76% of the land growing berries. While irrigation is important for crop production, it accounts for 38% of the total freshwater withdrawal in the United States, which equals freshwater withdrawal for thermoelectric power (Maupin et al., 2014). Approximately 75% of the total freshwater withdrawal in the seven western states is used for irrigation, and 67% of this is surface water (Maupin et al., 2014). Continued improvements in irrigation management, as well as agronomic management, are required to make the best use of limited water resources by applying the appropriate amount of water at the right time to all areas in a field.

Irrigation is generally categorized by the three main methods of applying water: surface, sprinkler, and microirrigation. Water flows over the soil by gravity for surface irrigation. Sprinkler irrigation applies water to soil by sprinkling or spraying water droplets from fixed or moving irrigation systems under pressure. Microirrigation applies frequent, small applications by dripping, bubbling, or spraying at low pressures and usually only wets a portion of the soil surface in the field. A fourth, and minor, irrigation method is subirrigation, where the water table is raised to or held near the plant root zone using ditches or subsurface drains to supply the water, used mostly in humid regions. According to the International Commission on Irrigation and Drainage (ICID), surface irrigation is used on about 85% of the 299 Mha of irrigated cropland in the world (ICID, 2013). India and China each irrigate >60 Mha of cropland (FAO, 2013) and ~95% of this land is surface irrigated (ICID, 2013). The United States and Pakistan each have about 20 Mha of irrigated land. These four counties account for 55% of the irrigated land in the world; all other countries each have <10 Mha of irrigated land. Of these four countries, only the United States uses sprinkler and microirrigation on a significant portion of the irrigated cropland. Sprinkler irrigation is used on 57% of the 22 Mha of cropland irrigated in the United States, while microirrigation is used on ~8% of the irrigated cropland, and these percentages continue to increase (Fig. 1). The amount of US farmland irrigated by center pivots has increased almost linearly since they were first marketed in the late 1960s. Center-pivot irrigation in the United States has increased ~240,000 ha yr^{-1} and is used on 80% of the sprinkler-irrigated land and 45% of all irrigated land (USDA-NASS, 2014b).

Available Water Supply

Irrigation water for ~25% of the US irrigated land is supplied by off-farm sources (USDA-NASS, 2014b). The percentage is likely higher in China, India, and Pakistan. Precision irrigation on farms in these irrigation projects must include allowances for water delivery limitations. Irrigation projects have rules for irrigation water delivery that govern how far in advance a request must be made and whether the request is for a volume or flow rate for a specified time. In general, irrigation projects are either supply oriented (rigid schedule) or demand oriented (flexible schedule). Some projects may shift from demand-based to supply-based operation as irrigation demand increases during the season to a point where irrigation demand exceeds the capacity of the canal system (Lozano and Mateos, 2008). Supply-oriented systems generally operate canals at constant flow rates to efficiently use the canal system that required large capital investment to construct (Merriam

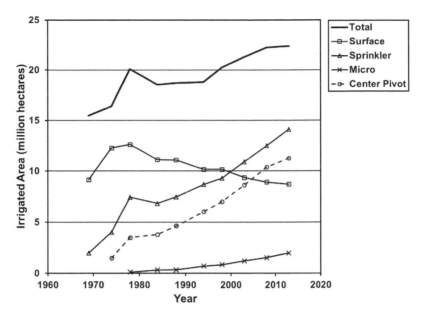

Fig. 1. Trends in US irrigated acreage for 1969 through 2013 (from USDA-NASS, 2014a).

et al., 2007). However, this operating method does not account for specific crop irrigation needs in specific fields, which led Merriam et al. (2007) to call supply-oriented delivery an engineer's dream and a farmer's nightmare. Clemmens (2006) referred to supply-oriented delivery as water disposal not irrigation water management. Supply-oriented systems tend to be hydraulically efficient but inefficient for crop water management.

Irrigation projects using supply-oriented schedules may allocate water on a flow rate per unit area, a volume per unit area, or flow rate for specified time interval. A specific example of a flow-rate allocation project is the Twin Falls Canal Company in southern Idaho. The Twin Falls Canal Company was established in 1900 and originally planned to divert 85 m³ s⁻¹ from the Snake River to irrigate 97,000 ha (Bjorneberg et al., 2008). Since the canal company has a natural-flow water right, irrigation water is continually available at 52 L m⁻¹ ha⁻¹ (total flow rate divided by total project area) during the entire irrigation season. If flow in the river is not adequate to meet this supply, the allocation per hectare is reduced proportionally to all users. This flow-rate allocation continues today even though water stored in upstream reservoirs supplements the natural river flow so the need to reduce allocations is much less. While a flow rate is available during the entire irrigation season, it is not practical to continually use this flow rate, especially in the spring and fall. In the summer however, the flow rate allocation of 52 L m⁻¹ ha⁻¹ is only 7.5 mm d⁻¹, which does not meet peak irrigation demand for many crops. Consequently, farmers grow a variety of crops to spread irrigation demand throughout the irrigation season to better balance irrigation supply with crop demand. The flow rate allocation is unlikely to change as long as the natural-flow water right exists, demonstrating how policies and regulations impact irrigation management.

Rigid water allocation policies like the previous example provide little incentive for growers to adopt advanced irrigation systems and management practices such as precision irrigation. On the other hand, flexible irrigation water delivery strives to match delivery with irrigation demand. This is complicated in large irrigation projects (e.g., >20,000 ha) because it may take one or more days for water to travel from the diversion point to the field. Changing the flow rate at the main diversion causes a sudden flow-rate change in the canal that dissipates downstream. Operators further downstream often need to make multiple adjustments to account for the gradual flow-rate change (Clemmens, 2006). Automated diversion gates can be used to maintain desired flow rates within the irrigation project (van Overloop et al., 2010). Bautista et al. (2006) demonstrated a control system that used pre-established canal hydraulic relationships to make upstream flow changes to meet downstream delivery changes. Automated control of canal systems, however, is complicated, requiring excellent measurement and control equipment, good communication systems, and algorithms and strategies for controlling the equipment. Implementation usually involves problems that require adjustments to algorithms and control equipment (Burt and Piao, 2004), but these problems should not prohibit irrigation projects from implementing advanced control systems.

Flexible water delivery requires adequate canal design to carry and deliver varied flow rates that meet irrigation demand throughout the irrigation system for the entire irrigation season. Canal systems that provide flexible delivery tend to be more costly to construct and operate than canals designed for supply-based delivery. Exceeding existing canal flow capacity is an obvious problem. Low irrigation demand can also be a problem if the canal system requires a minimum flow rate before water reaches a specific elevation required for water delivery gates. Additional check structures may be needed to allow water delivery at lower flow rates. Additional small water storage facilities distributed within the canal system may also be necessary to store excess water during low demand periods and to more quickly supply water as demand increases.

On-farm water storage also increases irrigation flexibility when off-farm or on-farm water supplies are limited by flow rate or time. Continuous off-farm water supply can be stored to enable daytime only irrigation or odd irrigation set times for better surface irrigation management. Pressurized irrigation systems may be able to use off-peak electric supplies by storing water from an off-farm supplier during peak electrical usage time. Farms with low-capacity wells can also benefit from on-farm storage. The low-capacity well can continuously pump water into storage even if the field is not being irrigated as a result of precipitation or field operations. A disadvantage of on-farm storage is that land is taken out of production for the reservoir, and some water will be lost to evaporation and seepage. Producers need to compare the lost production from the reservoir area with the increased yield from the remaining irrigated land on the farm.

Although applying irrigation when the crop needs water is the best practice, Schlegel et al. (2012) determined that many irrigation systems cannot meet peak crop water demands and must rely on stored soil water to meet crop water needs. They found that preseason irrigation was profitable for wells with 2.5 to 5 mm d^{-1} capacities in Kansas, and water use efficiency (crop yield per unit water applied) was not significantly affected by preseason irrigation. In this case, the soil is the on-farm storage reservoir.

Crop Water Needs

At the project, farm, and field scales, the first step for precision irrigation is to know how much water a crop needs at a given time. This is typically done by calculating evapotranspiration (ET), which is the amount of water that evaporates from the soil and plant surfaces and transpires from plants. Evapotranspiration can be estimated from meteorological data, remotely sensed information (e.g., temperature and reflectance), soil water depletion, or crop condition. Precision irrigation requires ET information sufficiently in advance so water can be delivered (if supplied from an off-farm source) and applied when needed by the crop. At the field scale, irrigation systems typically cannot instantaneously irrigate the entire field, so allowances must be made for the physical ability to apply water over several days on an individual field. Irrigation system limitations can result in certain parts of the field being overirrigated, while other parts may experience some level of water stress.

Evapotranspiration

Various methods are available to calculate ET from meteorological data. A common method is the American Society of Civil Engineers' standardized Penman–Monteith method (Allen et al., 2005), which is used to calculate alfalfa (*Medicago sativa* L.) reference ET. The reference ET is multiplied by a crop coefficient to estimate ET for a specific crop during the growing season. Crop coefficients can be determined by methods described in FAO Irrigation and Drainage Paper No. 56 (Allen et al., 1998). Since calculating reference ET and crop coefficients for daily use is cumbersome and time consuming, various ET networks provide crop water use information to users for irrigation scheduling. The US Bureau of Reclamation, for example, has AgriMet, a network of over 70 automatic agricultural weather stations covering irrigated areas in the Pacific Northwest (http://www.usbr.gov/pn/agrimet/). Similarly, many state networks exist such as California Irrigation Management Information System (CMIS, http://www.cimis.water.ca.gov/), Texas ET Network (http://texaset.tamu.edu/), and Colorado Agricultural Meteorological Network (CoAgMet, http://www.coagmet.colostate.edu). It is important to remember that these networks provide potential ET values for a crop; actual ET could be greater or less depending on local field and crop conditions such as soil, fertility, disease, and precipitation. Specific knowledge about each field, and locations within fields, is required to precisely irrigate that field. Evett et al. (2012a) noted that crop coefficient values are sensitive to local climate conditions and often are not transferable to other areas. Specific soil or plant monitoring within the field is necessary to determine if net applied irrigation water is meeting or exceeding actual crop use.

Evapotranspiration can be directly measured using micrometeorological methods such as eddy covariance (e.g., Swinbank, 1951; Twine et al., 2000), Bowen ratio (e.g., Sinclair et al., 1975; Payero et al., 2003), or surface renewal (Paw U et al., 1995; Snyder et al., 2008). Eddy covariance and Bowen ratio methods require careful attention to site conditions and instrument operation. The surface renewal method has a relatively lower cost than other micrometeorological methods because sensible heat flux is determined with high frequency air-temperature measurements with fine-wire thermocouples (Mengistu and Savage, 2010). The disadvantage of surface renewal is the need to calibrate a weighting factor for a particular plant canopy (Snyder et al., 2008). All three methods require that measurements must be made in areas that represent the field conditions where the data will be applied, so these methods are not readily applicable for identifying variable irrigation requirements within a field. Furthermore, Evett et al. (2012b) concluded that it was difficult to estimate ET well using eddy covariance and Bowen ratio even with the best equipment and expert operators.

Soil Water Depletion

A key aspect of precision irrigation is measuring soil water content throughout the crop root zone to quantify the amount of water added to soil by irrigation and precipitation and removed from the soil by crops, evaporation, or deep percolation. These measurements can be used to schedule irrigation by applying enough water to replace water used since the last irrigation. Soil water content may also be used to verify and reset irrigation scheduling models that are based on ET estimations. A variety of direct and indirect methods can be used to measure soil water content. Direct measurement involves collecting a soil sample of known volume, determining the mass of the sample, drying the sample, and determining the mass of the dry sample. Direct measurement is highly accurate but time consuming, so it is mainly used to verify or calibrate indirect measurement methods rather than monitor soil moisture conditions for irrigation scheduling.

Indirect methods include neutron moisture meter and various sensors that use electromagnetic methods. The neutron moisture meter is an accurate method for repeated measurements of soil profile water content (Evett and Steiner, 1995). However, measurements cannot be automated, and a permit is required for the radiation source in the meter so neutron moisture meters have seen limited use beyond research. Time domain reflectometry is another indirect method of measuring soil water content that has been shown to be accurate and repeatable (Evett, 1998; Herkelrath et al., 1991). A variety of other electromagnetic sensors are now available for measuring soil water content. Evett et al. (2009) compared three capacitance-type sensors with neutron moisture meter and gravimetric measurements. Spatial variations of measured soil water profiles were similar between the neutron moisture meter and gravimetric methods. The electromagnetic sensors, however, had greater variability and were relatively inaccurate even though soil-specific calibrations were used. Evett et al. (2012b) found that capacitance sensors used in access tubes had soil water content errors up to 0.05 m^3 m^{-3}, which implied errors in soil water flux estimation of up to 50 mm d^{-1}. They concluded that capacitance-based soil water sensing was not suited to accurate measurement primarily because the electromagnetic field does not uniformly permeate the soil around the access tube in structured soils. For accurate determination of ET and crop water production, they recommended using neutron probe, gravimetric sampling, or conventional time domain reflectometry methods.

Soil water potential can be measured in addition to soil water content. Soil water potential indicates the amount of work required for a plant to take up water from the soil. While plants directly respond to soil water potential, it does not directly tell the irrigator how much water to apply. Relationships can be developed between soil water content and water potential so water potential can be used for scheduling irrigation. Various instruments are available to directly or indirectly measure soil water potential. Tensiometers have been used for the last 100 yr (Or, 2001) to directly measure soil water potential. The original measurement device on tensiometers was a mercury manometer, then bourdon-type gauges, and now digital pressure sensors can be used to continuously and remotely measure soil water potential. Two common indirect measurement devices are gypsum blocks and granular matrix sensors, which measure soil water potential over a wider range than tensiometers. Both sensors have porous material that loses water as the soil dries and gains water as the soil becomes wetter. While gypsum blocks are relatively inexpensive, they are better suited to measure relative soil water changes to determine when to irrigate rather than determining irrigation amounts. If properly calibrated, granular matrix sensors indicate trends that can be used for irrigation scheduling.

Remote Sensing to Determine Crop Water Needs

Remote sensing can be used to determine crop water use at project, farm, and field scales. Satellite-based remote sensing has advanced to where submeter resolution and daily coverage are possible with many defined spectral indices available besides normalized difference vegetation index (Mulla, 2013). Sensors are also available to mount on tractors, sprayers, or moving irrigation systems to provide real-time information. The same remote sensing technologies and management can be applied on irrigated farms and nonirrigated farms with the obvious exception that remote sensing can inform and guide water application decisions for irrigated farms. Irrigation also allows fertilizer or other chemicals to be applied during the growing season when ground-based applicators will damage the crop.

Remote sensing is useful for systematic measurements with time over large areas (Bastiaanssen et al., 2000). Jackson et al. (1977) first related the difference between canopy temperature and air temperature to ET. Satellite or aerial remote sensing can provide repeated field observations for an entire irrigation project that can be used in models to guide diversions of irrigation water or identify variations within fields. Remote sensing can also identify the actual irrigated area of specific crops within an irrigation project so managers can better plan diversions to match crop water needs. Michael and Bastiaanssen (2000) developed a technique to define crop coefficient maps from satellite images. Total crop water use for an irrigated area can then be calculated by multiplying reference ET by the crop coefficient. SEBAL (Bastiaanssen et al., 1998), METRIC (Allen et al., 2007), and ALEXI (Anderson et al., 2012) are energy balance models that use thermal infrared imagery for estimating ET for large areas. The Surface Energy Balance Algorithm for Land (SEBAL) model describes the spatial variability of most micrometeorological variables with semiempirical functions. SEBAL can be applied for diverse agroecosystems and does not require ancillary information on land use or crop types (Bastiaanssen et al., 1998). Validation efforts have shown that the error at a 1-ha scale varies between 10 and 20% and that the uncertainty diminishes with increasing scale. For an area of 1000 ha, the error is reduced to 5% and for regions of 1 million ha of farmland, the error becomes negligibly small (Bastiaanssen et al., 2000). Mapping EvapoTranspiration at High Resolution and Internalized Calibration (METRIC) uses the same basic energy balance algorithms as SEBAL to calculate ET from satellite imagery (Allen et al., 2007, 2011). However, METRIC uses reference ET calculated from micrometeorological data to extrapolate ET between instantaneous satellite images. The Atmosphere-Land Exchange Inverse (ALEXI) model (Anderson et al., 1997; Mecikalski et al., 1999) was developed as an auxiliary means for estimating surface fluxes over large regions using primarily remote-sensing data. This flux model is unique in that no information regarding antecedent precipitation or moisture storage capacity is required because the surface moisture status is deduced from a radiometric temperature change signal. Therefore, ALEXI can provide independent information for updating soil moisture variables in more complex regional models. However, these models require significant processing time, so real-time daily ET estimates are not available.

Direct ET measurements like eddy covariance, Bowen ratio, and surface renewal do not provide information about spatial variability within a field. Satellite and air-based remote sensing provide spatial information, but measurement intervals are usually >1 d. Combining direct or reference ET measurements from micrometeorological methods with remotely sensed spatial information can provide daily, site-specific ET information within fields. Jackson et al. (1977) developed a technique to calculate daily ET from one-time-of-day measurements. Peters and Evett (2004) used one-time-of-day measurements with a reference temperature curve from a fixed location to model diurnal canopy temperature dynamics within a field. Ben Asher et al. (2013) combined the Penman–Monteith ET equations with infrared radiometers mounted on a linear-move irrigation system to calculate hourly ET to control an irrigation system without monitoring soil water. This system provided a bulk representation of the field not site-specific information within the field. Remotely sensed thermal and reflectance information from the crop canopy can be used to calculate site-specific water use. Colaizzi et al. (2012) used a two-source energy balance model to predict daily ET of corn, cotton (*Gossypium hirsutum* L.), grain sorghum [*Sorghum bicolor* (L.) Moench], and wheat. Their technique may allow the center pivot to be the remote sensing platform for collecting real-time information for calculating site-specific ET.

Field-Scale Precision Irrigation

For at least the last 50 yr, irrigation research has attempted to understand crop water needs and to apply water uniformly to fields. Good irrigation management should strive to optimize crop yield in a field. When irrigation is not uniform, some areas receive more water than required and some areas receive less. To reduce the assumed negative effects of too little irrigation, operators may apply additional water to the entire field, which means a larger portion of the field is overirrigated and may actually reduce total crop yield from the field. Soils are seldom uniform within a field, indicating that infiltration rates and potential crop yields will also vary within a field. Although modern center-pivot irrigation systems uniformly apply irrigation water, water will not uniformly infiltrate if soil properties are not uniform. Uniformly applied irrigation water, or rainfall, that runs off and infiltrates in another area in the field leads to nonuniform infiltration. This in-field runoff causes variations in stored soil water, which in turn causes variable irrigation requirements. Couple this with variable productivity and noncropped areas within fields and the justification for site-specific irrigation management (SSIM) becomes evident. However, there is little scientific information demonstrating its cost-effectiveness or documenting the capability of site-specific sprinkler irrigation to conserve water or energy on a field scale (Evans and King, 2012).

Center-pivot and linear-move sprinkler irrigation systems are well suited to precision irrigation management. These moving irrigation systems can be designed and controlled to apply different rates of water on different locations in the field. Solid set or permanent sprinkler irrigation systems, including microirrigation, can also be designed to differentially apply water. Since these systems do not move, sprinkler application rates can be permanently or seasonally adjusted by changing sprinklers or nozzles to meet water needs for specific areas. Varying the duration of irrigation events among management zones can also be used provided that the irrigation zones match the field variability. Set-move irrigation systems (e.g., side-roll systems) can apply variable amounts of water if sprinkler nozzles vary along the system or are changed between irrigation sets. At this time, this would require extensive manual labor. Varying application rates to precisely meet irrigation requirements within a field is nearly impossible with surface irrigation because water flows by gravity on the soil surface through the field. Individual basins, borders, or furrows can be irrigated for different time intervals, but application depth cannot be controlled spatially within these units. Even if surface irrigation could be controlled to spatially vary water application, the irrigator would need to know how much water will infiltrate in a specific time period (through measurement or models) to precisely apply the right amount of water in the right location.

Moving sprinkler irrigation systems, like center-pivot and linear-move irrigation systems, are probably the most appropriate platform for SSIM in crop fields. Center-pivot irrigation systems mechanically move through the entire field multiple times per year enabling irrigation amounts and rates to vary within the season and between seasons. Technology for controlling center-pivot operation is continually changing from the original on–off switch and speed control dial. Operators can now communicate with irrigation machines by cell phones, satellite radios, and internet-based systems (Kranz et al., 2012). Center-pivot manufactures now offer control panels that can change pivot speed in 1 to 10° increments (Evans et al., 2013). Controlling center-pivot speed allows the operator to change application depth in pie-shaped areas within fields; however, field variability seldom occurs in triangular-shaped parcels. Center-pivot manufacturers also offer variable-rate irrigation systems that can change application rates along the lateral by pulsing individual or groups of sprinklers on and off. Using variable speed and rate enables a field to be divided into several thousand parcels to accurately match the management zones within a field.

Although center-pivot irrigation systems are commercially available to precisely apply water to management zones within a field, user-friendly systems to manage SSIM are not. To fully implement SSIM, sensor systems are needed to measure crop, soil, and weather factors so the plants and soil control the amount and timing of irrigation in each zone in the field. Within-season temporal variation in rainfall, infiltration, and water use require periodic feedback of plant and soil status. Temporal feedback can be used for real-time irrigation management and improved decision support.

Most current irrigation scheduling programs calculate the timing and duration of water applications using algorithms based on historical weather patterns and predicted crop water use over a relatively short period (e.g., 3–14 d). Feedback to estimate crop water use is usually made by spot measurements of soil water or other data after the irrigation is completed, and adjustments are made for the following irrigation event. Specialized software programs need to be developed for SSIM so that data obtained from multiple sources can be used to determine the amount and timing of irrigation for each zone in a field.

Recent innovations in low-voltage sensors and wireless radio frequency data communications combined with advances in internet technologies offer tremendous opportunities for advancement of SSIM. Spatially distributed, within-field plant and soil sensors in combination with agroweather stations are potentially more accurate for controlling irrigation than historical or static map-based input projections. Sensor systems can be used to measure climatic, soil water, plant density, canopy temperature differences, irrigation application amounts, and other types of variability. Remote sensing by satellites can also provide synoptic crop feedback in both space and time to supplement real-time measurements. Real-time feedback from multiple sources, in combination with modern wireless communication and computerized control systems, are fundamental to the development and implementation of optimal site-specific irrigation management strategies.

Moving irrigation systems provide a unique platform to collect data from sensors in the field in addition to applying irrigation water. O'Shaughnessy and Evett (2010) demonstrated that wireless infrared thermometers functioned reliably when mounted on a center-pivot irrigation system. Peters and Evett (2008) used infrared temperature to automate irrigation with a center pivot. There was no difference in irrigation efficiency or water use efficiency compared with manual irrigation scheduling using a neutron probe to measure soil water content. This system used the temperature–time threshold method (Wanjura et al., 1995) to determine when irrigation was required. The center pivot traveled through the field each day to measure canopy temperature. If a threshold canopy temperature is exceeded for a predetermined threshold time, an irrigation event is scheduled.

Presently, there has been limited adoption of site-specific irrigation systems. One estimate is that there are <200 center-pivot systems with variable-rate technology in the United States (Evans et al., 2013). One reason for limited adoption is that variable-rate irrigation technology has only recently been commercially marketed by center-pivot system manufacturers. Another potential barrier to installing variable-rate irrigation technology is the current limited use of soil or plant sensing to schedule irrigation. The USDA Farm and Ranch Irrigation Survey showed that <10% of irrigated farms used soil or plant sensing or a scheduling service (Table 2). Furthermore, the trend is not increasing for any of these technologies. These results indicate that acceptable technology is currently not available, or the need for more sophisticated scheduling has not been realized by irrigators. In addition to demonstrating the benefits of SSIM, reliable, easy-to-use equipment is needed for large-scale adoption of SSIM. The common use of yield monitors in today's harvesting equipment allows farmers to see the variability in their fields and may increase interest in spatially varying irrigation.

Integrating information from various sensors and systems into a decision support program will be critical to highly managed, spatially varied irrigation (Evans and King, 2012). Advanced decision systems should integrate real-time monitoring with plant growth and pest models to seamlessly interface with the irrigation system to optimally manage crop production. For example, the best response to an identified diseased area in a field may be a pesticide treatment to limit the disease spread or may be eliminating further irrigation so a limited

Table 2. Percent of farms reporting a method of deciding when to irrigate. Respondents could select more than one method (USDA NASS, 2014b).

Method	2013	2008	2003	1998	1994	1988
	---------------------------- % ----------------------------					
Crop Condition	78	78	80	74	68	72
Feel of Soil	39	43	35	41	39	36
Calendar	21	25	19	17	17	15
Scheduled by water delivery organization	16	12	12	10	14	11
Other method	8	9	9	7	9	5
Soil moisture sensing device	10	9	7	8	10	7
Scheduling service	8	8	6	4	5	5
Neighbors	6	7	6	NA	NA	NA
Plant moisture sensing device	2	2	1	NA	NA	NA
Computer simulation model	1	1	1	1	2	0

water supply can be saved for healthy areas of the field. Advanced decision systems also need to consider the whole farm to best use water and other resources among multiple fields, crops, and even years. For example, if groundwater use for irrigation is limited to 400 mm yr^{-1} on a 3-yr average, using <400 mm in 1 yr allows an irrigator to use >400 mm in future years.

Summary

Precision irrigation involves applying the right amount of water at the right time and location. Precision irrigation management is needed on large irrigation projects and on individual fields to make the best use of irrigation water, especially in water-limited areas. On large irrigation projects, managers need to adjust irrigation delivery to match the irrigation needs of the crops in the project. On individual farms and fields, managers need specific information about crop water needs on their fields to apply the right amount of water at the right time and place.

Technology is commercially available to precisely apply water when and where it is needed by crops; however, user-friendly decision tools are still needed to quantify specific irrigation needs and control water application within fields. Researchers and managers need to remember that the technology to precisely apply irrigation water is wasted if the water does not infiltrate where it was applied. Irrigation system design must consider soil properties along with irrigation capacity.

Satellite imagery can be used to calculate actual crop water use in fields, but this information is not available in real-time for daily irrigation management. Combining direct or reference ET measurements from micrometeorological methods with remotely sensed spatial information can provide daily, site-specific ET information within fields. Unmanned aerial vehicles are a developing technology that may make it feasible to collect frequent, high-resolution aerial imagery for managing irrigation. An alternative to satellite-based remote sensing is in-field sensors that provide information to a decision support system to precisely manage field irrigation. Integrating information from various sensors into a decision support program is the next potential advancement to precisely apply the right amount of water at the right time to unique areas within a field.

Water rights and policies can be a barrier to adopting precision irrigation practices by restricting water delivery amounts or timing at the district or field level. However, irrigation managers and designers can partially account for these restrictions through on-farm storage, irrigation system design, or crop choice, allowing the right amount of water to be applied at the right time and place.

References

Allen, R.G., A. Irmak, R. Trezza, J.M.H. Hendrickx, W. Bastiaanssen, and J. Kjaersgaard. 2011. Satellite-based ET estimation in agriculture using SEBAL and METRIC. Hydrol. Processes 25:4011–4027. doi:10.1002/hyp.8408

Allen, R.G., L.S. Pereira, D. Raes, and M. Smith. 1998. Crop evapotranspiration: Guidelines for computing crop water requirements. FAO Irrigation and Drainage Paper No. 56. Food and Agric. Org. of the United Nations, Rome.

Allen, R.G., M. Tasumi, and R. Trezza. 2007. Satellite-based energy balance for mapping evapotranspiration with internalized calibration (METRIC)–model. J. Irrig. Drain. Div., Am. Soc. Civ. Eng. ASCE 133:380–394. doi:10.1061/(ASCE)0733-9437(2007)133:4(380)

Allen, R.G., I.A. Walter, R.L. Elliott, T.A. Howell, D. Itenfisu, M.E. Jensen, and R.L. Snyder. 2005. ASCE standardized reference evapotranspiration equation. Am. Soc. of Civil Engineers, Reston, VA.

Anderson, M.C., W.P. Kustas, J.G. Alfieri, F. Gao, C. Hain, J.H. Prueger, S. Evett, P. Colaizzi, T. Howell, and J.L. Chavez. 2012. Mapping daily evapotranspiration at Landsat spatial scales during the BEAREX'08 field campaign. Adv. Water Resour. 50:162–177. doi:10.1016/j.advwatres.2012.06.005

Anderson M.C., J.M.Norman, G.R. Diak, W.P. Kustas, and J.R. Mecikalski. 1997. A two-source time-integrated model for estimating surface fluxes using thermal infrared remote sensing. Remote Sens. Environ. 60:195–216. doi:10.1016/S0034-4257(96)00215-5

Bastiaanssen, W.G.M., M. Menenti, R.A. Feddes, and A.A.M. Holtslag. 1998. A remote sensing surface energy balance algorithm for land (SEBAL), part 1. Formulation. J. Hydrol. 212–213:198–213. doi:10.1016/S0022-1694(98)00253-4

Bastiaanssen, W.G.M., D.J. Molden, and I.W. Makin. 2000. Remote sensing for irrigated agriculture: Examples from research and possible applications. Agric. Water Manage. 46:137–155. doi:10.1016/S0378-3774(00)00080-9

Bautista, E., A.J. Clemmens, and R.J. Strand. 2006. Salt River Project canal automation pilot project: Simulation tests. J. Irrig. Drain. Eng. 132:143–152. doi:10.1061/(ASCE)0733-9437(2006)132:2(143)

Ben Asher, J., B.B. Yosef, and R. Volinsky. 2013. Ground-based remote sensing system for irrigation scheduling. Biosystems Eng. 114:444–453. doi:10.1016/j.biosystemseng.2012.09.002

Bjorneberg, D.L., D.T. Westermann, N.O. Nelson, and J.H. Kendrick. 2008. Conservation practice effectiveness in the irrigated Upper Snake/Rock Creek Watershed. J. Soil Water Conserv. 63:487–495. doi:10.2489/jswc.63.6.487

Burt, C.M., and X. Piao. 2004. Advances in PLC-based irrigation canal automation. Irrig. Drain. 53:29–37. doi:10.1002/ird.106

Clemmens, A.J. 2006. Improving irrigated agriculture performance through an understanding of the water delivery process. Irrig. Drain. 55:223–234. doi:10.1002/ird.236

Colaizzi, P.D., S.R. Evett, T.A. Howell, P.H. Gowda, S.A. O'Shaughnessy, J.A. Tolk, W.P. Kustas, and M.C. Anderson. 2012. Two-source energy balance model: Refinements and lysimeter tests in the Southern High Plains. Trans. ASABE 55:551–562. doi:10.13031/2013.41385

Evans, R.G., and B.A. King. 2012. Site-specific sprinkler irrigation in a water-limited future. Trans. ASABE 55:493–504. doi:10.13031/2013.41382

Evans, R.G., J. LaRue, K.C. Stone, and B.A. King. 2013. Adoption of site-specific variable rate sprinkler irrigation systems. Irrig. Sci. 31:871–887. doi:10.1007/s00271-012-0365-x

Evett, S.R. 1998. Coaxial multiplexer for time domain reflectometry measurement of soil water content and bulk electrical conductivity. Trans. ASAE 41:361–369. doi:10.13031/2013.17186

Evett, S.R., R.J. Lascano, T.A. Howell, J.A. Tolk, S.A. O'Shaughnessy, and P.D. Colaizzi. 2012a. Single- and dual-surface iterative energy balance solutions for reference ET. Trans. ASABE 55:533–541. doi:10.13031/2013.41388

Evett, S.R., R.C. Schwartz, J.J. Casanova, and L.K. Heng. 2012b. Soil water sensing for water balance ET and WUE. Agric. Water Manag. 104:1–9. doi:10.1016/j.agwat.2011.12.002

Evett, S.R., R.C. Schwartz, J.A. Tolk, and T.A. Howell. 2009. Soil profile water content determination: Spatiotemporal variability of electromagnetic and neutron probe sensors in access tubes. Vadose Zone J. 8:926–941. doi:10.2136/vzj2008.0146

Evett, S.R., and J.L. Steiner. 1995. Precision of neutron scattering and capacitance type soil water content gauges from field calibration. Soil Sci. Soc. Am. J. 59:961–968. doi:10.2136/sssaj1995.03615995005900040001x

FAO. 2013. AQUASTAT database. Food and Agriculture Organization of the United Nations (FAO), Rome. www.fao.org/nr/water/aquastat/main/index.stm (accessed 20 Apr. 2013).

Herkelrath, W.N., S.P. Hamburg, and F. Murphy. 1991. Automatic, real-time monitoring of soil moisture in a remote field area with time domain reflectometry. Water Resour. Res. 27:857–864. doi:10.1029/91WR00311

Holmen, H., C.W. Carlson, R.J. Lorenz, and M.E. Jensen. 1961. Evapotranspiration as affected by moisture level, nitrogen fertilization, and harvest method. Trans. ASAE 4:41–44. doi:10.13031/2013.41004

International Commission on Irrigation and Drainage (ICID). 2013. ICID databaase. ICID, Chanakyapuri, New Delhi. www.icid.org/icid_data.html (accessed 20 Apr. 2013)

Jackson, R.D., R.J. Reginato, and S.B. Idso. 1977. Wheat canopy temperature: A practical tool for evaluating water requirements. Water Resour. Res. 13:651–656. doi:10.1029/WR013i003p00651

Kranz, W.L., R.G. Evans, and F.R. Lamm. 2012. A review of center-pivot irrigation control and automation technologies. Applied Eng. in Agric. 28:389–397. doi:10.13031/2013.41494

Lenka, S., A.K. Singh, and N.K. Lenka. 2009. Water and nitrogen interaction on soil profile water extraction and ET in maize–wheat cropping system. Agric. Water Manage. 96:195–207. doi:10.1016/j.agwat.2008.06.014

Lozano, D., and L. Mateos. 2008. Usefulness and limitations of decision support systems for improving irrigation scheme management. Ag. Waste Manag. 95:409–418.

Maupin, M.A., J.F. Kenny, S.S. Hutson, J.K. Lovelace, N.L. Barber, and K.S. Linsey. 2014. Estimated use of water in the United States in 2010. U.S. Geological Survey Circular 1405. U.S. Dep. of the Interior, Washington, DC.

Mecikalski, J.R., G.R. Diak, M.C. Anderson, and J.M. Norman, 1999. Estimating fluxes on continental scales using remotely-sensed data in an atmospheric-land exchange model, J. Appl. Meteorol. 38:1352–1369. doi:10.1175/1520-0450(1999)038<1352:EFOCSU>2.0.CO;2

Mengistu, M.G., and M.J. Savage. 2010. Surface renewal method for estimating sensible heat flux. Water S.A. 36:9–17. doi:10.4314/wsa.v36i1.50902

Merriam, J.L., S.W. Styles, and B.J. Freeman. 2007. Flexible irrigation systems: Concept, design and application. J. Irrig. Drain. Eng. 133:2–11. doi:10.1061/(ASCE)0733-9437(2007)133:1(2)

Michael, M.G., and W.G.M. Bastiaanssen. 2000. A new simple method to determine crop coefficients for water allocation planning from satellites: results from Kenya. Irr. Drain. Sys. 14:237–256. doi:10.1023/A:1026507916353

Mulla, D.J. 2013. Twenty five years of remote sensing in precision agriculture: Key advances and remaining knowledge gaps. Biosystems Eng. 114:358–371. doi:10.1016/j.biosystemseng.2012.08.009

Or, D. 2001. Who invented the tensiometer? Soil Sci. Soc. Am. J. 65:1–3. doi:10.2136/sssaj2001.6511

O'Shaughnessy, S.A., and S.R. Evett. 2010. Developing wireless sensor networks for monitoring crop canopy temperature using a moving sprinkler system as a platform. Appl. Eng. Agric. 26:331–341. doi:10.13031/2013.29534

Pandey, R.K., J.W. Maranville, and A. Admou. 2000. Deficit irrigation and nitrogen effects on maize in a Sahelian environment I. Grain yield and yield components. Agric. Water Manage. 46:1–13. doi:10.1016/S0378-3774(00)00073-1

Paw U, K.T., J. Qiu, H.B. Su, T. Watanabe, and Y. Brunet. 1995. Surface renewal analysis: A new method to obtain scalar fluxes without velocity data. Agric. For. Meteorol. 74:119–137. doi:10.1016/0168-1923(94)02182-J

Payero, J.O., C.M.U. Neale, J.L. Wright, and R.G. Allen. 2003. Guidelines for validating Bowen ratio data. Trans. ASAE 46:1051–1060. doi:10.13031/2013.13967

Peters, R.T., and S.R. Evett. 2004. Modeling diurnal canopy temperature dynamics using one-time-of-day measurements and a reference temperature curve. Agron. J. 96:1553–1561. doi:10.2134/agronj2004.1553

Peters, R.T., and S.R. Evett. 2008. Automation of a center pivot using the temperature-time-threshold method of irrigation scheduling. J. Irrig. Drain. Eng. 134:286–291. doi:10.1061/(ASCE)0733-9437(2008)134:3(286)

Sadler, E.J., C.R. Camp, D.E. Evans, and J.A. Millen. 2002. Spatial variation of corn response to irrigation. Trans. ASAE 45:1869–1881. doi:10.13031/2013.11438

Schlegel, A.J., L.R. Stone, T.J. Dumler, and F.R. Lamm. 2012. Managing diminished irrigation capacity with reseason irrigation and plant density for corn production. Trans. ASABE 55:525–531. doi:10.13031/2013.41394

Sinclair, T.R., L.H. Allen, and E.R. Lemon. 1975. Analysis of errors in the calculation of energy flux densities above vegetation by a Bowen ratio profile method. Boundary-Layer Meteorol. 8:129–139. doi:10.1007/BF00241333

Snyder, R.L., D. Spano, P. Duce, K.T. Paw U, and M. Rivera. 2008. Surface renewal estimation of pasture evapotranspiration. J. Irrig. Drain. Eng. 134:716–721. doi:10.1061/(ASCE)0733-9437(2008)134:6(716)

Swinbank, W.C. 1951. The measurement of vertical transfer of heat and water vapor by eddies in the lower atmosphere. J. Meteorol. 8:135–145. doi:10.1175/1520-0469(1951)008<0135:TMOVTO>2.0.CO;2

Twine, T.E., W.P. Kustas, J.M. Norman, D.R. Cook, P.R. Houser, T.P. Meyers, J.H. Prueger, P.J. Starks, and M.L. Wesely. 2000. Correcting eddy-covariance flux underestimates over a grassland. Agric. For. Meteorol. 103:279–300. doi:10.1016/S0168-1923(00)00123-4

USDA-NASS. 2014a. Census of Agriculture. USDA National Agricultural Statistics Service. Washington, DC. http://www.agcensus.usda.gov/ (accessed 17 May 2016).

USDA-NASS. 2014b. Farm and ranch irrigation survey. USDA National Agricultural Statistics Service, Washington, DC. https://www.agcensus.usda.gov/Publications/2012/Online_Resources/Farm_and_Ranch_Irrigation_Survey/ (accessed 17 May 2016).

van Overloop, P.J., A.J. Clemmens, R.J. Strand, R.M.J. Wagemaker, and E. Bautista. 2010. Real-time implementation of model predictive control on Maricopa-Stanfield Irrigation and Drainage District's WM canal. J. Irrig. Drain. Eng. 136:747–756. doi:10.1061/(ASCE)IR.1943-4774.0000256

Wanjura, D.F., D.R. Upchurch, and J.R. Mahan. 1995. Control of irrigation scheduling using temperature–time thresholds. Trans. ASAE 38:403–409. doi:10.13031/2013.27846

GIS and GPS Applications for Planning, Design and Management of Drainage Systems

Vinayak S. Shedekar* and Larry C. Brown

Abstract

This chapter describes the application of various GIS and GPS tools available for the planning, design and management of surface and subsurface drainage systems. We discuss steps to acquire and process relevant GIS and GPS data, methods for applying such data to the planning and design of drainage systems, and guidelines for managing drainage systems. Readers will be introduced to existing software and online tools that can be used for GIS–GPS data, drainage design, and data visualization.

Agricultural drainage removes excess water from the soil profile and helps sustain crops by improving the aeration and trafficability of the soil. Among other factors that make drainage necessary are flat or depressional topography, restrictive geologic layers underlying the soil profile, and periods of excess precipitation (Zucker and Brown, 1998). Agricultural drainage systems may include surface drains, subsurface drains (commonly referred as "tile drains"), or a combination of both depending upon site conditions such as soil type, topography, location and elevation of natural outlets, and intended land use. The benefits of drainage include reduced risk of crop loss from excess water stress, improved control of pest and diseases, and consistent crop yields under climate variability (Zucker and Brown, 1998; Skaggs et al., 2012; King et al., 2015). In some soils, drainage is also known to improve crop yields by 5 to 25% and reduce long-term yield variability (Fausey et al., 1995; Skaggs et al., 2012). Along with its intended benefits, agricultural drainage is known to help reduce some of the nonpoint source pollution issues compared to croplands without drainage. Surface drainage systems reduce slope length causing a reduction in overland flow and hence erosion. Surface drainage ditches designed for adequate capacity, slope and appropriate vegetative cover may help control the velocity of discharge, reduce sediment transport, and reduce the loads of pollutants and nutrients downstream.

Abbreviations: DEM, Digital Elevation Models; DGPS, Differential Global Positioning System; DORIS, Doppler Orbitography and Radio-positioning Integrated by Satellite; DSM, Digital Surface Model; DTM, Digital Terrain Model; GLONASS, Global Navigation Satellite System; GPS, Global Positioning System; LiDAR, Light Detection and Ranging; NED, National Elevation Dataset; RTK, Real-time Kinematic; QZSS, Quasi-Zenith Satellite System; WSS, Web Soil Survey.

Ohio State University, Food, Agricultural and Biological Engineering Dept., Columbus, OH 43210. *Corresponding author (shedekar.1@osu.edu)

doi:10.2134/agronmonogr59.2014.0026

Precision Conservation: Geospatial Techniques for Agricultural and Natural Resources Conservation
J.A. Delgado, G.F. Sassenrath, and T. Mueller, editors. Agronomy Monograph 59.

Subsurface drainage can reduce the amount of surface runoff and reduce the peak rate of discharge through surface channels, thereby further reducing erosion and the associated off-site impacts of erosion (Fausey et al., 1995). In general, agricultural drainage has been found to reduce sediment loss due to water erosion, reduce particulate phosphorus loss and overall nutrient loss due to reduced soil loss, total runoff and peak runoff rate. Given these benefits, surface as well as subsurface drainage are considered conservation practices by the NRCS. However, research in recent decades suggests that drainage has adverse effects on water quality in the receiving streams. It has been established that agricultural drainage in the states of Minnesota, Iowa, Illinois, Indiana, and Ohio contribute to about 35% of total N discharged by the Mississippi River to the Gulf of Mexico (Goolsby et al., 2000). This contribution increases even more during wet years. Drainage water may also remove significant amounts of phosphorus in dissolved form, which may further contribute to eutrophication of fresh water bodies (King et al., 2015).

In spite of its potential adverse effects on water quality, agricultural drainage is essential to sustain crop production on productive but poorly drained soils. A suite of emerging technologies and methods that help reduce these adverse effects while ensuring the intended benefits of agricultural drainage are now available and are commonly referred to as "conservation drainage". These approaches include controlled drainage, shallow drainage, woodchip bioreactors, saturated buffers, rock inlets, alternative ditch design and various kinds of storage basins. Thus a well-planned, properly designed, and well-managed drainage system may help reduce erosion and improve uniformity of field conditions for timely and efficient management, which promotes adoption of conservation best management practices including no-till and cover crops. More recent studies also suggest that conservation drainage, along with other conservation practices may help increase adaptation to a changing climate and have the potential to reduce environmental impacts (Strock et al., 2010; Pease et al., 2017).

Although the primary purpose of drainage has remained unchanged since the 1800's, the field of drainage has evolved significantly with respect to design, installation and management. Unlike the old drainage tiles installed randomly, the modern drainage systems tend to be more intense (narrow spacing) and more systematically installed with the use of modern machinery and better precision. Drainage systems can be classified as Surface and Subsurface. Surface drainage involves installing a network of trenches and ditches to safely and quickly remove the surface runoff from the field, allowing less time for infiltration. Subsurface drainage involves installing a network of perforated pipes (laterals) with a certain grade; usually below the maximum root depth such that water from the root zone of crops flows into these laterals and discharges into nearby natural or artificial channels. A typical procedure of drainage system planning consists of the following steps:

Step 1: Conduct reconnaissance and identify evidence of need for drainage (erosion, wet spots, ponding, and yield maps illustrating areas with low yield).

Step 2: Conduct physical surveys to obtain and analyze a topographic map (determine possible location of surface and subsurface outlets, determine control points and grades).

Step 3: Assess environmental considerations (water quality, wetlands) and regulatory permit requirements for the project

Step 3: Design surface drainage infrastructure.

Step 4: Design subsurface drainage infrastructure.

In general, the design steps for surface drainage consist of (i)estimating maximum discharge capacity (drainage coefficient) of the system, (ii) determining size, shape, slope, depth and spacing of ditches and then (iii) planning the layout of the surface ditches network. Engineering design of subsurface drainage systems involves (i) estimating maximum discharge capacity (drainage coefficient) of the system, (ii) determining size, slope, depth and spacing of laterals and main pipes for the given drainage coefficient and then (iii) planning the layout of the pipe network. Based on the drainage theory developed over the years, various agencies have developed standard charts and nomographs that can assist in designing drainage systems. Depending upon the field topography, subsurface drainage systems can be installed in systematic, herringbone or random layout (Figure 1). The design of surface and subsurface drainage systems involves the following steps:

1. Conduct reconnaissance and topographic survey.

2. Create topographic map (e.g. contour map).

3. Determine drainage group for the existing soil type in the field.

4. Determine surface and/or subsurface drain spacing

5. Analyze topo map for drain layout alternatives, location of drainage outlet etc.

6. Determine layout and station main and laterals on topo map

7. Plot profile of existing ground along main

8. Plot control points for main on profile (e.g. low points, high points, outlet, lateral depth, minimum grade, minimum cover, obstructions).

9. Create gradelines and/or calculate grades.

10. Calculate total drainage area at the outlet.

11. Determine drainage coefficient.

12. Determine required surface and/or subsurface drai

13. For subsurface drainage, determine drainage pipe and lengths needed for outlet, main and laterals.

14. Prepare a detailed plan of project along with budge

The detailed theory and procedure for drainage des 16 of National Engineering Handbook and part 650 (C. Field Handbook (United States Department of Agricul Conservation Service, 2007).

GIS and GPS have made it easy to obtain and proces ral data relevant to agricultural drainage systems. In thi how these technologies can be used to facilitate design, and investigation of agricultural drainage systems.

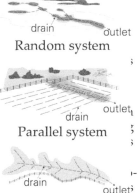

Fig. 1. Types and components of drainage systems (Source: United States Department of Agriculture - Natural Resources Conservation Service, 2007).

Surveying for Drainage Design

Field topography is the most important information required for planning the layout of any drainage system. Topographic surveys provide

Table 1. Current and future global and/or regional navigation systems

	Offered by	Number of satellites	Status
Global navigation systems			
GPS (Global Positioning System)	United States	32	Operational
GLONASS (Global Navigation Satellite System)	Russian Federation	24	Operational
BeiDou-2 (formerly COMPASS)	China	35 (16 launched)	In development. Partially operational in Asia-Pacific region since December 2012.
Galileo	European Union	30 (27 operational, 3 spares), 4 launched	In development. Expected to be operational by 2019
Regional navigation systems			
Beidou 1	China	4 (3 working, 1 spare)	Operational
DORIS Doppler Orbitography and Radio-positioning Integrated by Satellite	France	7	Operational. Uses a network of 50-60 ground stations.
IRNSS Indian Regional Navigational Satellite System	India	7	In development. Expected to be operational by 2016
QZSS Quasi-Zenith Satellite System	Japan	4	In development. Expected to be operational by 2017

detailed information about elevation changes within the field. Techniques such as differential leveling and profile leveling are typically used during topographic surveys for drainage design. This section gives a brief overview of existing surveying technology and compares some of the pros and cons of different surveying techniques.

Advances in Surveying Technology

Traditionally, laser surveying has been the most popular method for surveying and grade control in design and installation of drainage systems. The laser transmitter contains a laser light generator and columinator that generates a narrow beam of light, which is then rotated to produce a plane of laser light. The plane of light may be level and parallel to a datum or it may have grade by setting the laser transmitter as desired. The laser transmitter is usually set on a tripod or other solid base. A small laser receiver, which slides on the top section of a survey rod or Lenker rod, is then used to detect the laser plane. At each survey point, the surveyor typically aligns the receiver with the laser beam using an electronic display and sound signals. The receiver has a pointer at the numbers on the rod where the laser light beam would hit. The accuracy of laser surveying is in the range of 6 to 12 mm(0.02 to 0.04 ft).

With the advancements in space programs across the world, networks of satellites are being increasingly used for determination of location and time on and near the surface of the earth. Global Positioning System (GPS) is one such satellite navigation system that provides location and time information using a network of

US satellites. The GPS involves a constellation of more than 24 satellites (32 as of December 2013) orbiting in six orbits 20,350 km above the surface of the earth. Each satellite completes two orbits in a day. The system was first developed by the U.S. Government for military purposes and then made available for civilian services. A GPS receiver is used to receive the radio signals from at least four satellites in its field of view and determine the spatial location as well as altitude of the point of interest. The receiver uses distance from each satellite, geometry based on position of satellites in the sky and precise time signal sent by satellites to determine the precise location and time on the surface of the earth. Such GPS devices are now becoming common on navigation systems used in all modes of transport.

Similar to the GPS, additional services are available from other satellite navigation systems. Table 1 summarizes a list of currently available and future navigation systems at regional and international scales.

Depending upon the level of accuracy required, different types of GPS devices are available. The accuracy of GPS devices ranges from a few centimeters to a few meters. The horizontal accuracy of GPS devices is usually greater than the vertical accuracy. The errors associated with GPS devices depend upon several factors, such as clock limitations, ephemeris (orbital position) variation, satellite configuration, atmospheric interference, and multipath signal. Some of these errors are random and others are systematic. Accuracy of GPS devices can be improved by correcting for these systematic and random errors. Differential GPS (DGPS) is an example of such a system.

Application of Differential GPS (DGPS) in Drainage Design

Differential GPS (DGPS) systems provide a horizontal accuracy of about one to five meters. Examples of DGPS systems are Beacon and Wide Area Augmentation System (WAAS). Dual Frequency GPS devices usually have an accuracy of 5 to 10 cm. A Real Time Kinematic (RTK) GPS is a DGPS technique that offers a positional accuracy of up to one to two centimeters. With enhanced storage capacities, automation and ease of use, the RTK GPS devices are becoming popular among surveyors and contractors (Langley, 1998; Gakstatter, 2014).

A typical RTK GPS system consists of a base station GPS set at a known fixed location. The base station generates a correction signal based on its known position and calculated position. These corrections are transmitted to another rover GPS by FM radio signal. The rover GPS uses the signal to correct its calculated position. More recently, correction signals are becoming available from networks of permanent RTK base stations established in different states by different entities. Subscription to such networks eliminates the need to maintain and set up your own base station for surveying and/or equipment guidance.

Guidelines for Conducting a GPS Survey

Conducting a topographic survey using GPS is usually a multi-step process that involves pre-survey preparations, initialization and calibration of equipment, field measurements and post-processing of data. GPS equipment consists of multiple battery operated devices. It is important to know the status of batteries of all the devices and keep them fully charged before conducting surveys. Topographic surveying involving multiple sites often requires that the devices be used for several hours of the day without access to charging station. A portable charging station or back up batteries may be useful in such situations. It is also important to keep the equipment software and firmware updated on a regular basis. For RTK–network-based surveys, it is recommended that the users check the status and signal coverage at the site of interest.

After arriving at the survey site, selecting locations for the base station and local bench marks are the first steps. A base station should be set up at a location that is easily accessible, has a clear view to satellites without any major physical obstructions, and is permanent or semi-permanent (to allow resetting the base station during repeated visits). Benchmarks may or may not be necessary depending upon the need to geo-reference the GPS survey data with other spatial data for the site. GPS units usually take a few minutes to initiate and determine the final coordinates and elevation of a point. Once the base station and rover GPS are initiated and have established radio links, the topographic survey can be conducted either by moving the rover unit manually or installing it on a vehicle that can moved across the field. Similar to a conventional topographic survey, it may be useful to take repeated measurements at a few known points to ensure the reliability and accuracy of GPS. After the GPS survey is complete, the survey data from GPS units can be downloaded using specialized software in various formats. A GIS-based shape or vector file and a grid based ACSII (text) file that lists the X, Y and Z coordinates of survey points are most common formats of the survey data. Once the survey data are downloaded in these formats, they can be further customized and/or transformed for use on several platforms.

Web-based Tools for Deriving Site-specific Information

The National Map Viewer (www.viewer.nationalmap.gov) is a one-stop shop for site-specific information. It provides a map-based interface and tools to download a variety of data and create maps such as aerial imagery, Digital Elevation Models (DEMs), stream network, and soil maps. Web Soil Survey (WSS) is another such tool made available by the NRCS specifically for accessing soils information (http://websoilsurvey.sc.egov.usda.gov/). It provides map-based access to two soils databases, SSURGO and STATSGO. Users can view, download and print soil properties for an area of interest or for a county. In addition to the National Map Viewer, several online tools are available that provide similar data for certain states, watersheds and/or counties. State level LiDAR (Light Detection and Ranging) data sets are one such example discussed in detail in the subsequent sections. Airborne LiDAR imaging technology was used to collect elevation data in several states and is available for public use (Dollison, 2010; El-Rabbany, 2002).

Using Digital Elevation Models for Planning of Drainage Systems

Topography is one of the most important variables required for planning and design of drainage systems. Traditional methods of obtaining topographic information are usually time consuming, labor intensive and coarse in resolution (for example laser surveying). Advanced remote sensing techniques now allow us to obtain elevation information for large areas instantly, with a greater resolution. A typical remote sensing approach involves aerial imaging using different radiation spectrums (typically visible, near infrared and shortwave radiations) through satellites and airplanes. Different objects on the earth's surface reflect and absorb radiations differently, causing a unique "spectral signature". Such spectral signatures can be associated with properties (e.g., elevation, land use, crop growth stage, water bodies) of objects in a remotely sensed image. A Digital Elevation Model (DEM) is one of the several products of remote sensing.

What is a Digital Elevation Model?

In scientific literature, three terms, namely DSM, DTM and DEM, are used interchangeably to represent elevation models. A Digital Surface Model (DSM) refers to elevations of earth's surface including the natural and man-made objects such as vegetation and buildings. A Digital Terrain Model (DTM) refers to elevation values for bare earth and excludes objects like vegetation and buildings. Digital Elevation Model (DEM) is commonly used as a generic term to represent either DSM or DTM. As described by the National Digital Elevation Program (NDEP, 2004) Digital Elevation Models (DEMs) have at least three different meanings:

1. "DEM" is a generic term for digital topographic and/or bathymetric data in all its various forms. The generic DEM normally implies elevations of the terrain (bare earth z-values) void of vegetation and manmade features.

2. As used by the U.S. Geological Survey (USGS), a DEM is the digital cartographic representation of the elevation of the land at regularly spaced intervals in x and y directions, using z-values referenced to a common vertical datum.

3. As used by other users in the U.S. and elsewhere, a DEM has bare earth z-values at regularly spaced intervals in x and y, but normally following alternative specifications, with narrower grid spacing and State Plane coordinates for example.

Typically, DEMs are obtained using remote sensing techniques such as photogrammetry, LiDAR, and ifSAR (Interferometric synthetic aperture radar). In rare cases, DEMs can also be produced from existing land survey data using map digitization techniques (Source: NDEP, 2004).

Sources of Digital Elevation Models

Digital Elevation Models can be obtained from several sources, mostly free of cost (Gesch, 2007). The primary source of DEMs in the United States is the National Elevation Dataset (NED), provided by the USGS. The NED offers DEMs at resolutions of one arc-second (about 30 meters) and 1/3 arc-second (about 10 m), nationally and at 1/9 arc-second (about three meters) in limited areas. The NED is constantly being updated with more accurate DEMs at finer resolutions. In future, it will be possible to obtain DEMs with resolutions as low as 30 cm for the entire

United States. The readers are advised to visit the NED website (www.ned.usgs.
gov) for latest availability of these data products.

The second and more recent source of DEMs is statewide LiDAR Imagery.
LiDAR is an optical remote sensing technique that measures the distance to
(or other properties of) a terrestrial object by illuminating the target with light
pulses from a laser. The DEMs from LiDAR are slightly different from those available at NED. Unlike NED images with continuous elevation data, LiDAR typically
involves point data representing the X–Y location and Z elevation of any target
on earth's surface that reflects the laser pulse. Thus, a DEM created from LiDAR
involves interpolation between grid-points and represents location and elevations of terrestrial objects (similar to a DSM).

LiDAR data are currently available for only a few states. However, the USGS
has taken the initiative to build a National LiDAR Dataset that would be available
through the National Elevation Dataset using a National Map Viewer (Stoker et
al., 2008; USGS, 2017). As of the publication date, this project is still in its inception
phase. The users are advised to look up the USGS website for latest updates for
the same. Appendix 1 summarizes the current status and availability of LiDAR
data in the United States (Maune, 2001; NDEP, 2004).

Applications of Digital Elevation Models
for Drainage Planning and Design

There is a wide range of options available for selecting DEMs from the sources
mentioned above. For planning and design of drainage systems, moderate to high
resolution DEMs are recommended (ground sample distance of 10 meters or less).
In fields with relatively rough topography, high resolution would be necessary to
capture all but the abrupt changes in elevations within the field. For such fields,
DEMs with a spatial resolution of 1 meter or less should be used. DEMs with
such high resolution can be obtained from LiDAR data. However, LiDAR data
are currently available for only a limited number of regions in the United States
(Please refer to Appendix-1 for availability of LiDAR data in the United States).
The remaining discussion in this chapter will be limited to using high resolution
DEMs for design and planning of drainage systems.

A DEM with appropriate resolution can be used to draw a contour map for
the field and to identify low spots, high spots, possible drainage outlets, existing
waterways and gullies in the field. Although high resolution DEMs can provide
very accurate elevation information, it is very important to note that this information was collected at least a few years ago (imagine looking at your own picture
from 2 to 5 years ago). It is strongly advised that you conduct a field investigation
(typically using a GPS or laser survey) in order to verify location and elevation of
at least a few key features or spots in the field. For example, you can verify if the
DEM is accurately showing the location and elevation of an existing waterway or
a distinct low or high spot within the field, with respect to a benchmark. Depending upon the source of the DEM, areas adjacent to large water bodies and/or state
borders are likely to inherit some errors in their respective DEMs. Furthermore,
if the field has undergone a major reshaping (e.g., land leveling, construction of
waterways, etc.) recently, any DEM based on data collected before that reshaping
happened is not useful for our analysis. In such cases (when the most recent DEM
is not available), it becomes necessary to conduct a field survey using one of the
traditional methods (e.g., GPS Survey, Laser Survey).

How to Use Digital Elevation Models

Downloading Digital Elevation Models

Digital Elevation Models can be obtained from several different sources. In general, the agency that makes these DEMs available offers them through various electronic media. Most common forms are web servers and compact discs or DVDs. Since these media and their interfaces are constantly evolving, it is practically impossible to describe each and every source and its interface. Therefore, a generic description is given in the subsequent text. The readers are advised to refer to the appendices of this chapter for more detailed and up-to-date information.

In general, the source agency (e.g. USGS) offers DEMs in different formats compatible with different image processing software tools. The most common formats are Raster (elevation data stored in an image file with each pixel representing an elevation value), ASCII (elevation data stored in a text file separated by a certain delimiter such as a comma) and LAS (a special file format for LiDAR data). The DEM data can be downloaded at different scales, ranging from square tiles of uniform size (e.g. 1000 m by 1000 m), to counties and states. For advanced users, these DEM tiles are usually made available in a table format. The same data are also made available through a map-viewer web interface, for users who may need DEMs for custom-shaped areas. A typical map-viewer web interface is shown in Figure 2. The interface is usually designed to be self-explanatory, and consists of some GIS processing tools, similar to those in existing GIS software tools. You can use these tools to go through the following general steps:

1. Zoom into your area of interest on the map-viewer (common tools: Zoom-in, Zoom-out, Pan, Center Map, Full Extent, Zoom to Previous/Next Scale)

Fig. 2. Example of online map viewer application for downloading DEM.

2. Use "selection" tools to select the area of interest by drawing a regular or irregular shaped polygon around the same (common tools: Select, Add to Selection, Remove from Selection, Clear Selection).

3. Overlay relevant supporting layers, that may facilitate your selection process (common layers: aerial image, roads, county boundaries, Hydrography)

4. Once you have selected the area of interest, the map-viewer offers additional tools to download the DEMs and relevant data layers, in different formats mentioned above.

5. It is common for the downloaded files to be in a compressed folder format. In such cases you need to extract the actual data files from the compressed folder using a file extraction utility program (e.g. WinZip, WinRAR, 7-Zip).

Preparing Digital Elevation Models for Use

The first step after downloading a DEM is to read the metadata file for the same. Metadata is data that describes the data. Most agencies provide a "Read Me" document along with data files, which consists of all the relevant information for the data files. The metadata typically consists of information such as: source of DEM, date of image capture, the geographic coordinate system used, projection system used, units used and so on. Once the required information is obtained from the metadata, all the data files associated with the DEM should be stored in a folder or the project directory. If the DEM data is in ASCII format, additional steps would be required to convert the data into a GIS file format (usually a raster file). If the rest of the project files are in a different coordinate or projection system than that of the DEM data file, then the DEM data file needs to be transformed using a 'projection' command. If DEM data are downloaded in the form of multiple tiles that need to be combined to represent the area of interest, a GIS tool can be used to mosaic (or stitch together) the respective tiles, so that there is one single DEM that contains the entire field.

Processing Digital Elevation Models for Obtaining Information

Once the DEM is ready for processing, GIS tools can be used to create a contour map (Shedekar and Brown, 2011). The GIS tools and steps required to process DEMs may vary depending upon type of DEM. The following section describes the common tools and steps used for processing a DEM. Several GIS tools are available to process the LiDAR data and/or DEMs.

    1.   ESRI – ArcMAP – Arcview license

      a. Spatial Analyst

         i. Hydrologic modeling, Raster analysis

      b. 3-D Analyst (ArcScene, ArcGlobe)

         i. Cut/Fill, Line of Sight, 3-D Perspective

         ii. LAS Reader

      c. ArcMAP extension

         i. LiDAR Analyst, LP360

    2.   Stand-Alone Programs

      a. MARS

      b. AutoCAD

         i. Civil 3D

c. Stand-Alone Viewers

   i. Qcoherent (www.qcoherent.com, verified 21 Aug. 2017)

   ii. Fugro EarthData (https://www.fugro.com/about-fugro/our-expertise/technology/fugroviewer, verified 21 Aug. 2017)

   iii. Sanborn (http://www.sanborn.com/aerial-lidar/, verified 21 Aug. 2017)

   iv. Global Mapper (www.bluemarblegeo.com/products/global-mapper.php, verified 21 Aug. 2017)

   v. ALDPAT - Airborne LIDAR Data Processing and Analysis Tools (http://lidar.ihrc.fiu.edu/lidartool.html)

One or a combination of the tools mentioned above can be used to further process the DEM. Depending upon the software and its capabilities, a DEM can be processed using one or all of the following steps:

1. Cut the DEM in the shape of field boundary.

2. Select appropriate symbology (color scheme) that facilitates the representation of the DEM on a map.

3. Derive a contour map for the field of interest.

4. Derive other elevation information such as slopes, waterways, and surface depressions.

5. Convert the DEM into an interactive 3-D map for visual interpretations. For example, for hill shade effect, use Triangular Irregular Networks (TIN).

Obtaining Information from Digital Elevation Models

A certain level of professional experience is required to be able to summarize all the information from a DEM into a presentable and/or usable report format. However, a semiskilled user can use GIS software to prepare such reports using pre-existing templates. Such a report may be comprised of various maps, layers, and numeric value tables derived from the DEM. A DEM can provide information that can be useful for planning and design of drainage systems. A topographic map showing contour lines can be used for planning the layout of the drainage system based on land slope, high points and depressions, and pre-existing drainage ways.. The same DEM can be used to determine the cross-sectional profiles of proposed drainage main(s) and laterals. Land slopes and profiles can be derived from contour maps, either manually or by using GIS software to process the contour map layers. The methods to derive these maps and layers are described in detail in the next section.

GIS and Web-based Drainage System Design (Approach, Tools, Software)

How to Use GIS for Designing Drainage Systems

Determine Average Slope Across the Field

The average slope across the field is the most representative slope for design purposes. The average slope can be derived using a "slope" tool in most GIS software. The algorithms used by most GIS software tools compute the difference in elevation of each cell with its adjacent cells and selects the highest value of difference to calculate a local value of slope between two corresponding cells. The slope values for all the cells are then averaged over the entire area to derive an average slope. The field's average slope provides an indication of the overall topography of the field and relates more to guidelines for design of surface drainage system. However, more specific field topographic information is required for the design of subsurface drainage systems. For a systematic drainage system, it is essential to derive the specific slopes in the parts of the field where drainage mains and laterals are proposed.

Determining Slope Along a Proposed Main or Lateral (Profile)

Slope value along a straight line is simply the elevation difference per unit of horizontal distance along that line. Thus, a slope may be computed as the elevation difference between upstream (inlet) and downstream (outlet) ends of the proposed drainage main or lateral divided by the horizontal distance between the two points. GIS software provides a method to derive a profile for the proposed drainage main and/or lateral. A profile is a graph showing elevation versus length, or distance, along a proposed line. A typical procedure involves drawing the proposed line on the map at the proposed location, then combining the line layer with the DEM layer to derive elevation information along the proposed line, and finally displaying the ground elevations vs the length of the proposed line. The 3D Analyst in ArcGIS offers a tool to create a profile graph for proposed lines using a DEM or a contour map as the source layer.

Special Features in the Field

A contour map can help identify special features in a field. Closely spaced contour lines indicate steeper slopes, while those spread farther apart indicate relatively flatter terrain. A low spot (depression) can be identified by contour lines that form a circle, with lower elevations towards the center of the circle. On the other hand, high spots, small hills or knobs can be identified by circular contour lines with higher elevations towards the center of the circle. Contour lines that form a V-pattern may indicate a ridge or drainage way. For example, a series of "Vs" pointing downslope, from higher to lower elevations indicates a ridge line. Conversely, a drainage way (valley) may be illustrated by a series of "Vs" pointing from lower to higher elevations.

Identifying the Drainage Outlet

A contour map of a field and its surrounding area may help in identifying possible locations of the drainage outlet. If the contour map is based on a recent DEM or topographic survey, it can be used to determine the elevation of a proposed outlet. If the contour map is based on a lower resolution DEM or if field conditions have changed recently, it is essential to check the actual conditions in the field by conducting a survey.

Estimating Cut and Fill Volumes for Surface Drainage

Surface drainage usually involves reshaping the field by moving soil from higher positions in the field (hills, ridges, knobs) to lower positions (depressions), a process called land leveling. The goal is to provide a more uniform field elevation that allows surface water to flow off the field without causing localized ponding, or wet spots, and without causing erosion. Simple GIS tools using high quality topographic data can be used to make precise estimates of the amount of earthwork required in terms of cuts and fills. Most GIS-based software programs provide special tools to estimate cut and fill volumes. The typical procedure involves selecting the area to be shaped, selecting the required slope and/or depth of cut in a particular direction, and assigning parameters such as soil bulk density. The software will calculate the volume of cut and/or fill, and the weight of material to be transported.

Identifying Spatial Variation of Slope
Most GIS programs allow users to select different classes of slopes and add colorful labels to represent these slope classes. Users can easily identify steeper slopes with respect to flat grounds, grade breaks and lengths of different slopes.

Flow Paths and Watershed Analysis
One of the key benefits of GIS and DEMs is the ability to derive watersheds, their boundaries, and flow paths within the field. Such information is vital for identifying drainage zones within a field. A drainage design layout that extends across watershed boundaries may result in deeper or shallower than normal drain depths, and sometimes may increase the installation complexity and costs. Drainage layouts that follow natural slopes and/or natural contour lines tend to be less complex to install and more cost effective. Tools to derive watersheds and flow paths vary with the GIS software, as some programs have tools that can instantly generate watershed boundaries and flow lines. However, some programs may require you to perform some data pre-processing before watershed boundaries can be derived. For example, high resolution DEMs for large fields may result in several small watersheds with very minor elevation differences. In such cases, a user may use a generalization or resampling tool that helps remove minor elevation changes (roughness) within the field, essentially making the topography of the field surface smoother.

Importing and Processing GPS Survey Data
If GPS survey data are available, users can import it into GIS software. If the GPS survey was conducted at a finer resolution than that available with the DEM, it is advisable to use the GPS data for developing the topographic map. Typically, GPS data can be imported in shape format or in X-Y-Z coordinate format. These data points can then be processed to derive the contour map, 3-D surface, watersheds and boundaries, flow directions, etc.

Importing and Analyzing Soil Layer Information
Soil properties guide the determination of drainage system depths and spacing. Once the topographic layers are processed, they can be overlaid by a soils map. In some cases, drainage-related properties of soils are less likely to vary much within a relatively small field with less complex slopes compared to larger fields with more complex slopes. In either case soil information can help determine the optimum drain depth and spacing for the site. Soils information is usually available in the form of a shape file, with polygons representing different soil types within the field. The soils layer can be further processed and used as a background image while presenting the final design map.

Overlaying
The ability to overlay different informational layers within the same map is a very useful functionality of GIS. Designers can refer to the available information (e.g., soils, topography, special features, outlet, etc.) all at once and produce

a suitable drainage design. Overlaying maps also helps create more professional and interactive maps. Once the baseline information is organized, the next step is to design the drainage system, either with design software or manually using standard design procedures.

Tools Available for Drainage Design

One of the recent advancements in the science of drainage design is availability of several computer tools and software. These software tools range from small individual programs relating to one of the drainage design steps all the way up to highly sophisticated, fully automated programs that can carry out several design steps in a matter of minutes. Since, this is a fast-evolving field, it is impossible to capture all of the drainage design tools and/or their latest versions. This section, however will introduce you to some of the leading software tools and their developers. Readers are encouraged to explore them further based on information given in this section. Please note that the authors do not intend to endorse any of these products and/or developers. All the description is based on scholarly evaluation of these tools by the authors.

Appendix 2 summarizes the existing tools available for drainage design. Although several small tools are available for individual steps in the design process, only a handful of software packages are available for comprehensive planning and design of drainage systems, at present. A few programs are described in detail below. Considering the pace at which these programs are evolving, the users are strongly encouraged to explore further for latest developments of these tools.

1. WM-DRAIN: Developed by Trimble, this package provides tools for survey, analysis, design, installation and mapping of surface and subsurface drainage. A suite of tools and programs is available from Trimble® and Farmworks. Trimble offers WM-Topo, a topographic data collection system that can be combined with WM-DRAIN. Farmworks, a division of Trimble, offers Farmworks Surface, an analysis and design tool that is intended to be used with WM-DRAIN. Farmworks Surface ensures the optimal placement of tile and surface drains in both surface and subsurface drainage water management projects.

2. LANDRAIN: Developed by A.B. Consulting Company, this package allows users to layout drainage network, determine pipe grades, pipe sizes and analyze design options. Other programs offered by A.B. Consulting are LIContour, LANDIMPROVE and Discover. LIContour provides the capability to import topographic data from GPS and other inputs, and to conduct topographic analysis (e.g. contour maps).

3. SMS: This is a program developed by Ag Leader Technology, primarily as a farm management software. The recently added "water management" module allows users to perform a few of the drainage design steps. Users can import topographic survey data from GPS and other sources and overlay multiple reference layers (such as soil map, planting and harvesting imagery). The module consists of tools for drawing layout of drainage network, determining design parameters (minimum grade, depth and sizes of pipes), and comparing different designs plans for optimization. The module is also capable of generating design summaries, inventory of supplies and printable maps. Ag Leader is in the process of adding more features to this module. This type of integration may gain more popularity in the future, as it would provide one-stop-shop for the storage, management,

and analysis of all the farm-related data as well as tools for planning and designing of on-farm systems.

4. AGPS-Pipe Pro: calculates the slopes and grade breaks for an entire main or lateral using Vertical Curve Technology and controls the machine grade during installation. The program also captures lines during installation to easily make CAD maps showing drain layout.

5. Illinois Drainage Guide: Provides some simple calculators based on Drainage Theory to determine drainage depth, spacing and outflows using soils information. It provides detailed guidelines for design and management of drainage systems. It provides several programs useful for design of drainage systems. List of important programs includes drain spacing, county specific simulations, sizing drainage pipes, lateral specifications, region of influence, flow profiles in laterals, sizing ditch, Manning's equation, flow under submerged outlet, minimum curvature, design of riprap lined channel etc.

Integrating GIS with CAD and Other Tools

Computer Aided Design (CAD) software can be used to design the layout and plan of the drainage system. AutoCAD is one of the commonly used design software. It offers tools to design systems in 2-D as well as 3-D modes. Most of the layers from GIS can be imported into CAD software. Conversely, the design created in CAD can be exported into a format that is compatible with GIS softwares. Most common interchangeable formats are shape (.shp), drawing (.dwg), and raster (.rst).

An example of application of GIS, GPS, and precision technology for a case study is shown in Figure 3. The maps show various steps in the planning and design process for a subsurface drainage system.

Precision Installation of Drainage Systems

With the advancement of RTK–GPS and machine control, precision installation of drainage systems is possible. Such systems involve a drainage installation machine and/or plow that is mounted with an RTK–GPS unit and grade-control device. A central computer controls the location and depth of installation based on a predesigned drainage layout of the system. Systems that integrate laser technology for grade control and GPS for horizontal location are also being used (Chieng et al, 1980).

Geographic Information System–based Management of Drainage Systems

The first step of successful management of drainage systems is to ensure that the installed system layout is georeferenced to the existing GIS data for the same field. Overlaying georeferenced system layout over a DEM or contour map can help delineate zones of water table elevations for the field. This is particularly important for practices that involve water table management such as drainage water management (controlled drainage) and subirrigation. The practice of drainage water management (DWM), also referred to as controlled drainage involves raising the elevation of a drainage system outlet during the non-growing and in some cases growing season (typically 30 to 45 cm below ground surface), thereby regulating the discharge from the system outlet (United States Department of Agriculture - Natural Resources Conservation Service, 2007). This reduction in outflow helps reduce nutrient loading into the natural water ways. Subirrigation is a further modification of controlled drainage, and involves adding water through the system outlet

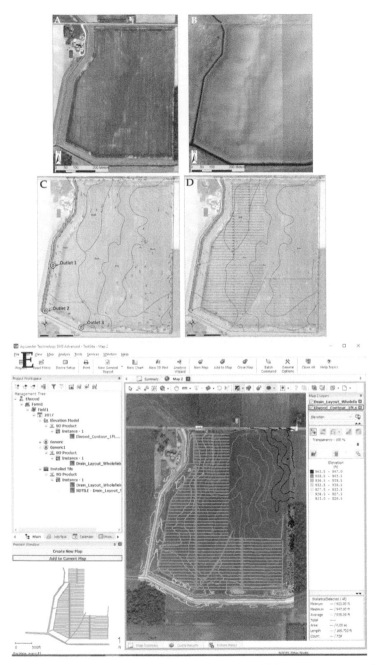

Fig. 3. Example case study showing application of GIS, GPS, and precision technology in planning and design of a subsurface drainage system. A. An areal ortho image. B. Digital Elevation Model derived from LiDAR data. C. Overlay of soil map, contour map, and planned outlets on areal ortho-image. D. A layout of a planned drainage system. E. An example of drainage design software (SMS Advanced, Ag Leader Technology).

to maintain the water table at a desired stage in order to provide capillary water for plant use during the growing season (United States Department of Agriculture-Natural Resources Conservation Service, 2007). Delineation of zones of water table elevations helps estimate the spatial extent of effectiveness of the water table management practices. Another application of such water table zones is to assess the vulnerability of crops to excess water stress (flooding damage).

Subirrigation, controlled drainage and drainage water management involve maintaining the water table at a certain depth below the surface in the field. In most of the cases, the drainage system outlet is located further away from the network of laterals and submains. A georeferenced outlet elevation can help determine the board and/or dam heights to be set at the outlet for achieving the target water table depths in the field. This demonstrated clearly in the example shown in Figure 4. In this case, the outlet is about 40 m away from the edge of the field and about 13 cm lower than the drain depth at the edge of the field. If the outlet elevation is set at a certain depth above the drain pipe or below the ground surface, it is going to be 13 cm lower or higher at the edge of the field. This elevation difference should be factored in when one decides to manage the water table in the field by controlling the outlet elevation. Thus, if the target water table depth is 30 cm below the surface at the first lateral in the field, then the water table at the outlet needs to be set at about 80 cm above the outlet elevation (285.15 cm). Zones of influence can also be delineated based on this information as shown in Figure 5. However, it is important to note that in reality, the water table does not move as a flat plane in the field as shown in the illustrations. Generally, following a rainfall event causing complete saturation of the soil profile, the areas above the laterals drain faster than those midway between laterals. In steady state conditions, this phenomenon results in a seemingly elliptical shape of the water table in a cross-section

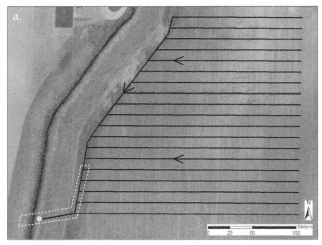

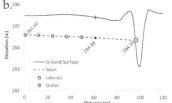

Fig.4 (a) Drainage layout for a field, and (b) Profile chart for part of main line highlighted by dotted area.

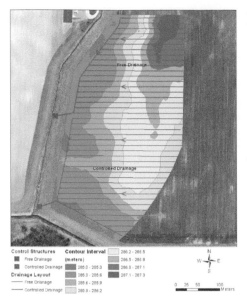

Figure 5. Zones of influence delineated for a drainage system for different depths of water table control.

of soil profile between two laterals. Delineating the water table based on a theoretical understanding of these shapes is rather complicated. For practical considerations, the delineation of the zone of influence is based on a maximum elevation of water table in the field that can be achieved with the given elevation set at the outlet.

With the evolution of conservation drainage, georeferencing and precision installation of drainage systems are becoming more important. Advances in surveying technology, GIS, GPS, and access to precision design and installation software can greatly enhance the implementation of the suite of practices under the realm of conservation drainage.

Conclusions

GPS, remote sensing, DEM, and new software and/or technologies can be used to integrate spatial and temporal information that accounts for soil type, hydrology and crop management information at a given site. These new technologies can be used to implement very precise conservation practices to improve management of drainage systems. We now have the potential to apply this science to implement precision conservation in the field to increase soil and water conservation and the effectiveness of soil and water conservation practices. Improving drainage management could also contribute to increased yields (Skaggs et al., 2012). Additionally, we could integrate all of this precision information to improve drainage management as part of precision conservation efforts. Other precision conservation practices, together with this improved management, will have the potential to reduce environmental impacts (Strock et al., 2010). Additionally, we need to improve drainage management practices to increase adaptation to a changing climate. Although some regions of the United States are projected to experience increased drought periods, others are projected to have increased precipitation, and these types of improved drainage management systems resulting from precision conservation may contribute to adaptation to a changing climate. Recent advances of technology are allowing the frequent use of precision information to implement precision conservation for management of drainage systems making this an important tool for higher productivity during the twenty-first century.

References

Chieng, S.T., R.S. Broughton, and R. Kok. 1980. Computer-aided drafting of subsurface drainage plans. Canadian Agricultural Engineering 22(2):137–143.

Dollison, R.M. 2010. The national map: New viewer, services, and data download. U.S. Geological Survey Fact Sheet. United States Geological Survey. 2010–3055, p. 2. http://pubs.usgs.gov/fs/2010/3055/ (verified 21 Aug. 2017).

El-Rabbany, A. 2002. Introduction to GPS: The Global Positioning System. Mobile Communication Series. Artech House Publications, Norwood, MA.

Fausey, N.R., L.C. Brown, H.W. Belcher, and R.S. Kanwar. 1995. Drainage and water quality in Great Lakes and Cornbelt states. J. Irrig. Drain. Eng. 121(4):283–288. doi:10.1061/(ASCE)0733-9437(1995)121:4(283)

Gakstatter E. 2014. Centimeter-Level RTK accuracy more and more available – for less and less. GPS World. March 2014. http://gpsworld.com/centimeter-level-rtk-accuracy-more-and-more-available-for-less-and-less/ (verified 21 Aug. 2017).

Gesch, D.B. 2007. The national elevation dataset. In: D. Maune, editor, Digital Elevation Model Technologies and Applications: The DEM User's Manual. 2nd ed. American Society for Photogrammetry and Remote Sensing, Bethesda, Maryland. p. 99–118.

Goolsby, D.A., W.A. Battaglin, B.T. Aulenbach, and R.P. Hooper. 2000. Nitrogen flux and sources in the Mississippi River Basin. Sci. Total Environ. 248(2):75–86. doi:10.1016/S0048-9697(99)00532-X

King, K.W., M.R. Williams, M.L. Macrae, N.R. Fausey, J. Frankenberger, D.R. Smith, P.J. Kleinman, and L.C. Brown. 2015. Phosphorus transport in agricultural subsurface drainage: A review. J. Environ. Qual. 44(2):467–485. doi:10.2134/jeq2014.04.0163

Langley, R. B. 1998. RTK GPS. GPS World 9:70-76.

Maune, D.F. 2001, Digital elevation model technologies and applications: The DEM user's manual. American Society for Photogrammetry and Remote Sensing. Bethesda, MD. p. 539.

National Digital Elevation Program. 2004. Guidelines for digital elevation data. NDEP. Reston, VA.

Pease, L.A., N.R. Fausey, J.F. Martin, and L.C. Brown. 2017. Projected climate change effects on subsurface drainage and the performance of controlled drainage in the Western Lake Erie Basin. J. Soil Water Conserv. 72(3):240–250. doi:10.2489/jswc.72.3.240

Shedekar, V.S., and L.C. Brown. 2011. Using LiDAR data for creating contour maps. Overholt Drainage School, 21-25 March 2011, Wauseon, OH. The Ohio State University. Columbus, OH.

Skaggs, R.W., N.R. Fausey, and R.O. Evans. 2012. Drainage water management. J. Soil Water Conserv. 67(6):167A–172A. doi:10.2489/jswc.67.6.167A

Stoker, D., D. Harding, and J. Parrish. 2008. The need for a national Lidar dataset. Photogramm. Eng. Remote Sensing 74(9):1066–1068.

Strock, J.S., P.J.A. Kleinman, K.W. King, and J.A. Delgado. 2010. Drainage water management for water quality protection. J. Soil Water Conserv. 65(6):131A–136A. doi:10.2489/jswc.65.6.131A

United States Department of Agriculture-Natural Resources Conservation Service. 2007. Chapter 14: Water management (Drainage). Part 650: Engineering field handbook. In: USDA-NRCS, National engineering handbook. USDA-NRCS, Washington, D.C.

USGS. 2017. CLICK: The USGS Center for LIDAR information coordination and knowledge. USGS. http://lidar.cr.usgs.gov [2017 is year accessed.]

Zucker, L.A., and L.C. Brown. 1998. Agricultural drainage: Water quality impacts and subsurface drainage studies in the Midwest. Ohio State University Extension Bulletin 871. The Ohio State University. Columbus, OH.

Appendix A: LiDAR Data Availability and Sources in United States

State	Dataset Status	Source
Connecticut	Complete	Connecticut Lidar 2010 DEM
Delaware	Complete	Delaware Spatial Data Framework
Florida	Partial	FL Coastline Project
Idaho	Partial	Idaho LiDAR Consortium
Illinois	Partial	Illinois Height Modernization Program
Indiana	Complete	Indiana Spatial Data Portal
Iowa	Complete	Iowa GeoTree Lidar Mapping Project
Kansas	Partial	Kansas GIS/DASC
Louisiana	Complete	Louisiana Atlas
Maryland	Partial	Maryland DNR, LiDAR, or NOAA
Massachusetts	Partial - Boston area only	Available for purchase from MassGIS
Michigan	Partial	2010 Southeast Michigan Imagery Project
Minnesota	Partial	Minnesota Lidar Status
New Hampshire	Partial	GRANIT/Coastal data
New Jersey	Partial	New Jersey Lidar Status October 2007
New York	Partial	NYS Lidar Coverage
North Carolina	Complete	North Carolina Floodmapping Program NCDOT Elevation Data
North Dakota	Partial	North Dakota LIDAR Dissemination Mapservice
Ohio	Complete	Ohio Statewide Imagery Program
Oregon	Partial	Oregon LiDAR Consortium
Pennsylvania	Complete	PAMAP Program LiDAR
South Carolina	Partial	SC Lidar Consortium
Texas	Partial	TNRIS/Texas
Utah	Partial	Utah 2m Lidar datasets
Virginia	Partial	William and Mary CGA State LiDAR datasets
Wisconsin	In progress	WisconsinView Data Portal
Wyoming	In progress	Wyoming Statewide LiDAR Effort

Appendix B: Software and Hardware Tools
Available for Surveying and Drainage Design

Tool	Source or Manufacturer	Remarks
Discover	A B Consulting Co	A program designed to supplement the mapping capabilities of LI Contour (V+). Allows 2-D and 3-D viewing, localized flow pattern, slope, watershed and contour analysis.
LANDIMPROVE	A B Consulting Co	A topographical surface modeling, analysis, and design program. Uses triangulated surface models for contour mapping, volumetric calculation, and surface model design capabilities. Capable of providing both 2-D and 3-D graphics during data editing and mapping.
LANDRAIN	A B Consulting Co	A drainage design software that provides tools to layout drainage network, bury the network, determine the required pipe grades and size the pipes.
LI Contour	A B Consulting Co	A program designed to generate contour maps from coordinate data. X, Y, Z coordinates may be input either by hand or loaded from coordinate files previously stored on disk. The field surface is modeled using the triangulation method.
LI Contour V+	A B Consulting Co	This includes all capabilities of LI Contour, and adds the ability to calculate the volumetric difference between two surfaces. Output includes the total volumes of cut and fill, and a print out of the cut/fill amount at each data point.
SMS	Ag Leader Technology	Farm management software with "water management" module that provides tools for tile drainage design.
AGPS Pipe Design	AGPS	This is a new software for use in conjunction with earlier drainage software solutions, including Pipe Pro, Pipe FM. Notable features include: • Draw mains with designed depth – including profile view. • Automatically draw a group of laterals within a watershed boundary or other break lines. • Export to AGPS-Pipe software for automatic blade control in the field. • Follow exact design or make use of flexibility to do changes in the field. • Compatibility of background imaging enhancing design capabilities. • Other features; pipe sizing, reports, etc. coming in future releases.
AGPS-Pipe Pro	AGPS	Calculates the slopes and grade breaks for an entire main or lateral using Vertical Curve Technology™ and controls the machine grade during installation. Also captures lines during install to easily make CAD maps showing drain layout.
AGPS-Shape Pro	AGPS	This unique patent pending program lets you create custom shaped surfaces and achieve drainage by moving less soil using Vertical Curve Technology.
AGREM LLC Agriculture Water Management Software	AGREM	This software can be used to produce detailed topographic maps, design of AGREM water management systems, mapping of drainage and irrigation networks and design of drainage systems.

AgForm-3D	TopCon	A field design software that lets you view and design the existing field. View existing field contours and profiles, and then design the field as a single plane or break it into as many sections as needed.
System 210	TopCon	Incorporates Topcon's AGS land leveling, surveying, and field design capability. Integrated system consisting of an Interactive Display (X20), the MC-R3 receiver (with GPS receivers, radios, and controllers) and MC-A1 antenna.
FmX integrated display	Trimble	An advanced, full-featured guidance display for precision farming applications. It provides advanced precision agriculture functionalities, including: • Guidance and mapping • Planting and nutrient and/or pest management • Harvesting • Water management • Information exchange and farm software
Trimble WM-Drain	Trimble	Tools for survey, analysis, design, installation, and mapping steps of surface and subsurface drainage
WM-Topo Survey System	Trimble	Topographic data collection device that can be taken into hard-to-reach areas such as ditches, steep terrain, muddy fields, or fields with mature crop cover. Includes a complete GPS survey system with software program to record and process data on a handheld computer.
Farm Works Software	Trimble (Farmworks)	Farmworks is a division of Trimble® that offers a range of solutions for field and farm office—including mapping, accounting, water management, and more. Available programs: • Farm Works View • Farm Works Mapping • Farm Works Accounting • Farm Works Surface • Farm Works Stock • Farm Works Mobile • Farm Works Stock Mobile
Farm Works Surface	Trimble (Farmworks)	Analysis and design tool for use with the Trimble® WM-Drain farm drainage solution. Surface ensures the optimal placement of tile and surface drains in both surface and subsurface drainage water management projects.
FieldLevel II system	Trimble (Farmworks)	Tools for surveying, designing, and leveling steps required for land leveling projects. Also provides two methods for installing rice levees.
Illinois drainage guide	University of Illinois	Provides detailed guidelines for design and management of drainage systems. Provides several programs useful for design of drainage systems. List of important programs: General recommendations, county specific simulations, sizing drainage pipes, lateral specifications, region of influence, flow profiles in laterals, sizing ditch, Manning's equation, flow under submerged outlet, minimum curvature, and design of riprap lined channel.

Calculating Soil Organic Turnover at Different Landscape Position in Precision Conservation

David E. Clay,* Jiyul Chang, Graig Reicks, Sharon A. Clay, and Cheryl Reese

Abstract

An important aspect of precision conservation is assessing changes in soil health and soil organic carbon (SOC) across landscapes. Carbon (C) cycling can be determined using carbon flux towers, modeling, and by experimentally measuring budgets. Once the carbon budgets are understood, this information can be used to assess the value of implementing precision conservation and the potential impacts of targeted residue harvesting on soil health. This chapter provides a review of methods to determine carbon budgets, and the potential impacts of crop residue harvesting on SOC maintenance across landscapes. The chapter also provides examples on how to convert point carbon budget measurements into a precision conservation assessment.

Precision conservation is designed to target conservation activities to landscape areas that disproportionally impact the environment and the fields' long-term resilience (Ortega et al., 2002; Wilhelm et al., 2004; Varvel, 2006; VandenBygaart et al., 2003; Sisti et al., 2004; Russell et al., 2005; Clay et al., 2012b). Outcomes of precision conservation might include increased soil resilience, reduced erosion, and decreased need to convert grazing lands to annual crops (Clay et al., 2014). Examples of precision conservation include identifying and planting adapted plants in waterways, identifying and using no-tillage or reduced tillage techniques in highly erodible areas, and matching rotations, tillage, residue harvesting, N rates, and cover crops to landscape problems. A critical component of precision conservation is the maintenance of the soil organic carbon (SOC). SOC maintenance requires that the amount of C added to the system equals the amount of relic C

Abbreviations: DPM, decomposable plant material; NHC, nonharvested carbon; NHC_a, amount of nonharvested C applied; NHC_m, nonharvested C maintenance requirement; PCR, plant carbon retained; PCR_{incorp}, new plant carbon incorporated into SOC; RothC, Rothamsted Carbon Model; RPM, resistant plant materials; SOC, soil organic carbon; SOC_e, soil organic carbon at equilibrium; SOC_i, initial soil organic carbon; SOM, soil organic matter. South Dakota State University, Brookings, SD 57007. *Corresponding author (david.clay@sdstate.edu).

doi:10.2134/agronmonogr59.2013.0028

© ASA, CSSA, and SSSA, 5585 Guilford Road, Madison, WI 53711, USA.
Precision Conservation: Geospatial Techniques for Agricultural and Natural Resources Conservation
J.A. Delgado, G.F. Sassenrath, and T. Mueller, editors. Agronomy Monograph 59.

mineralized (Barber 1971, 1979; Anderson 1982; Barber and Martin, 1976; Balesdent et al., 1988; Allmaras et al., 2004; Bradford et al., 2005; Clapp et al., 2000; Causarano et al., 2006; Clay et al., 2010, 2012a). Precision conservation can increase SOC by reducing tillage intensity in erodible areas, which in turn can lead to reduced erosion and increased soil C levels.

Today's precision tools make precision conservation feasible. For example, GPS controlled sprayers and fertilizer applicators can be turned off when they drive over water ways and planters can seed different cultivars at different landscape positions. Ultimately, precision conservation can result in cool season grasses being seeded into summit or shoulder areas, traditional row crops being no-tillage seeded into backslope positions, and salt tolerant native plants being seeded into footslope positions. When successfully implemented, precision conservation should decrease erosion and increase SOC.

Developing a C Budget

Carbon inputs and outputs

The carbon cycle is driven by photosynthesis, which produces organic biomass that is respired by microorganisms (Frye and Blevins, 1997). Developing a C budget starts with accurate C output and input measurements. In carbon budgets, outputs are the amount of carbon dioxide (CO_2) released to the atmosphere, while inputs are the amount of organic carbon added to the soil. CO_2 releases can be measured directly or indirectly. For direct measurements, CO_2 is measured continuously. For indirect measures, changes in SOC at the beginning and end of the study are determined. Input measurements require accurate accounting of above and belowground biomass. Obtaining good measures of aboveground biomass is relatively easy and it is typically accomplished by weighing the amount of biomass returned to soil or estimating nonharvested C (NHC) from the harvest index (grain divided by all above-ground biomass). However, obtaining accurate measures of belowground biomass and exudates is very difficult (Dwyer et al., 1996; Rochette and Flanagan., 1997; Rochette et al., 1999; Kuzyakov and Domanski, 2000; Huggins et al., 1998; Amos and Walters, 2006; Mamani-Pati et al., 2010; Bolinder et al., 2007; Ehleringer et al., 2000). The three basic approaches used to estimate belowground biomass are modeling (Gilmanov et al., 1997; Molina et al., 2001), estimation, and measurement.

Modeling belowground biomass

Models have the capacity to account for climate, soil, and vegetation differences. The Century model estimates cereal root growth as a function of precipitation and plant age (Parton et al., (1987). For example, at emergence, 60% of the fixed carbon is allocated to roots, which decreases linearly with time and 3 months after planting the amount allocated to roots is 10%. If 60% of the fixed carbon is allocated to roots the associated root-to-shoot ratio would be 1.5, and if 10% of fixed carbon is allocated to roots the associated root-to-shoot ratio would be 0.11. The Rothamsted Carbon Model (RothC) model does not estimate plant growth.

Care must be used when comparing modeling results with field studies because experiments may define roots differently. For example, the Century model, allocates C to roots as a function of time and of total C fixed, while many field studies report roots as a ratio between roots and shoots (roots/shoots). In addition, some studies include the root crown in the root sample, while other studies do not. For carbon budgets, roots must be clearly defined and all sources of C must be considered. Plant growth simulation models may improve the accuracy of C input estimates by improving belowground biomass estimates.

Measurement of belowground biomass

Measuring roots is complicated by roots not equally distributed below the plant. Follett et al. (1974) collected 10 monoliths (91 cm deep) centered over the corn row at silking. In these monoliths, 73% of the root biomass, not including crowns, were contained in the surface 20.3 cm. Laboski et al. (1998) reported that over a three-year period, in soil with a subsurface (15–60 cm) bulk density of 1.57 Mg m^{-3}, 94% of the roots were within the surface 60 cm with 85% of the roots were within the upper 30 cm. Dwyer et al. (1996) reported that root mass density with depth could be explained by the equation, $Y = Y_0 e^{-BX} + C$, where Y is the root mass density (mg cm^{-3}), Y_0 is the root mass density extrapolated to $X = 0$, B is the shape coefficient, C represents the root mass in the deepest increment, and X is soil depth (cm).

Within the plant, C is allocated to many different parts including roots, exudates, crown, stover, grain, and cobs (Lloyd and Taylor, 1994; Hébert et al., 2001; Hanson et al., 2000; Kuzyakov and Cheng, 2001; Kuzyakov and Larionova, 2005, 2006; Melnitchouck et al., 2005; Wichern et al., 2008). Nearly all belowground biomass estimates do not consider exudates. Exudates, may represent 5 to 21% of all C fixed by the plant (Marschner, 2012), and they have a range of roles. They can be used to help regulate microbial communities, change the chemical and physical conditions of the root, and inhibit the growth of competing plants. Exudates need to be included in C budget calculations. In Table 1, it was assumed that 50% of the roots were exudates. When exudates were included in the calculations, the root to shoot ratios [(stover + grain + cob)/(roots + exudates + crown)] ranged from 0.47 to 0.61 (Table 1). Following harvest, most of the below ground (crown, and roots 0–15 cm), stover, and cobs would be added back to the surface soil. In experiments in South Dakota, 67% of the roots + exudates + crown were in the surface 15 cm, 83% of all biomass returned to the soil (roots, exudates, crown, stover, and cob) was in the surface 15 cm, and 12% of the ear was cob (Chang et al., 2014; Clay et al., 2015).

Differences between roots and shoots

Stover, crowns, and roots may have different mineralization rates (Wilts et al., 2004). Gale and Cambardella (2000) reported that in no-tillage, 75% of the new C incorporated into SOC was root derived, while a large percentage of surface residue was released as CO_2. Barber and Martin (1976) had similar results and reported that 50% of the root derived C was retained in SOC while only 13% of shoot derived C was retained in SOC. Differences between surface and root residue may be related to greater biochemical recalcitrance of root biomass, physical protection of root biomass, and lower O_2 concentrations with increasing soil depth. Soil depth also influences the rate at which the roots are mineralized. Clay et al. (2015) reported that: i) the first order rate constant for root in the 0- to 15-cm depth was much higher than the rate constants for the 15- to 30- or 30- to 60-cm soil depths; and ii) that harvesting surface residue influenced the mineralization belowground C.

Measurement of soil organic C

The ability to accurately determine C turnover depends on an accurate measurement of the SOC content at the beginning and end of the experiment. To convert gravimetric SOC values to volumetric values the concentration is multiplied by the bulk density. Changes in management resulting from precision conservation are likely to change bulk densities, which in turn will impact the relative depth sampled. To minimize changes in bulk density-derived sampling error, two approaches can be used. The first approach is to use a bulk density adjusted sampling protocol. In this approach, the depth at the final sampling is adjusted to account for changes in bulk density. The ratio, Sampling depth 2/Sampling depth 1 = Bulk density 1/Bulk density 2, defines the relationship between depth and bulk density. This ratio can be rearranged into Sampling depth 2 = (Bulk density 1 × sampling depth 1)/(Bulk density 2). For example, if the bulk density of the surface 15-cm initial sample is 1.1 g cm^{-3} and the soil bulk density 2 yr later is 1.25 g cm^{-1}, then to sample the same soil, the sample should be collected from the surface 13.2 cm (= 15 × 1.1/1.25).

In the second approach, the C budget in the entire soil column is determined at the beginning and end of the experiment. In these calculations, the entire soil column is sampled and the associated bulk densities of each zone and C concentration determined. Techniques for measuring bulk density are available at Grossman and Reinsch (2002).

Using models to estimate SOC turnover

Carbon turnover can be estimated using a simulation model or point or calculated using the input and output measurement described above. For simulations, the Century or RothC models are routinely used. The Century model (http://www.nrel.colostate.edu/projects/century5/) divides plant material, based on the lignin to N ratio into structural or metabolic pools and SOC into active, slow, and passive soil C pools. The active pool represents microbial biomass with a turnover time of days to years. The slow pool represents more recalcitrant material with turnover times in years to decades. The passive pool is humified C stabilized on mineral surfaces with turnover times of hundreds to thousands of years. Each pool has different rate constants. It is difficult to estimate the size of different pools based on only field-derived information.

The RothC uses a five pool structure, decomposable plant material (DPM), resistant plant materials (RPM), microbial biomass, humified organic matter, and inert organic matter (Coleman and Jenkinson, 1996; Skjemstad et al., 2004; Guo et al., 2007). The first 4 pools decompose by first order kinetics. RothC does not include a plant growth submodule, and therefore NHC inputs must be known, estimated, or calculated by inverse modeling.

Many scientists have attempted to use chemical extraction techniques to quantify the sizes of these C pools used by the Century and RothC models (Wolf et al., 1994; Olk and Gregorich, 2006; Olk, 2006; Zimmermann et al., 2007). These efforts have had mixed results. Olk and Gregorich (2006) stated that, "Each procedure has its strengths and weaknesses; each is capable to some degree of distinguishing labile SOM (soil organic matter) fractions from nonlabile fractions for studying soil processes, such as the cycling of a specific soil nutrient or anthropogenic compound, and each is based on an agent for SOM stabilization". In general, low-density soluble SOC material turns over faster (i.e., has a higher k value) than high density mineral-associated SOC, and hydrolyzable SOC turns over faster than non-hydrolyzable SOC (Martel and Paul, 1974; Six and Jastrow, 2002).

Direct measurement of C turnover

Direct calculation of C turnover can be based on non-isotopic and isotopic techniques. In these calculations, average SOC and NHC mineralization rates, based on field-derived information are mathematically defined. In nonisotopic and isotopic experiments, C inputs are modified and the temporal changes in SOC are measured (Larson et al., 1972). In the past, direct measurement has been underutilized, and many studies have reported the SOC at the end of the study and have failed to report the initial conditions. If the initial and final values are known, the SOC rate constants can be determined.

Determining rate constants is complicated by changes in method to measure SOC content. For example, in the 1950s many experiments SOC was determined using the Walkley–Black method, whereas in 2010 many experiments determined SOC by combusting the samples at 1000 C°. To convert SOC values from one method to another, an appropriate constant should be used. For example, soil organic matter can be converted into SOC by dividing by 1.72, and Walkley–Black C can be converted to SOC by multiplying Walkley–Black C by 1.34 (Clay et al., 2010). However, it should be noted that the conversion factors are sensitive to soil type and revisions in the method.

Soil organic C budgets can be based on the relational diagram shown in Fig. 1. In this diagram, nonharvested crop residues (NHC) represent the annual additions of organic C added to soil. The rate constants (kNHC and kSOC) shown in Fig. 1 represent the annual rate that C is transformed from NHC into SOC or SOC to CO_2. Based on the relational diagram, several equations can be defined. The first equation, $\delta SOC/\delta t = 0$, defines the maintenance requirement. At this point,

$$k_{SOC} \times SOC_e = k_{NHC} \times NHC_m \qquad [1]$$

where SOCe is the amount of SOC at equilibrium, NHCm is the nonharvested C maintenance requirement, and kSOC and kNHC are first order rate constants. The second equation is,

$$\frac{dSOC}{dt} = k_{NHC}\left[NHC_a - NHC_m\right] \qquad [2]$$

where NHC$_a$ is the amount of nonharvested C applied. This equation can be rearranged into the form,

$$\frac{dSOC}{dt} = k_{NHC}NHC_a - k_{NHC}NHC_m \qquad [3]$$

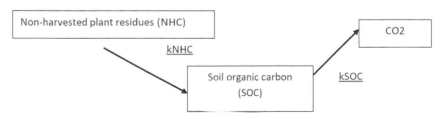

Fig. 1. A relational diagram showing the relationship between three carbon pools and the associated rate constants.

which is converted into a linear equation, $y = mX - b$, by defining $dSOC/dt$ as y, NHC_a as x, and k_{NHC} as m (Fig. 2). This derivation provides the theoretical basis for Johnson et al. (2006), and shows that the y-intercept is the product of the nonharvested carbon (NHC) first order mineralization rate constant (k_{NHC}) and the nonharvested carbon (NHC_m) maintenance requirement, whereas the slope is the nonharvested carbon rate constant (k_{NHC}). The maintenance requirement is calculated by dividing the y-intercept by the slope.

Clay et al. (2006) proposed an alternative approach where both k_{NHC} and k_{SOC} are calculated. The derivation of this approach is as follows. First, equations 1 and 2 were integrated into the equation, $NHC_a = NHC_m + (NHC_a - NHC_m)$, which resulted in,

$$\frac{NHC_a}{SOC_e} = \frac{k_{SOC}}{k_{NHC}} \bullet \frac{SOC_e}{SOC_e} + \left[\frac{dSOC}{dt} \bullet \frac{1}{k_{NHC} \bullet SOC_e} \right] \qquad [4]$$

This was accomplished by replacing $(NHC_a - NHC_m)$ with $dSOC/dt \times 1/k_{NHC}$ and NHC_m with $k_{SOC}/k_{NHC} \times SOC_e$, and dividing both sides by SOC_e. After cancelling units, the equation

$$\frac{NHC_a}{SOC_e} = \frac{k_{SOC}}{k_{NHC}} + \frac{dSOC}{dt} \left[\frac{1}{k_{NHC} \bullet SOC_e} \right] \qquad [5]$$

was derived. This equation was solved by defining SOC_i (initial SOC) as SOC_e, NHC_a/SOC_i as y, and $dSOC/dt$ as x. SOC_e was replaced with SOC_i because as time approaches infinity, SOC_i approaches SOC_e, and the assumption is that when the experiment was initiated, SOC_i approached SOC_e. The resulting y-intercept is k_{SOC}/k_{NHC} and the slope is $1/k_{NHC} \times SOC_i$. Based on these values, maintenance requirement and first order rate constants are determined with the equations,

$$NHC_m = b \cdot SOC_i \qquad [6]$$

$$k_{NHC} = 1/(m \cdot SOC_i) \qquad [7]$$

$$k_{SOC} = b/(m \cdot SOC_i) \qquad [8]$$

The advantages of this solution are that site-specific rate constants are calculated which can be used to calculate the impact of management on C turnover. For example, based on Eq. [5], if $k_{SOC} = 0.011$ g SOC/ (g × SOC year), $k_{NHC} = 0.13$ g NHC/(g NHC × year), and NHC = 4000 kg C (ha yr)[-1], then SOC_e is 47,300 kg C ha[-1] [47,300 = (0.13/0.011)(4000)]. If NHC is reduced to 2000 kg C (ha yr)[-1] then SOC decreases to 23,600 kg ha[-1]. The disadvantages with this approach are that it assumes that: (i) above and belowground biomass make equal contributions to SOC; (ii) the amount of belowground biomass is known; (iii) SOC is near the equilibrium point; and (iv) the rate constants are constant. Based on this solution, rate constants were calculated for a large number of historical studies (Table 1).

SOC maintenance calculations are very sensitive to the root-to-shoot ratio (Fig. 3), which is impacted by crop type. Johnson et al. (2006) used root to shoot ratios of 0.82, 0.55, and 0.62 for wheat (*Triticum aestivum* L.), corn, and soybean (*Glycine max* L. Merr.), respectively, whereas Amos and Walters (2006) reported

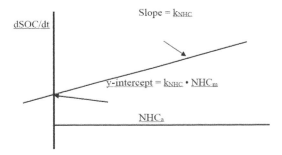

Fig. 2. Graphical representation showing the relationship between the change in soil organic carbon with time (dsoc/dt) and amount of nonharvested carbon returned to soil.

Table 1. The calculated percentages of soil organic carbon (SOC), using the nonisotopic approach that must be returned annually to maintain SOC. Root to shoot ratios for corn, soybean, and wheat were identical to the values reported in Johnson et al. (2006). The initial soil organic C (SOC_i) was from the 0- to 15-cm soil depth.

Location	Tillage	Landscape Position of soil	SOC_i†	% SOC Returned	References
			kg ha^{-1}		
Rosemont, MN	no-till		55,600	5.35	Allmaras et al. (2004)
	Chisel			9.41	
	Plow			23.31	
Lafayette, IN	Plow		34,900	18.3	Barber (1979)
Morris, MN	Plow		48,400	15.9	Wilts et al. (2004)
Iowa	Plow		26,750	16	Larson et al. (1972)
Kentucky	Plow		28,270	17.7	Frye and Blevins (1997)
Moody, SD	Strip till	Footslope	47,100	9.9	Clay et al. (2005)
		Lower backslope	46,700	9.9	
		Backslope	44,000	9.9	
		Upper backslope	43,600	9.9	
		Shoulder or summit	46,700	9.9	
Sterling, CO	No-tillage	Footslope	14,210	8.72	Peterson and Westfall (1997)
		Backslope	12,980	8.26	
		Summit	18,530	8.54	
		Backslope	14,500	9.26	
		Summit	13,570	9.66	
		Footslope	6180	8.54	
		Summit	16,810	8.04	
Southern Brazil	No-tillage or plowed	Oxisol	40,180	5.4	Sisti et al. (2004)

†SOC_i estimated from the 0- to 15-cm depth were calculated by multiplying the 0 to 5 depth by 2.

that root to shoot ratios increased with N and P deficiencies and decreased with increasing water stress, population, shade, and soil compaction. Sensitivity analysis showed that the amount of corn stover that could be harvested increased with the root to shoot ratio. If roots were not considered in the NHC value, then the estimated amount of aboveground biomass that could be safely harvested was 35%, whereas if the root to shoot ratio was 1.00 then 70% of the aboveground biomass could be harvested. Drainage class, tile drainage, soil characteristics, tillage, and initial SOC levels can also impact SOC maintenance requirements (Arrouays and Pelissier, 1994; Zach et al., 2006; Clay et al., 2007). Tillage and installing tile drainage generally increase SOC mineralization (West and Post, 2002; Clay et al., 2006, 2012a). Landscape position also impacts C turnover by influencing the amount of C contained in the soil, soil water contents, productivity, and the relative age of the SOC (Campbell et al., 2005; Clay et al., 2005). Campbell et al. (2005) reported that in Colorado, soil organic C gains increased with cropping intensity and tended to be highest in the lowest evaporation sites. Others have reported

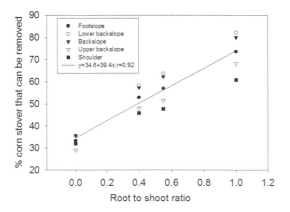

Fig. 3. Relationship between root to shoot ratio and the amount of above ground biomass that can be harvested and still maintain the soil organic carbon (SOC) level at the current level.

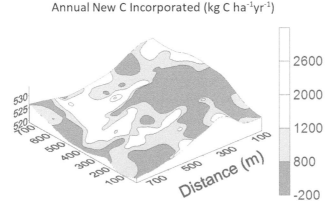

Fig. 4. Landscape position influence on annual carbon additions from 1995 to 2003 (Clay et al., 2005).

that footslope areas may have higher turnover rates than summit shoulder areas (Campbell et al., 2005, Soon and Malhi, 2005; Clay et al., 2005). Landscape differences can result from two interrelated factors, higher soil water contents and larger SOC concentrations in footslope than summit or shoulder areas (Clay et al., 2001). In addition, C contained in summit or shoulder areas may be more resistant to mineralization than C from footslope areas (Clay et al., 2005, Fig. 4). Similar contour maps can be developed for mineralized C and the amount of relic carbon remaining in the soil after mineralization. Based on these maps, the data can be aggregated into landscape positions (Table 2) and precision conservation management recommendations implemented.

Using ^{13}C fractionation to assess turnover

A common technique for assessing C turnover in production fields is the ^{13}C natural abundance approach (Ågren et al., 1996; Balesdent and Mariotti, 1996; Natelhoffer and Fry, 1988; Collins et al., 1999; Kuzyakov, 2001; Baisden et al., 2002; Ekblad et al., 2002; Fernandez and Cadisch, 2003; Griebler et al., 2004; Cheng et al., 2005). The ^{13}C isotopic approach is based on soil, C_4, and C_3 plants having different δ^{13}C signatures. This approach can be used to determine the amount of NHC remaining in soil, SOC half-lives, and SOC turnover because relic SOC and new plant material additions have different isotopic values. When making these calculations, it is important to consider that aboveground and belowground carbon inputs may have different isotopic signatures. For example, plant roots are often ^{13}C-enriched compared to plant leaves (Badeck et al., 2005, Bowling et al., 2008). Furthermore, mycorrhizal fungi are frequently ^{13}C-enriched compared to host plant leaves, probably because mycorrhizal fungi receive ^{13}C-enriched carbon from the host plant.

The ^{13}C natural abundance isotopic carbon budget approach is based on C_3 plants having lower δ^{13}C value than C_4 plants (Ehleringer, 1991; Clay et al., 2006) and that the signatures can be tracked by placing C_3 plant residue into a soil derived from C_4 plants or vice versa. In these calculations, the delta ^{13}C value is calculated with the equation,

$$\delta^{13}C = [R(sample)/R(standard)-1] \times 1000‰ \qquad [9]$$

where, R (sample) is the ratio of ^{13}C and ^{12}C in the sample, and R(standard) is the reference materials, that should be specified (Craig, 1957). Using mass balance relationships, the δ^{13}C values in a soil sample and total carbon in soil can be defined by the equations,

Table 2. The influence of landscape position and ^{13}C fractionation on calculated half-lives of soil organic carbon (SOC) at the Moody field (modified from Clay et al., 2007).

Landscape position	^{13}C fraction considered	
	No	Yes
	————yr————	
Footslope	49.8	89.1
Lower backslope	56.1	87.8
Backslope	113.1	232
Upper backslope	181	341
Shoulder or summit	78.9	151

$$\delta^{13}C_{\text{soil final}} = \frac{\left[PCR_{\text{incorp}} \left(\delta^{13}C_{PCR} \right) + SOC_{\text{retained}} \left(\delta^{13}C_{\text{SOC retained}} \right) \right]}{\left(PCR_{\text{incorp}} + SOC_{\text{retained}} \right)} \qquad [10]$$

$$SOC_{\text{final}} = PCR_{\text{incorp}} + SOC_{\text{retained}} \qquad [11]$$

$$SOC_{\text{initial}} = SOC_{\text{retained}} + SOC_{\text{lost}} \qquad [12]$$

In these equations, SOC_{initial} was the SOC in the soil at the beginning of the experiment, SOC_{final} was SOC at the end of the study, $\delta^{13}C_{\text{soil final}}$ was the $\delta^{13}C$ value of SOC when the experiment was completed, PCR_{incorp} was the new plant carbon incorporated into SOC, $\delta^{13}C_{PCR}$ was the $\delta^{13}C$ value of the plant material retained in the soil after mineralization, SOC_{retained} was the amount of relic C (SOC_{initial}) retained in the soil at the end of the study, and $\delta^{13}C_{\text{SOC retained}}$ was the associated $\delta^{13}C$ value. The equations,

$$SOC_{\text{retained}} = \frac{\left[SOC_{\text{final}} \left(\delta^{13}C_{\text{soil final}} - \delta^{13}C_{PCR} \right) \right]}{\left(\delta^{13}C_{\text{SOC retained}} - \delta^{13}C_{PCR} \right)} \qquad [13]$$

$$PCR_{\text{incorp}} = \frac{SOC_{\text{final}} \left(\delta^{13}C_{\text{soil final}} - \delta^{13}C_{\text{SOC retained}} \right)}{\left(\delta^{13}C_{PCR} - \delta^{13}C_{\text{SOC retained}} \right)} \qquad [14]$$

were derived by simultaneously solving these equations. Equations 13 and 14 have been modified by making a variety of assumptions. The first assumption is that ^{13}C fractionation during SOC and plant carbon retained (PCR) mineralization is minimal, that is, $\delta^{13}C$ SOC retained = $\delta^{13}C$ soil initial and $\delta^{13}C$ PCR = $\delta^{13}C$ plant. Based on these assumptions, the equation

$$PCR_{\text{incorp}} = \frac{SOC_{\text{final}} \left(\delta^{13}C_{\text{soil final}} - \delta^{13}C_{\text{soil initial}} \right)}{\left(\delta^{13}C_{\text{plant}} - \delta^{13}C_{\text{soil initial}} \right)} \qquad [15]$$

was derived. Solving Eq. [15] required that soil and plant material be collected at time zero ($\delta^{13}C_{\text{soil initial}}$ and $\delta^{13}C_{\text{plant}}$) and at the end of the experiment (SOC_{final} and $\delta^{13}C_{\text{soil final}}$). Equation [15] can be reorganized into,

$$\frac{PCR_{\text{incorp}}}{SOC_{\text{final}}} = \frac{\left(\delta^{13}C_{\text{soil final}} - \delta^{13}C_{\text{soil initial}} \right)}{\left(\delta^{13}C_{\text{plant}} - \delta^{13}C_{\text{soil initial}} \right)} \qquad [16]$$

Equation [16] can be modified by replacing $\delta^{13}C_{\text{soil initial}}$ with δ_{c3}, $\delta^{13}C_{\text{plant}}$ with δc_4, and $\delta^{13}C_{\text{soil final}}$ with δ. Based on these modifications, the equations,

$$p = \frac{\delta - \delta_{c3}}{\delta_{c4} - \delta_{c3}} \qquad\qquad [17]$$

$$\delta = p\,\delta_{c4} - (1-p)\,\delta_{c3} \qquad\qquad [18]$$

reported by Wolf et al. (1994) were derived. The p and δ equations are based on the assumption that ^{13}C discrimination during SOC and nonharvested biomass mineralization is minimal. Extreme care must be used when using equations 16, 17, and 18 because they assume that ^{13}C fractionation during SOC and PCR mineralization is insignificant (Ehleringer et al., 2000; Clay et al., 2007). Research has shown that these assumptions can produce large errors (Clay et al., 2007, 2008). This error increases with the length of the experiment, and is the result of biological processes discriminating against the ^{13}C isotope.

The assumption of minimal ^{13}C fractionation in fresh biomass may be appropriate. Boutton (1996) stated that, "Direct measurements indicate that the $\delta^{13}C_{\text{PDB}}$ of plant tissue remains relatively constant during the early stages of decomposition (1–7 yr)." The apparent lack of ^{13}C enrichment during the early stages of nonharvested biomass mineralization may result from two independent processes cancelling each other out. The first factor is that many SOC consumers tend to accumulate ^{13}C. The second factor is that resistant materials (waxes and lignin) tend to be depleted in ^{13}C (Lichtfouse et al., 1995; Boutton, 1996; Huang et al., 1999; Conte et al., 2003).

If ^{13}C fractionation during SOC mineralization occurs, not accounting for fractionation when C_4 plant materials are added to soil can result in overestimating the importance of C_4 plant material and underestimating the half-life of the relic carbon. For C_3 plants, the reverse was true. There are three approaches to account for ^{13}C isotopic fractionation. The first approach is to use a model such as Century to calculate the amount of fractionation (http://nrel.colostate.edu/projects/century5/reference/html/Century/labeled-C.htm).

The second approach is to include no-plant control areas where the fractionation is measured and ultimately integrated into the calculations. The third approach is to estimate fractionation based on previously measured values (Clay et al., 2006). ^{13}C fractionation during SOC mineralization can be integrated into equations 13 and 14 using the equation,

$$\delta^{13}C_{\text{SOC retained}} = \delta^{13}C_{\text{soil initial}} + \varepsilon_{\text{SOC}} \cdot \ln(SOC_{\text{retained}}/SOC_{\text{initial}}) \qquad [19]$$

where, ε_{SOC} was the Rayleigh fractionation constant for the relic SOC. If fractionation occurs during fresh biomass mineralization, a similar equation can be used. The Rayleigh fractionation constant of the soil organic carbon (ε_{SOC}) is calculated from plots where plant growth is prevented. This equation has been used to explain ^{13}C fractionation in a variety of systems (Balesdent and Mariotti, 1996; Accoe et al., 2002; Fukada et al., 2003; Spence et al., 2005; Wynn et al., 2005). Once ^{13}C isotopic fractionation is determined, carbon budgets are determined by using

Eq. [19] to calculate $\delta^{13}C_{SOC\ retained}$, which is then used to calculate plant C remaining (PCR) in the soil (Eq. [15]).

After the temporal changes in the size of different soil organic C components are determined, the first order rate constant (k), half-life, and mean residence time can be calculated using the equations,

$$k = -\frac{\ln\left(SOC_{remaining} / SOC_{initial}\right)}{number\ of\ years} \qquad [20]$$

$$t_{half\text{-}life} = -\frac{\ln(2)}{k} \qquad [21]$$

$$Mean\ residence\ time = \frac{1}{k} \qquad [22]$$

It is important to remember that assuming that isotopic fractionation does not occur, can result in large errors in the calculated amount of plant carbon retained in the soil (PCR) and the amount of relic carbon retained in the soil (Table 3). Clay et al. (2015) used the isotopic approach to determine soil organic C turnover in the 0- to 15-, 15- to 30-, and 30- to 60-cm soil depths. This work showed that C turnover slowed with increasing soil depth and that harvesting surface residue influenced belowground C turnover. Sample calculations for this approach are available in Clay et al. (2016b).

Precision Conservation and Crop Residue Harvesting

Crop residues and associated food products have been harvested for a multitude of reasons ranging from providing food, livestock feed, and energy for heating homes or cooking food (Garcia and Kalscheur, 2006; Clay et al., 2016a; Carlson et al., 2010). The removal of these materials can result in increased erosion and reduced productivity (Wilhelm et al., 2004; West and Post, 2002; Qin et al., 2015). For example, when the U.S. Great Plains was homesteaded, prairies were plowed and the grains and associated crop residues were harvested. The net result was extensive erosion and a 40 to 50% reduction in the soil organic matter in the 20th century (Clay et al., 2012a, 2012b, 2015, 2016a). Associated with this loss was the mining of soil nutrients, and the increased use of inorganic fertilizer. Of the

Table 3. The potential impact of not considering ^{13}C fractionation during soil organic carbon (SOC) mineralization on amount of relic carbon remaining in the soil and the incorporation of plant carbon (PCR) into the SOC. The soil contained at the end of the experiment contained 50,000 kg SOC ha^{-1}.

	PCR	soil initial	soil final	SOC retained	PCR/SOC	PCR incorporation	Relic carbon
			‰			kg ha^{-1}	
no-fract	-13	-16.50	-15.50	-16.25	0.29	14,286	35,714
fract	-13	-16.50	-15.50	-16.25	0.23	11,538	38,462

numerous studies that have investigated the impacts of residue harvesting on soil resilience and adaptability, few have provided the information needed to calculate soil organic C and nonharvested C rate kinetics (West and Post, 2002; Clay et al., 2010). Soil organic carbon rate kinetics information is needed to directly calculate SOC maintenance requirements.

The maintenance or the enhancement of SOC is a critical component of building resilience to perturbations such as drought, reducing erosion, and increasing yield. The impact of crop residue harvesting on productivity and soil health is influenced by many factors including the length of time that the field was cultivated, the crop rotation, soil texture, climatic conditions, initial carbon content of the soil, and crop yields. Because mineralization of crop residue to CO_2 is a biologically mediated process, factors that stimulate soil microbial activity reduce carbon sequestration.

The process of converting crop residues to soil organic matter helps the soil's living organisms recycle critical nutrients, builds and enhances soil structure, increases water infiltration, and produces conditions needed to sustain the production of food, feed, fiber, and energy. As with all living organisms, soil organisms require food (organic matter) to sustain their growth and development. Because soil organic C degradation is a biological process, the degradation process is influenced by complex interactions between the numerous factors impacting living organisms including temperature, moisture, quality of the material added to the soil, management, and texture.

The SOC maintenance requirement is dependent on many factors including climate, soil properties, and management. Clay et al. (2015) showed that yield goal and tillage both impacted the SOC maintenance requirements. In the moderate yield (between 8.9 to 11 Mg ha^{-1} or 150 and 180 bu per acre) continuous corn tilled system the amount of carbon contained in the soil could only be maintained if all of the carbon was returned, whereas in a high yield (> 11 Mg ha^{-1} or > 180 bu per acre) continuous corn rotation, 60% of the stover could be removed without producing a loss of soil organic matter. Across landscapes, the moderate yield zone would be associated water stress that reduces yields in summit or shoulder areas, whereas high yield zones would be associated with lower backslope areas. The mineralization of corn residue in the surface and subsurface soils were very different. In the surface soil, between 15 and 20% of the fresh biomass remained in the soil after 1 yr, whereas in the 15- to 30-cm soil depth approximately 68% of the roots remained in the soil after 1 yr. In landscapes with differential yield goals, these findings indicate that yields have a direct impact on C sequestration, and that C sequestration can be accelerated by increasing the yield. In addition, Clay et al. (2015) showed that: i) residue harvesting increased the conversion of fresh biomass to CO_2 and did not influence the conversion of SOC to CO_2, and ii) less SOC was mineralized in the high yield zone than the low yield zone, which in turn reduced the maintenance requirement.

Higher SOC levels have many benefits including increased productivity, higher cation exchange capacities, and more plant available water (Morachan et al., 1972; Fig. 5). In soil, where yields are controlled by available water, Fig. 5 suggests that there is a feedback loop between soil organic C and productivity. As SOC increases, the soil productivity increases, which increases SOC and the yield potential.

Carbon budget information can be used to predict the impact of a range of crop residue harvesting on soil health (Clay et al., 2005). Across the United States crop residues are being harvested for either livestock feed or bedding. In addition to these uses, crop residues can be harvested for ethanol production or fuel. The wide scale harvesting of crop residues, can have detrimental impact on soil health and result in increased erosion (Delgado and Berry, 2008; Cruse and Herndl, 2009; Peterson and Westfall, 1997). Crop residue protects the soil by increasing water infiltration and aggregate stability, while reducing erosion. The impact of residue removal on soil productivity and erosion is related to many factors including soil erodibility, slope, and landscape position (Thomas et al., 2011; Meki et al., 2011). Predicting the impact of any specific management practice on erosion and carbon sequestration is complicated by spatial variability.

In many fields, SOC has high spatial dependency with shoulder soils containing less SOC than lower footslope soils (Ritchie et al., 2007). SOC patterns can be produced by differential erosion, different amounts of carbon returned to the soil annually, management, and differential mineralization rates (Ritchie et al., 2007). For example, Clay et al. (2006, 2007) reported that in a South Dakota field soils in the lower backslope position (523.4–527.3 m) have higher SOC and shorter SOC half live than soil from the summit area (Table 4). Differences between the landscape positions were attributed to drainage that was installed in the footslope area and a lower SOC mineralization rate constant in shoulder than lower backslope soils.

Similar results were reported by Clay et al. (2011) for long-term studies conducted in Colorado (Sherrod et al., 2003). At Stratton, soil from the summit had a half-life of 26.8 yr, while SOC from the footslope had a half-life of 20.1 yr. Differences in SOC contents and k values resulted in higher maintenance requirements in the toeslope (2597 kg ha^{-1} yr) than the summit (1373 kg ha^{-1} yr). Landscape differences in SOC and mineralization rates complicate the development of sustainable management plans across watersheds and suggest that the SOC maintenance requirement is higher in footslope than lower backslope areas. We believe that to improve sustainability both erosion and SOC maintenance must be considered. In many fields, this means that the SOC level from soil with a high erosion risk must be increased. This is accomplished by applying more NHC than the SOC maintenance requirement.

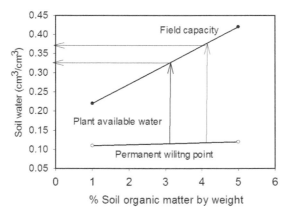

Fig. 5. Relationship between soil carbon content and plant available water for a silt loam soil (Clay et al., 2014).

Table 4. The impact of landscape position on soil organic carbon (SOC) and associated half-life (modified from Clay et al., 2007). These values were calculated using the ^{13}C isotopic fractionation approach. The 523.4- to 527.3-m zone was the lower backslope while the 532.2- to 532.2-m zone was the shoulder area.

Elevation zone	SOC	Half-life	k
m	Mg ha^{-1}	yr	g: g yr
523.4–527.3	52.2	97.8	0.007087394
532.2–532.2	49.1	341	0.00203269

Examples Estimating SOC Turnover

Example 1: Estimating SOC grains making several assumptions

The rate of SOC increase can be estimated by measuring the current SOC level and making several assumptions. For example, if SOC is 25 Mg ha^{-1} and k_{SOC} is 0.02 g (g SOC· year)$^{-1}$ and k_{NHC} is 0.2 g (g SOC· year)$^{-1}$, and the amount of non-harvested carbon returned to the field annually is 5000 kg ha^{-1}, then 500 kg C [(5000·0.2)– (25,000·0.02)] will be added to the SOC. Applying less NHC than the maintenance requirement results in a decrease in SOC.

Example 2: Using precision conservation to maintain SOC

Prior research has shown that landscape position influences SOC turnover. In this example, calculate SOC maintenance for shoulder and lower backslope positions based on the following data. The maintenance requirement can be calculated with the equation, $NHC_m = (SOC) \times (k_{SOC}/k_{NHC})$.

Shoulder lower backslope

k_{SOC} 0.014 g (g×year)$^{-1}$ 0.017 g (g ×year)

k_{NHC} 0.20 g (g × year)$^{-1}$ 0.20 g (g × year)$^{-1}$

SOC 20,000 kg SOC ha^{-1} 25,000 kg SOC ha^{-1}

$SOC_{maintenance}$ = 20,000 × (0.014/0.2) = 1400 = 25,000 × (0.017/0.20) = 2125

These calculations suggest that the maintenance requirements for the two landscape positions are different. However, this calculation does not consider the importance of increasing SOC in the shoulder area. The failure to protect the soil surface and maintain the SOC can produce severe consequences. Additional examples are provided in Clay et al. (2012b, 2016b).

Summary

Even though sensitivity analysis of carbon budget maintenance equations shows that belowground biomass estimates influence SOC maintenance rate calculations, most experiments do a poor job of estimating belowground biomass. The minimum data required to estimate SOC maintenance requirements is $SOC_{initial}$, SOC_{final}, and the amount of nonharvested carbon returned to the soil during the study period. This minimum data set is not available for most studies. Belowground biomass is generally estimated using root to shoot ratios. The impact of the root to shoot ratios on calculated maintenance requirements is important because root to shoot ratios are highly variable and almost always underestimate belowground biomass. Amos and Walters (2006) reported that the net belowground C deposition in corn at physiological maturity was 29 ± 13% of the shoot biomass (leaves, stems, husks) in 41 studies. The use of these values is further complicated by the use of different definitions for root to shoot ratio (Johnson et al., 2006).

For ^{13}C isotopic approaches, accurate measurements of ^{13}C fractionation in the bulk soil are needed. Carbon-13 isotopic discrimination can be measured in no-plant control areas. These plots are used to measure ^{13}C enrichment of the relic carbon during the experimental timeframe. Once the carbon budgets and rate constants are known, this information can combined in measured spatial variability to estimate the impact of precision conservation on changes in SOC.

Acknowledgments

Support for this project was provided by the South Dakota Corn Utilization Council, South Dakota Soybean Research and Promotion Council, National Resources Conservation Service, National Institute of Food and Agriculture seed technology grant, South Dakota 2010 initiative, USDA Sustainable Agriculture Research and Education (enc07-095, 2007-47001-03883), and NASA.

References

Accoe, F., P. Boeckx, O. Van-Cleemput, G. Hofman, Y. Zhang, R. Li, and C. Guanxiong. 2002. Evolution of the delta ^{13}C signature related to total carbon content and carbon decomposition rate constants in a soil profile under grassland. Rapid Commun. Mass Spectrom. 16:2184–2189. doi:10.1002/rcm.767

Ågren, G.I., E. Bosatta, and J. Balesdent. 1996. Isotope discrimination during decomposition of organic matter: A theoretical analysis. Soil Sci. Soc. Am. J. 60:1121–1126. doi:10.2136/sssaj1996.03615995006000040023x

Allmaras, R.R., D.R. Linden, and C.E. Clapp. 2004. Corn-residue transformation into root and soil carbon as related to nitrogen, tillage, and stover management. Soil Sci. Soc. Am. J. 68:1366–1375. doi:10.2136/sssaj2004.1366

Amos, B., and D.T. Walters. 2006. Maize root biomass and net rhizodeposited carbon: An analysis of the literature. Soil Sci. Soc. Am. J. 70:1489–1503. doi:10.2136/sssaj2005.0216

Anderson, J.P.E. 1982. Soil respiration. In: A.L. Page, R. Miller, and D.R. Kenny, editors, Methods of soil analysis, chemical and microbiological properties, part II. 2nd ed. Agron. Monogr. 9.2. SSSA, Madison, WI. p. 831-871.

Arrouays, D., and P. Pelissier. 1994. Changes in carbon storage in temperate humic loamy soils after forest clearing and continuous cropping in France. Plant Soil 160:215–223. doi:10.1007/BF00010147

Badeck, F.W., G. Tcherkez, S. Nogués, C. Piel, and J. Ghashghaie. 2005. Post-photosynthetic fractionation of stable carbon isotopes between plant organs: A widespread phenomenon. Rapid Commun. Mass Spectrom. 19:1381–1391. doi:10.1002/rcm.1912

Baisden, W. T., R. Amundson, D.L. Brenner, A.C. Cook, C. Kendall, and J.W. Harden. 2002. A multi-isotope C and N modeling analysis of soil organic matter turnover and transport as a function of soil depth in a California annual grassland soil chronosequence. Global Biogeochem. Cycles 16:82-1-82-26 doi:10.1029/2001GB001823.

Balesdent, J., and A. Mariotti. 1996. Measurement of soil organic matter turnover using 13C natural abundance. In: T.W. Boutton and S. Yamasaki, editors, Mass spectrometry of soils. Marcel Dekker, Inc., New York. p. 83–112.

Balesdent, J., G.H. Wagner, and A. Mariotti. 1988. Soil organic matter turnover in long-term experiments as revealed by carbon-13 natural abundance. Soil Sci. Soc. Am. J. 52:118–124. doi:10.2136/sssaj1988.03615995005200010021x

Barber, S.A. 1971. Effect of tillage practice on corn (Zea mays L.) root distribution and morphology. Agron. J. 63:724–726. doi:10.2134/agronj1971.00021962006300050020x

Barber, S.A. 1979. Corn residue management and soil organic matter. Agron. J. 71:625–627. doi:10.2134/agronj1979.00021962007100040025x

Barber, S.A., and J.K. Martin. 1976. The release of organic substance by cereal roots in the soil. New Phytol. 76:69–80. doi:10.1111/j.1469-8137.1976.tb01439.x

Bolinder, M.A., H.H. Janzen, E.G. Gregorich, D.A. Angers, and A.J. VandenBygaart. 2007. An approach for estimating net primary productivity and annual carbon inputs to soil for common agricultural crops in Canada. Agric. Ecosyst. Environ. 118:29–42. doi:10.1016/j.agee.2006.05.013

Boutton, T.W. 1996. Stable carbon isotopes ratios of soil organic matter and their use as indicators of vegetation and climate change. In: T.W. Boutton and S. Yamasaki, editors, Mass spectrometry of soils. Marcel Dekker, Inc., New York. p. 47–82.

Bowling, D.R., D.E. Pataki, and J.T. Randerson. 2008. Carbon isotopes in terrestrial ecosystem pools and CO_2 fluxes. New Phytol. 178:24–40. doi:10.1111/j.1469-8137.2007.02342.x

Bradford, J.B., W.K. Lauenroth, and I.C. Burke. 2005. The impact of cropping on net primary production in the U. S. Great Plains. Ecology 86:1863–1872. doi:10.1890/04-0493

Campbell, C.A., H.H. Janzen, K. Paultian, E.G. Gregorich, L. Sherrod, B.C. Liang, and R.P. Zentner. 2005. Carbon storage in soil of the North American Great Plains: Effect of cropping frequency. Agron. J. 97:349–363. doi:10.2134/agronj2005.0349

Carlson, C.G., D.E. Clay, C. Wright, and K.D. Reitsma. 2010. Potential impacts of linking ethanol, crop production, and backgrounding calves on economics, carbon, and nutrient budgets. SDSU Extension Publication. Brookings, SD.

Causarano, H.J., A.J. Franzluebbers, D.W. Reeves, and J.N. Shaw. 2006. Soil organic carbon sequestration in cotton production systems of southeastern United States: A review. J. Environ. Qual. 35:1374–1383. doi:10.2134/jeq2005.0150

Chang, J., D.E. Clay, S.A. Hansen, S.A. Clay, and T. Schumacher. 2014. Water stress impacts on transgenic corn in the northern Great Plains. Agron. J. 106:125–130. doi:10.2134/agronj2013.0076

Cheng, W., S. Fu, R.B. Susfalk, and R.J. Mitchell. 2005. Measuring tree root respiration using 13C natural abundance: Rooting medium matters. New Phytol. 167:297–307. doi:10.1111/j.1469-8137.2005.01427.x

Clapp, C.E., R.R. Allmaras, M.F. Layese, D.R. Linden, and R.H. Dowdy. 2000. Soil organic carbon and 13C abundance as related to tillage, crop residue, and nitrogen fertilizer under continuous corn management in Minnesota. Soil Tillage Res. 55:127–142. doi:10.1016/S0167-1987(00)00110-0

Clay, D.E., C.G. Carlson, S.A. Clay, J. Chang, and D.D. Malo. 2005. Soil organic C maintenance in a corn (*Zea mays* L.) and soybean (*Glycine max* L.) as influenced by elevation zone. J. Soil Water Conserv. 60:342–348.

Clay, D.E., C.G. Carlson, T.E. Schumacher, V. Owens, and F. Mamani Pati. 2010. Biomass estimation approach impacts on calculated SOC maintenance requirements and associated mineralization rate constants. J. Environ. Qual. 39:784–790. doi:10.2134/jeq2009.0321

Clay, D.E., C.G. Carlson, S.A. Clay, C. Reese, Z. Liu, and M.M. Ellsbury. 2006. Theoretical derivation of new stable and non-isotopic approaches for assessing soil organic C turnover. Agron. J. 98:443–450. doi:10.2134/agronj2005.0066

Clay, D.E., C.G. Carlson, and S.A. Clay. 2011. Maximizing nutrient efficiency through the adoption of management practices that maintain soil organic carbon, calculating carbon turnover kinetics. In: D.E. Clay and J. Shanahan, editors, GIS in agriculture: Nutrient management for improved energy efficiency. CRC Press, Boca Raton, FL. p. 191–208.

Clay, D.E., C.G. Carlson, and S.A. Clay. 2008. Calculating site-specific carbon budgets: Carbon footprints and implications of different assumptions. Proceedings of 9th International Conference of Precision Farming. Denver CO. July 20-23.

Clay, D.E., C.E. Clapp, C. Reese, Z. Liu, C.G. Carlson, H. Woodard, and A. Bly. 2007. 13C fractionation of relic soil organic C during mineralization effects calculated half-lives. Soil Sci. Soc. Am. J. 71:1003–1009. doi:10.2136/sssaj2006.0193

Clay, D.E., J. Chang, S.A. Clay, J.J. Stone, R.H. Gelderman, C.G. Carlson, K. Reitsma, M. Jones, L. Janssen, and T. Schumacher. 2012a. Corn yield increases and no-tillage affects carbon sequestration and carbon footprint. Agron. J. 104:763–770. doi:10.2134/agronj2011.0353

Clay, D.E., S.A. Clay, C.G. Carlson, and S. Murrell. 2012b. Mathematics and calculations for agronomists and soil scientists. International Plant Nutrition Institute, Peachtree Corners, GA.

Clay, D.E., S.A. Clay, Z. Lui, and C. Reese. 2001. Spatial variability of C-13 isotopic discrimination in corn (Zea mays). Commun. Soil Sci. Plant Anal. 32:1813–1827. doi:10.1081/CSS-120000252

Clay, D., S. Clay, K. Reitsma, B. Dunn, A. Smart, G. Carlson, D. Horvath, and J. Stone. 2014. Does the conversion of grasslands to row crop production in semi-arid areas threaten global food security? Glob. Food Secur. 3:22–30. doi:10.1016/j.gfs.2013.12.002

Clay, D.E., T. M. DeSutter, S.A. Clay, and C. Reese. 2016a. From plows, horses, and harnesses to precision technologies in the north American Great Plains. Oxford University Press, Oxford, United Kingdom.

Clay, D.E., G. Reicks, C.G. Carlson, J. Miller, J.J. Stone, and S.A. Clay. 2015. Residue harvesting and yield zone impacts C storage in a continuous corn rotation. J. Environ. Qual. doi:102134/jeq2014.070322.

Clay, D.E., C. Reese, S.A.H. Bruggeman, and J. Moriles-Miller. 2016b. The use of enriched and natural abundance nitrogen and carbon isotopes in soil fertility research. In: A. Chatterjee and D. Clay, editors, Soil fertility management in agroecosystems. ASA, CSSA, SSSA, Madison, WI. p. 1-13.

Coleman, K., and D.S. Jenkinson. 1996. RothC-26.3: A model for the turnover of carbon in soil. In: D.S. Powlson, P. Smith, and J.U. Smith, editors, Evaluation of soil organic matter models using existing long-term datasets, NATO ASI Series I, 38:237-246.

Collins, H.P., R.L. Blevins, L.G. Bundy, D.R. Christenson, W.A. Dick, D.R. Huggins, and E.A. Paul. 1999. Soil carbon dynamics in corn-based agroecosystems: Results from carbon-13 natural abundance. Soil Sci. Soc. Am. J. 63:584–591. doi:10.2136/sssaj1999.03615995006300030022x

Conte, M.H., J.C. Weber, P.J. Carlson, and L. B. Flanagan. 2003. Molecular and carbon isotope composition of leaf waste in vegetation and aerosols in northern prairie ecosystem. Oecologia 135:67–77. doi:10.1007/s00442-002-1157-4

Craig, H. 1957. Isotopic standards for carbon and oxygen and correction factors for mass spectrometric analysis of carbon dioxide. Geochim. Cosmochim. Acta 12:133–149. doi:10.1016/0016-7037(57)90024-8

Cruse, R.M., and C.G. Herndl. 2009. Balancing corn stover harvest for biofuels with soil and water conservation. J. Soil Water Conserv. 64:286–291. doi:10.2489/jswc.64.4.286

Delgado, J.A., and J.K. Berry. 2008. Advances in precision conservation. Adv. Agron. 98:1–44. doi:10.1016/S0065-2113(08)00201-0

Dwyer, L.M., B.L. Ma, D.W. Stewart, N.N. Hayhoe, D. Balchin, J.L.B. Culley, and M. McGovern. 1996. Root mass distribution under conventional and conservation tillage. Can. J. Soil Sci. 76:23–28. doi:10.4141/cjss96-004

Ehleringer, J.R. 1991. 13C/12C fractionation and its utility in terrestrial plant studies. In: D. Coleman and B. Fry, editors, Carbon isotopic techniques. Academic Press, New York. p. 187-200.

Ehleringer, J.R., N. Buchmann, and L.B. Flanagan. 2000. Carbon isotope ratios in belowground carbon cycle processes. Ecol. Appl. 10:412–422. doi:10.1890/1051-0761(2000)010[0412:CIRIBC]2.0.CO;2

Ekblad, A., G. Nyberg, and P. Högberg. 2002. 13C-discrimination during microbial respiration of added C_3-, C_4- and ^{13}C-labelled sugars to a C_3-forest soil. Oecologia 131:245–249. doi:10.1007/s00442-002-0869-9

Fernandez, I., and G. Cadisch. 2003. Discrimination against ^{13}C during degradation of simple and complex substrates of two white rot fungi. Rapid Commun. Mass Spectrom. 17:2614–2620. doi:10.1002/rcm.1234

Follett, R.F., R.R. Allmaras, and G.A. Reichman. 1974. Distribution of corn roots in a sandy soil with a declining water table. Agron. J. 66:288–292. doi:10.2134/agronj1974.0002196 2006600020030x

Frye, W.W., and R.L. Blevins. 1997. Soil organic matter under long-term no-tillage and conventional tillage corn production in Kentucky. In: E.A. Paul, K. Paustian, E.T. Elliott, and C.V. Cole, editors, Soil organic matter in temperate agroecosystems. CRC Press, Boca Raton, FL. p. 227–234.

Fukada, T., K.M. Hiscock, P.F. Dennis, and T. Grischek. 2003. A dual isotope approach to identify denitrification at a river-band infiltration site. Water Res. 37:3070–3078. doi:10.1016/S0043-1354(03)00176-3

Gale, W.J., and C.A. Cambardella. 2000. Carbon dynamics of surface residue- and root-derived organic matter under simulated no-till. Soil Sci. Soc. Am. J. 64:190–195. doi:10.2136/sssaj2000.641190x

Garcia, A.D., and K.F. Kalscheur. 2006. Ensiling wet distillers grain with other feeds. South Dakota Cooperative Extension, Extension Extra 4029, South Dakota State University, Brookings, SD.

Gilmanov, T.G., W.J. Parton, and D.S. Ojima. 1997. Testing the CENTURY ecosystem level model on data sets from eight grassland sites in the former USSR representing a wide climatic/soil gradient. Ecol. Modell. 96:191–210. doi:10.1016/S0304-3800(96)00067-1

Griebler, C., L. Adrian, R.U. Meckenstock, and H.H. Richnow. 2004. Stable carbon isotope fractionation during aerobic and anaerobic transformation of trichlorbenzene. FEMS Micro. Ecol. 48:313-321.

Grossman, R.B., and T.G. Reinsch. 2002. Bulk density and linear extensibility. In: J.H. Dane and C.G. Topp, editors, Methods of soil analysis: Part 4 physical methods. SSSA, Madison, WI. p. 201-228.

Guo, L., P. Falloon, K. Coleman, B. Zhou, Y. Li, E. Lin, and F. Zhang. 2007. Application of the RothC model to the results of long-term experiments on typical upland soils in northern China. Soil Use Manage. 23:63–70. doi:10.1111/j.1475-2743.2006.00056.x

Hanson, P.J., N.T. Edwards, C.T. Garten, and J.A. Andrews. 2000. Separating root and soil microbial contributions to soil respiration: A review of methods and observations. Biogeochemistry 48:115–146. doi:10.1023/A:1006244819642

Hébert, Y., E. Guigo, and O. Loulet. 2001. The response of root/shoot partitioning and root morphology to light reductions in maize genotypes. Crop Sci. 41:363–371. doi:10.2135/cropsci2001.412363x

Huang, Y., G. Eglinton, P. Ineson, R. Bol, and D.D. Harkness. 1999. The effects of nitrogen fertilisation and elevated CO_2 on the lipid biosynthesis and carbon isotopic discrimination in birch seedlings (Betula pendula). Plant Soil 216:35–45. doi:10.1023/A:1004771431093

Huggins, D.R., C.E. Clapp, R.R. Allmaras, J.A. Lamb, and M.F. Layese. 1998. Carbon dynamics in corn-soybean sequences as estimated from natural carbon-13 abundance. Soil Sci. Soc. Am. J. 62:195–203. doi:10.2136/sssaj1998.03615995006200010026x

Johnson, J.M.F., R.R. Allmaras, and D.C. Reicosky. 2006. Estimating source carbon from crop residues, roots, and rhizodeposits using the national grain-yield data-base. Agron. J. 98:622–636. doi:10.2134/agronj2005.0179

Kuzyakov, Y.V. 2001. Tracer studies of carbon translocation by plants from the atmosphere into the soil (A review). Eurasian Soil Sci. 34:28–42.

Kuzyakov, Y., and W. Cheng. 2001. Photosynthesis controls of rhizosphere respiration and organic matter decomposition. Soil Biol. Biochem. 33:1915–1925. doi:10.1016/S0038-0717(01)00117-1<jrn>

Kuzyakov, Y., and G. Domanski. 2000. Review Carbon inputs by plants into the soil. J. Plant Nutr. Soil Sci. 163:421–431. doi:10.1002/1522-2624(200008)163:4<421::AID-JPLN421>3.0.CO;2-R

Kuzyakov, Y., and A.A. Larionova. 2005. Root and rhizomicrobial respiration: A review of approaches to estimate respiration by autotrophic and heterotrophic organisms in soil. J. Plant Nutr. Soil Sci. 168:503–520. doi:10.1002/jpln.200421703

Kuzyakov, Y., and A.A. Larionova. 2006. Contribution of rhizomicrobial and root respiration to CO_2 emission from soil (A review). Eurasian Soil Sci. 39:753–764. doi:10.1134/S106422930607009X

Laboski, C.A.M., R.H. Dowdy, R.R. Allmaras, and J.A. Lamb. 1998. Soil strength and water content influences on corn root distribution in a sandy soil. Plant Soil 203:239–247. doi:10.1023/A:1004391104778

Larson, W.E., C.E. Clapp, W.H. Pierre, and Y.B. Morachan. 1972. Effects of increasing amounts of organic residues on continuous corn: II. Organic carbon, nitrogen, phosphorous, and sulfur. Agron. J. 64:204–208. doi:10.2134/agronj1972.00021962006400020023x

Lichtfouse, E., S. Dou, C. Girardin, M. Grably, J. Balesdent, F. Behar, and M. Vanden-broucke. 1995. Unexpected ^{13}C-enrichment of organic components from wheat crop soils: Evidence for the in situ origin of soil organic matter. Org. Geochem. 23:865–868. doi:10.1016/0146-6380(95)80009-G

Lloyd, J., and J.A. Taylor. 1994. On the temperature dependence of soil respiration. Funct. Ecol. 8:315–323. doi:10.2307/2389824

Mamani-Pati, E.M., D.E. Clay, C.G. Carlson, and S.A. Clay. 2010. Calculating soil organic carbon maintenance using stable and isotopic approaches: A review. In: E. Licht-fouse, editor, Sustainable agricultural reviews: Sociology, organic farming, climate change and soil science. Springer Netherlands, Dordrecht, the Netherlands. p. 189-216. doi:10.1007/978-90-481-3333

Meki, M.N., J.P. Marcos, J.D. Atwood, L.M. Norfleet, E.M. Steglich, J.R. Williams, and T.J. Gerik, 2011. Effects of site-specific factors on corn stover removal thresholds and sub-sequent environmental impacts in the Upper Mississippi River Basin. J. Soil and Water Cons. 66:386-399.

Martel, Y.A., and P.A. Paul. 1974. The use of radiocarbon dating of organic mat-ter in the study of soil genesis. Soil Sci. Soc. Am. J. 38:501–506. doi:10.2136/sssaj1974.03615995003800030033x

Marschner, P. 2012 Mineral Nutrition of Higher Plants. 3rd ed. Academic Press, London, United Kingdom.

Melnitchouck, A., P. Leinweber, K.U. Eckhardt, and R. Beese. 2005. Qualitative differences between day- and night-time rhizodeposition in maize (*Zea mays* L.) as investigated by pyrolysis-field ionization mass spectrometry. Soil Biol. Biochem. 37:155–162. doi:10.1016/j.soilbio.2004.06.017

Molina, J.A.E., C.E. Clapp, D.R. Linden, R.R. Allmaras, M.F. Layese, R.H. Dowdy, and H.H. Cheng. 2001. Modeling of incorporation of corn (*Zea mays*) carbon from roots and rhizodeposition into soil organic matter. Soil Biol. Biochem. 33:83–92. doi:10.1016/S0038-0717(00)00117-6

Morachan, Y.B., W.C. Modenhauer, and W.E. Larson. 1972. Effects of increasing amount of organic residues on continuous corn: 1. Yields and soil physical properties. Agron. J. 64:199–203. doi:10.2134/agronj1972.00021962006400020022x

Natelhoffer, K.J., and B. Fry. 1988. Controls on natural nitrogen-15 and carbon-13 abun-dances in forest soil organic matter. Soil Sci. Soc. Am. J. 52:1633–1640. doi:10.2136/sssaj1988.03615995005200060024x

Olk, D.C., and E.G. Gregorich. 2006. Overview of the symposium proceedings, meaningful pools in determining soil carbon and nitrogen dynamics. Soil Sci. Soc. Am. J. 70:967–974. doi:10.2136/sssaj2005.0111

Olk, D.C. 2006. A chemical fractionation for chemical structure-function relations of soil organic matter in nutrient cycling. Soil Sci. Soc. Am. J. 70:1013–1022. doi:10.2136/sssaj2005.0108

Ortega, R., G.A. Peterson, and D.G. Westfall. 2002. Residue accumulation and changes in soil organic matter as affected by cropping intensity in no-till dryland agroecosys-tems. Agron. J. 94:944–954. doi:10.2134/agronj2002.9440

Parton, W.J., D.S. Schimel, C.V. Cole, and D.S. Ojima. 1987. Analysis of factors controlling soil organic matter levels in Great Plains grasslands. Soil Sci. Soc. Am. J. 51:1173-117.

Peterson, G.A., and D.G. Westfall. 1997. Management of dryland agroecosystems in the central Great Plains of Colorado. In: E.A. Paul, C.V. Cole, K.H. Paustian, and E.T. Elliot, editors, Soil organic matter in temperate agroecosystems. CRC Press. New York. p. 371-380.

Qin, Z., C.E. Canter, J.B. Dunn, S. Mueller, H.Y. Kwon, J. Han, M.M. Wander, and M. Wang. 2015. Incorporating agricultural management practices into the assessment of soil carbon change and life-cycle greenhouse gas emissions of corn stover ethanol pro-duction (No. ANL/ESD-15/26). Argonne National Laboratory (ANL), Argonne, IL.

Ritchie, J.C., G.W. McCatrty, E.R. Venteris, and T.C. Kaspar. 2007. Soil and soil organic carbon redistribution on the landscape. Geomorphology 89:163–171. doi:10.1016/j.geomorph.2006.07.021

Rochette, P., and L.B. Flanagan. 1997. Quantifying rhizosphere respiration in a corn crop under field conditions. Soil Sci. Soc. Am. J. 61:466–474. doi:10.2136/sssaj1997.03615995006100020014x

Rochette, P., D.A. Angers, and L.B. Flanagan. 1999. Maize residue decomposition measurements using soil surface carbon dioxide fluxes and natural abundance of carbon-13. Soil Sci. Soc. Am. J. 63:1385–1396. doi:10.2136/sssaj1999.6351385x

Russell, A.E., D.A. Laird, T.B. Parkin, and A.P. Mallarino. 2005. Impact of nitrogen fertilization and cropping systems on carbon sequestration in Midwestern Mollisols. Soil Sci. Soc. Am. J. 69:413–422. doi:10.2136/sssaj2005.0413

Sherrod, L.A., G.A. Peterson, D.G. Westfall, and L.R. Ahuja. 2003. Cropping intensity enhances soil organic carbon and nitrogen in no-tillage agroecosystems. Soil Sci. Soc. Am. J. 67:1533–1543. doi:10.2136/sssaj2003.1533

Sisti, C.P.J., H.P. dos Santos, R. Kohhann, B.J.R. Alves, S. Urquiaga, and R.M. Boddey. 2004. Change in carbon and nitrogen stocks in soil under 13 years of conventional or zero tillage in Southern Brazil. Soil Tillage Res. 76:39–58. doi:10.1016/j.still.2003.08.007

Six, J., and J.D. Jastrow. 2002. Organic matter turnover. In: R. Lal, editor, Encyclopedia of Soil Science. Marcel Dekker, New York. p. 936–942.

Skjemstad, J.O., L.R. Spouncer, B. Cowie, and R.S. Swift. 2004. Calibration of the Rothamsted organic carbon turnover model (RothCver 26.3) using measurable soil organic carbon pools. Aust. J. Soil Res. 42:79–88. doi:10.1071/SR03013

Soon, Y.K., and S.S. Malhi. 2005. Soil nitrogen dynamics as affected by landscape position and nitrogen fertilizer. Can. J. Soil Sci. 85:579–587. doi:10.4141/S04-072

Spence, M.J., S.H. Bottrell, S.F. Thornton, H.H. Richnow, and K.H. Spence. 2005. Hydrochemical and isotopic effects associated with petroleum fuel biodegradation pathways in a chalk aquifer. J. Contam. Hydrol. 79:67–88. doi:10.1016/j.jconhyd.2005.06.003

Thomas, M.A., B.A. Engel, and I. Chaubey. 2011. Multiple corn stover removal rates for cellulosic biofuels and long-term water quality impacts. J. Soil Water Conserv. 66:431–444. doi:10.2489/jswc.66.6.431

VandenBygaart, B., E.G. Gregorich, and D.A. Angers. 2003. Influence of agricultural management on soil organic carbon: A compendium and assessment of Canadian studies. Can. J. Soil Sci. 83:363–380. doi:10.4141/S03-009

Varvel, G.E. 2006. Soil organic carbon changes in diversified rotations of the western Corn Belt. Agron. J. 70:426–433.

West, T.O., and W.M. Post. 2002. Soil organic carbon sequestration rates by tillage and crop rotation: A global data analysis. Soil Sci. Soc. Am. J. 66:1930–1946. doi:10.2136/sssaj2002.1930

Wichern, F., E. Eberhardt, J. Mayer, R.G. Joergensen, and T. Müller. 2008. Nitrogen rhizodeposition in agricultural crops methods, estimates and future prospects. Soil Biochemistry. 40:30–48. doi:10.1016/j.soilbio.2007.08.010

Wilhelm, W.W., J.M.F. Johnson, J.L. Hatfield, W.B. Voorhees, and D.R. Linden. 2004. Crop and soil productivity response to corn residue removal: A literature review. Agron. J. 96:1–17. doi:10.2134/agronj2004.0001

Wilts, A.R., D.C. Reicosky, R.R. Allmaras, and C.E. Clapp. 2004. Long-term corn residue effects: Harvest alternatives, soil carbon turnover, and root-derived carbon. Soil Sci. Soc. Am. J. 68:1342–1351. doi:10.2136/sssaj2004.1342

Wolf, D.C., J.O. Legg, and T.W. Boutton. 1994. Isotopic methods for the study of soil organic matter. In: P.S. Bottomley, J.S. Angle, and R.W. Weaver, editors, Methods of soil analysis. Part 2: Microbial and biochemical properties. SSSA Book Ser. 5.2. SSSA, Madison, WI. p. 865-908.

Wynn, J.G., M.I. Bird, and V.N.L. Wong. 2005. Rayleigh distillation and the depth profile of 13C/12C ratios of soil organic carbon from soils of disparate texture in Iron range National Park, Far North Queensland, Australia. Geochim. Cosmochim. Acta 69:1961–1973. doi:10.1016/j.gca.2004.09.003

Zach, A., H. Tiessen, and E. Noellemeyer. 2006. Carbon turnover and carbon-13 abundance under land use change in semi-savanna soils of La Pampa, Argentina. Soil Sci. Soc. Am. J. 70:1541–1546. doi:10.2136/sssaj2005.0119

Zimmermann, M., J. Leifeld, M.W.I. Schmidt, P. Smith, and J. Fuhrer. 2007. Measured soil organic matter fractions can be related to pools in the RothC model. Eur. J. Soil Sci. 58:658–667. doi:10.1111/j.1365-2389.2006.00855.x

Precision Conservation
for Biofuel Production

Indrajeet Chaubey,* Raj Cibin, and Qingyu Feng

Abstract

It is expected that large amounts of agricultural lands will be needed to meet desired bioenergy production goals. Considerable land-use changes associated with the biofuel production presents an unprecedented opportunity to develop sustainable landscape design plans that meet biofuel, environmental, and economic goals. Multiple sustainability indicators related to water availability, water quality, soil health, ecosystem services, and profitability are available that can be used to develop sustainable biofuel production strategies. Selection and placement of various biofuel crops considering these sustainability indicators will require either targeting of specific landscape positions or optimization in a watershed. While targeting bioenergy crops in specific landscape positions (e.g. perennial grasses grown in vegetated filter strips or highly erodible lands) provide a simple-to-use recommendation, targeting methods may not always result in optimal improvement in water quality. Complex landscape optimizations can be performed to meet one or more objective functions of land management and biofuel production; however, these methods are currently difficult to implement. There is a need to develop targeting methods that are derived from complex spatial optimization so that results are easy to implement and meet crop production and environmental and economic decision criteria. There is a need to develop precision conservation plans that meet desired sustainability and biofuel production goals.

In recent years, growing energy demands, combined with a desire to attain energy independence from fossil fuels and concerns over global climate change, have generated considerable interest in biofuel production. The United States and Brazil are the two largest producers of ethanol and biofuel, accounting for 86% of the global biofuel production in 2010 (Renewable Fuels Association, 2011). Various US policies and mandates implemented since 1992 have resulted in considerable increase in biofuel production (Fig. 1). For example, a goal to double

Abbreviations: BMP, best management practice; CRP, Conservation Reserve Program; EISA, Energy Independence and Security Act.

I. Chaubey, Professor, Dep. of Earth, Atmospheric, and Planetary Sciences, Dep. of Agricultural and Biological Engineering, Purdue Univ., West Lafayette, IN 47907. R. Cibin, Assistant Professor, 319 Forest Resource Lab, Penn State University, University Park, PA 16802 (craj@psu.edu). Q. Feng, Research Assistant, Dep. of Agricultural and Biological Engineering, Purdue Univ., West Lafayette, IN 47907 (feng37@purdue.edu). *Corresponding author (ichaubey@purdue.edu).

doi:10.2134/agronmonogr59.2013.0030

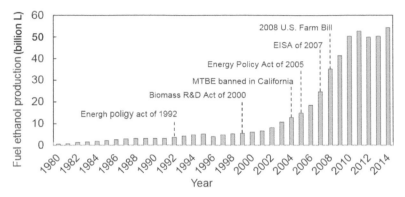

Fig. 1. Historic production of ethanol in the United States and key policies and mandates (1980–2010) based on data from the Renewable Fuels Association (http://ethanolrfa.org/pages/statistics#A).

previous years' production of ethanol to 28 billion L before 2012 was set by the Energy Policy Act of 2005. The actual production of ethanol exceeded this goal in 2008 with a production of 34 billion L. A goal of producing 136 billion L of biofuel by 2022 was set by the Energy Independence and Security Act (EISA) of 2007. Currently, corn (*Zea mays* L.) grain is the primary feedstock of ethanol and soybean [*Glycine max* (L.) Merr.] is the primary feedstock of biodiesel in the United States; however, numerous studies have indicated that grain-based ethanol production may induce competition among food, feed, and fuel (Nonhebel, 2005; Trostle, 2011) and may lead to substantial increase in losses of sediment and nutrients from production fields with unintended negative environmental impacts (Thomas et al., 2009). To reduce this competition, a production limit of 56.8 billion L of ethanol derived from grain-based feedstock has been set by the EISA with the remaining 79.5 billion L from low-input production systems.

Agriculture is expected to play a significant role in producing biofeedstocks to meet EISA goals for biofuel production. The USDA estimated that ~11 million ha of cropland would be required to achieve the EISA biofuel target (USDA, 2010). From these agricultural areas, corn production for ethanol is expected to be at 88 Tg and is expected to meet the EISA mandate by 2017. Approximately 18 Tg of soybean is expected to be the primary feedstock to produce biodiesel. In addition, cellulosic ethanol from crop residues is considered as a viable short-term source of biofuel because of its low cost, immediate availability, and relatively greater availability from the major crop producing areas (US Department of Energy, 2011). Stover from corn is considered to be the largest single source of crop residue. Straw and stubble from wheat (*Triticum aestivum* L.), barley (*Hordeum vulgare* L.), oat (*Avena sativa* L.), and sorghum [*Sorghum bicolor* (L.) Moench] is additional residue available in the United States for bioenergy production. Crop residue ranging from 80 to 180 Tg are expected to be available by 2030 depending on the price of the crop residue at the farm-gate (US Department of Energy, 2011). This availability can be potentially doubled with increase in crop yield. Secondary cropland residue and waste resources available for bioenergy production include sugarcane trash and bagasse, cotton gin trash and residues, soybean hulls, rice hulls and field residues, wheat dust and chaff, orchard and vineyard pruning residues, animal fats, and animal manure, among others. However, these secondary sources are expected to play a relatively minor role in meeting the feedstock requirements to achieve EISA goals.

Dedicated energy crops, such as perennial grasses (e.g., *Miscanthus ×giganteus* J. M. Greef & Deuter ex Hodk. & Renvoize, switchgrass [*Panicum virgatum* L.], and native prairie grasses), fast growing trees (e.g., hybrid poplar [*Populus trichocarpa* Torr. & A. Gray]), and selected annual crops (e.g., dual purpose sorghum and cover crops) can be grown in existing agricultural areas to supply large volumes of feedstock. The US Department of Energy estimates that as many as 6 to 11 million ha of current cropland and pastureland could be replaced by these dedicated energy crops with ~150 to 380 Tg of biofeedstock produced each year.

Land-use and land-management changes associated with bioenergy feedstock production can have positive and negative impacts depending on land management and crop production scenarios. Some researchers have argued that the expansion of energy crop production is likely to increase pollution, impair soils with intensive agricultural practices, and increase water use (Berndes, 2002; Groom et al., 2008). For example, harvesting of crop residue can result in excess erosion from poor surface cover, low soil nutrients including organic matter, and may require additional fertilizer input to compensate for nutrients removed with the crop residue. Gramig et al. (2013) and Cibin et al. (2012) indicated that removal of corn stover can generally decrease stream flow and losses of nitrate and dissolved P. However, greenhouse gas and sediment losses generally increase with removal of crop residue. The relative impacts are expected to vary with the residue removal rates and biophysical characteristics of the agricultural fields such as slope and soil types. Anticipated benefits of dedicated bioenergy crops include potential to reduce greenhouse gas emissions and improved water quality from low fertilizer input required for the perennial bioenergy crops (Solomon, 2010), sequestration of atmospheric C, creation of wildlife habitat, increased landscape diversity, return of agricultural marginal lands to production, and potential increase in farm revenue (US Department of Energy, 2011). Energy from biofuel is considered as C neutral, where C released with energy production is assumed to be used for energy crop production.

Coproduction of food, feed, and bioenergy crops within an agricultural landscape will require careful evaluation of economic, social, and environmental sustainability issues. A careful selection of energy crops and placement at suitable locations can minimize potential negative impacts and enhance positive impacts (Parish et al., 2012; Robertson et al., 2008). For example, precision conservation considerations, such as growing perennial grasses in environmentally critical areas having relatively greater rates of erosion or developing crop residue removal rates that consider risk of soil and nutrient losses, can reduce potential negative effects of biofuel production. Similarly, perennial grasses also can be used as best management practices (BMPs) to minimize negative hydrologic and water quality impacts from high-input, intensive row-crop production. For example, vegetated filter strips, grassed waterways, and riparian areas that can reduce pollutant load from agricultural fields to water bodies and increase biomass productivity of the BMP areas (Sahu and Gu, 2009).

Precision Conservation Planning
for Sustainable Biofuel Production

Environmental impacts and benefits of different biofuel production systems may vary considerably depending in the type of biofeedstock source and the biophysical characteristics of land area under production. Precision conservation planning can reduce the negative impacts and enhance the environmental and economic benefits. Since large-scale production of biofeedstock is still in early stages, there is an unprecedented opportunity to develop precision conservation planning for designing agricultural landscapes that can meet food, feed, and bioenergy energy demands while enhancing environmental benefits. At the same time, the precision planning for the futuristic scenario is highly challenging, as many of the attributes essential in conservation planning are heavily uncertain such as the location and size of ethanol conversion plant, type of conversion facility, demand and type of biofeedstock needed at the refinery gate, storage and transportation logistics, and environmental and economic targets of biofuel production.

Figure 2 provides a flowchart for precision conservation planning that could be adopted for biofeedstock production system. The first task in precision conservation planning is to define the criteria or goal of the conservation plan; this becomes crucial in biofuel production system planning, as the development is futuristic. The criteria or goals can be defined by (i) sustainability indicators including environmental, ecological, economical, food and fuel production, and societal sustainability indicators or (ii) ecosystem services including provisioning, regulating, and cultural ecosystem services. A detailed discussion on precision planning criteria is provided in the Conservation Planning Criteria for Biofuel Production section.

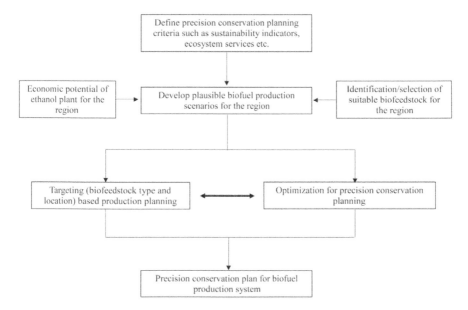

Fig. 2. Flow chart for precision conservation planning for biofuel production

Development of plausible biofuel production scenarios is vital in developing successful precision conservation plans. A key aspect in developing a plausible production scenario is an understanding of watershed or fuelshed biophysical and climate characteristics under consideration. The stakeholder should have a clear understanding of the biofeedstocks that can be produced with relatively low nutrient, pesticide, or water input in the region; the economic, environmental; ecological and societal constraints in the region for specific biofeedstock production; and biofeedstock requirements of the ethanol production plant. A good understanding of the opportunities and constraints can reduce the uncertainties in the futuristic precision planning. The criteria defined for precision planning can be evaluated using the plausible scenarios with targeting-based evaluation techniques. The targeting-based production planning can be developed by defining biofuel production in specific type of land areas such as dedicated perennial grasses on marginal lands, highly erodible areas, low productive areas, and crop residue removal from conservation tilled low-slope areas. A targeting-based evaluation gives stakeholders opportunity to develop simple and easy-to-communicate conservation plans. The targeting-based conservation planning may not provide an optimum (best) conservation plan, since impacts of different biofeedstock crops vary from soil, slope and climatic conditions, and general rules of thumb, a targeting-based approach may not lead to an optimum conservation plan. Optimization-based techniques can be used to obtain optimum (best) conservation plan for biofuel production for the region when there are multiple target criteria and multiple biofeedstocks available for selection. The targeting-based analysis can guide the optimization problem definition in optimization-conservation planning. Similarly the optimization results needs to be further analyzed to derive simpler rules of thumb for targeting the conservation plan to make more general plans and to increase adoptability. A detailed discussion on targeting-based and optimization-based evaluation is given in the Targeting Techniques to Evaluate Sustainable Biofuel Production Opportunities and Optimization Techniques for Sustainable Biofuel Production sections.

Conservation Planning Criteria for Biofuel Production

Precision conservation planning requires one or many decision criteria or objective functions to choose from many available options. The decision criteria can be defined in many ways and may range from a single criterion (e.g., minimum cost of production) to multiple conflicting criteria (e.g., minimum cost of production, minimum environmental impacts, and maximum biomass yield). The conservation plan developed will be specific to the decision criteria. The optimum conservation plan created with one set of decision criteria may be suboptimal for another set.

The decision criteria for conservation planning should be carefully selected and may become complex when multiple objectives and stakeholders with conflicting interests are involved. Sustainability indicators, including environmental, ecological, and socioeconomic indicators recognizing potential impacts and benefits of biofuel production system, could be used as decision criteria for conservation planning. Dale et al. (2015) proposed a framework for selecting and evaluating sustainability indicators for biofuel production systems. The proposed framework (Fig. 3) emphasizes the fact that the selection of appropriate criteria and indicators is driven by the specific purpose of the biofuel production

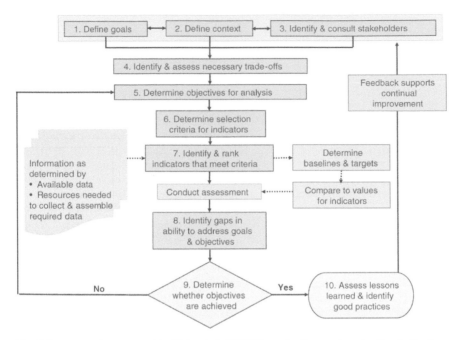

Fig. 3. Framework proposed by Dale et al. (2015) for selecting and evaluating indicators for bioenergy sustainability. Steps for the framework shown in blue; supporting components of the assessment process shown in green. Steps 1, 2, and 3 interact and occur concurrently. Adopted from Dale et al. (2015).

system. The framework starts with defining the sustainability goals and other goals for a bioenergy production project, gaining an understanding of the context, and identifying the values of stakeholders. The objectives for analysis and criteria for indicator selection can be developed after identifying production goals and stakeholders (Dale et al., 2015).

Selection of measurable indicators is critical in selection of evaluation criteria. McBride et al. (2011) proposed 19 measurable environmental sustainability indicators for biofuel production systems (Table 1). The indicators represented measure of soil quality, water quality and quantity, greenhouse gases, biodiversity, air quality, and productivity. Dale et al. (2013) proposed 16 socioeconomic sustainability indicators that include social well-being, energy security, trade, profitability, resource conservation, and social acceptability (Table 2). Indicators selected must be practical so that they can be easily measured in a timely and cost-effective way and should be unambiguous in terms of measurement methods. They should be responsive to both natural and anthropogenic stresses so that impacts of bioenergy production can quantified. There are numerous published articles evaluating impacts of biofuel crop production at various spatial and temporal scales; however, so far, no study has evaluated all the sustainability indicators listed by McBride et al. (2011) and Dale et al. (2013).

Ecosystem services, defined as the benefits people obtain from ecosystem (Millennium Ecosystem Assessment, 2005), are another set of criteria for evaluation of biofuel production system (Fieldsend and Singh, 2013). The ecosystem

Table 1. List of recommended environmental indicators proposed by McBride et al. (2011) for bioenergy sustainability along with associated management pressures and environmental effects expected to be captured by each indicator. (Adapted from McBride et al., 2011).

Category	Indicator	Units	Related management pressures	Potential related environmental effects
Soil quality	1. Total organic carbon (TOC)	Mg ha^{-1}	Crop choice, tillage	Climate change, N mineralization, humification, water holding capacity, infiltration, cation exchange capacity
	2. Total N	Mg ha^{-1}	Crop choice, tillage, N fertilizer application, harvesting practices	Eutrophication potential, N availability
	3. Extractable P	Mg ha^{-1}	Crop choice, tillage, P fertilizer application, harvesting practices	Eutrophication potential, P availability
	4. Bulk density	g cm^{-3}	Harvesting practices, tillage, crop choice	Water holding capacity, infiltration, crop nutrient availability
Water quality and quantity	5. Nitrate concentration in streams (and export)	Concentration: mg L^{-1}; export: kg ha^{-1} yr^{-1}	Crop choice, percentage of residue harvested, tillage, N fertilizer application	Eutrophication, hypoxia, potability
	6. Total P concentration in streams (and export)	Concentration: mg L^{-1}; export: kg ha^{-1} yr^{-1}	Crop choice, percentage of residue harvested, tillage, P fertilizer application	Eutrophication, hypoxia
	7. Suspended sediment concentration in streams (and export)	Concentration: mg L^{-1}; export: kg ha^{-1} yr^{-1}	Crop choice, percentage of residue harvested, tillage	Benthic habitat degradation through siltation, clogging of gills and filters
	8. Herbicide concentration in streams (and export)	Concentration: mg L^{-1}; export: kg ha^{-1} yr^{-1}	Crop choice, herbicide application, tillage	Habitat degradation through toxicity, potability
	9. Peak storm flow	L s^{-1}	Crop choice, percentage of residue harvested, tillage	Erosion, sediment loading, infiltration
	10. Minimum base flow	L s^{-1}	Crop choice, percentage reside harvested, tillage	Habitat degradation, lack of dissolved oxygen
	11. Consumptive water use (incorporates base flow)	Feedstock production: m^3 ha^{-1} d^{-1}; biorefinery: m^3 d^{-1}	Crop choice, irrigation practices, downstream biomass processing	Availability of water for other uses
Greenhouse gases	12. CO$_2$ equivalent emissions (CO$_2$ and N$_2$O)	kg ceq GJ^{-1}	N fertilizer production and use, crop choice, tillage, liming, fossil fuel use throughout supply chains	Climate change, plant growth

Table continued.

Table 1. Continued.

Category	Indicator	Units	Related management pressures	Potential related environmental effects
Biodiversity	13. Presence of taxa of special concern	Presence	Crop choice, regional land uses, management practices	Biodiversity
	14. Habitat area of taxa of special concern	ha	Crop choice, regional land uses	Biodiversity
Air quality	15. Tropospheric ozone	ppb	Fossil fuel use in production and processing, quality and mode of combustion of biofuel	Human health, plant health
	16. Carbon monoxide	ppm	Fossil fuel use in production and processing, mode of biofuel combustion	Human health
	17. Total particulate matter <2.5 μm diameter (PM2.5)	μg m^{-3}	N fertilizer application, fossil fuel use in production and processing, mode of biofuel combustion	Visibility, human health
	18. Total particulate matter <10 μm diameter (PM10)	μg m^{-3}	Fossil fuel use in production and processing, other agricultural activities, solid biomass combustion	Visibility, human health
Productivity	19. Aboveground net primary productivity (ANPP)/yield	g C m^{-2} yr^{-1}	Crop choice, management practices	Climate change, soil fertility, cycling of carbon and other nutrients

Table 2. List of recommended indicators for socioeconomic aspects of sustainable biofuel production with conditions related to each indicator proposed by Dale et al. (2013). Evaluation of each of these indicators should consider the attribution from the biofuel system being assessed. Food security, energy security premium, effective stakeholder participation, and risk of catastrophe require relatively more effort to develop data and measurement tools than the other indicators. Ten indicators in bold font are proposed to be the minimum list of practical measures of socioeconomic aspects of bioenergy sustainability. (Adapted from Dale et al., 2013.)

Category	Indicator	Units	Potential related conditions
Social well- being	**Employment**	Number of full-time equivalent (FTE) jobs†	Hiring of local people; rural development; capacity building; food security
	Household income	Dollars per day	Food security, employment, health, energy security, social acceptance
	Work days lost due to injury	Average number of work days lost per worker per year	Employment conditions, risk of catastrophe, social conditions, education and training
	Food security	Percentage change in food price volatility‡	Household income, employment, energy security
Energy security	Energy security premium	Dollars per gallon of biofuel	Crop failures, oil or bioenergy price shocks; macroeconomic losses; shifts in policy, geo-politics or cartel behavior; exposure to import costs; new discoveries and technologies affecting stock/demand ratio
	Fuel price volatility	Standard deviation of monthly percentage price changes over 1 yr	
External trade	Terms of trade	Ratio (price of exports/price of imports)	Energy security, profitability
	Trade volume	Dollars (net exports or balance of payments)	Energy security, profitability
Profitability	Return on investment (ROI)	Percentage (net investment/initial investment)	Soil properties and management practices; sustainability certification requirements; global market price, terms of trade
	Net present value (NPV)§¶	Dollars (present value of benefits minus present value of costs)	
Resource conservation	**Depletion of nonrenewable energy resources**	Amount of petroleum extracted per year (Tg)‡	Total stocks maintained; other critical resources depleted and monitored depending on context (e.g. water, forest, ecosystem services)
	Fossil energy return on investment (fossil EROI)	Ratio of amount of fossil energy inputs to amount of useful energy output (MJ) (adjusted for energy quality)	Petroleum share of fossil energy; imported share of fossil energy; energy quality factors; total petroleum consumed

Table continued.

Table 2. Continued.

Category	Indicator	Units	Potential related conditions
	Public opinion	Percentage favorable opinion	Aspects of social well-being, environment, energy security, equity, trust, work days lost, stakeholder participation and communication, familiarity with technology, catastrophic risk
	Transparency	Percentage of indicators for which timely and relevant performance data are reported#	Identification of a complete suite of appropriate environmental and socio–economic indicators
	Effective stakeholder participation	Percentage of documented responses addressing stakeholder concerns and suggestions, reported on an annual basis††	Public concerns and perceptions; responsiveness of decision-makers or project authorities to stakeholders; full suite of environmental and socio–economic indicators
Social acceptability	Risk of catastrophe‡‡	Annual probability of catastrophic event	Health, including days lost to injury; environmental conditions

† FTE employment includes net new jobs created, plus jobs maintained that otherwise would have been lost, as a result of the system being assessed.

‡ The inherent complexity of establishing and measuring an indicator of food security implies that significant time, cost, and analytical effort will be needed to reach agreement on its definition, methodology, and application. In the meantime, we propose that the previous indicators for employment and household income serve as practical proxy measures for food security.

§ Conventional economic models can address long-term sustainability issues by extending the planning horizon (e.g., projecting as an infinite geometric series) or calculating with a low discount rate.

¶ Can be expanded to include nonmarket externalities (e.g., water quality, GHG emissions).

This percentage could be based on the total number of social, economic, and environmental indicators identified via stakeholder consultation or on the indicators listed here and in McBride et al. (2011) for which relevant baseline, target, and performance data are reported and made available to the public on a timely basis (at least annually).

†† This indicator is relatively simple but may be difficult to interpret (e.g., whether an issue is effectively addressed is a subjective determination, and measurement is influenced by the ease with which stakeholder concerns and suggestions can be submitted, their comfort level in doing so, and how these inputs are tabulated).

‡‡ A catastrophic event can be defined as an event or accident that has >10 human fatalities, affects an area >1000 ha, or leads to extinction or extirpation of a species.

services include provisioning, regulating, cultural services, and other supporting services. Provisioning services—services people get from the ecosystem directly—include food, fuel, fiber, fresh water, and genetic resources. Regulating services include natural ecosystem processes such as erosion control, water and air quality purification, climate regulation, and human diseases regulation. Through cultural services, humans derive many nonmaterial benefits such as spiritual enrichment, cognitive development, reflection, recreation, and aesthetic experiences. Examples of supporting services include primary production, including oxygen production and soil formation, and directly support all other ecosystem services. Changes in these services affect human well-being through impacts on security, the basic material for a good life, health, and social and cultural relations. Fieldsend and Singh (2013) provided a detailed discussion on ecosystem services related to biofuel production systems. A major difficulty with ecosystem-services-based evaluation criteria is that many of the ecosystem services are not directly quantifiable. For example, the cultural services are often very subjective to quantify effectively. Cibin et al. (2015) have quantified impacts of biofuel production scenarios on five ecosystem services for two watersheds in US Midwest watersheds. The study used ecosystem services evaluation methodology proposed by Logsdon and Chaubey (2013). The study results (Fig. 4) indicated

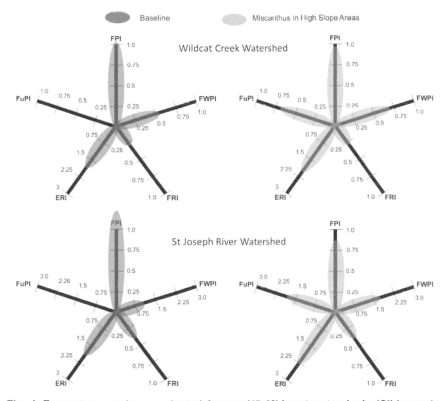

Fig. 4. Ecosystem services evaluated for two US Midwest watersheds (Cibin et al., 2015): comparison of base line and *Miscanthus* in high-slope (>2% slope) areas in Wildcat Creek watershed (top) and St Joseph River watershed (bottom).

cellulosic-perennial-grass-based biofuel production improves fuel provision, fresh water provision, erosion regulation, and flood regulation while reducing the food provision.

Targeting Techniques to Evaluate
Sustainable Biofuel Production Opportunities

Precision conservation planning requires plausible biofuel production scenarios to meet specified sustainability criteria. Plausible biofuel production design depends on targeting biofeedstock production at specific locations within an agricultural landscape, biofeedstock crop types, and management practices. This section provides general guidelines for each choice of targeting in terms of their benefits and constraints for biofeedstock production in developing conservation plans. Choice of land for biofeedstock production may vary geographically. Generally, the land that could be targeted includes cropland, grassland, and marginal land. Biofeedstock crop production will involve a series of management practices including planting, tillage, fertilizer and pesticide application, and harvesting. Table 3 shows potential hydrologic and water quality impacts of targeting various agricultural lands for biofuel production. Relationship of these practices for corresponding biofeedstock production and conservation planning should be carefully evaluated.

Targeting at Productive Agricultural Land

Biofeedstock production could be embed into current highly productive agricultural system mainly in three ways: (i) by producing more first-generation biofeedstock crops, such as corn grain to meet both biofuel development and the demand of existing agricultural products; (ii) by removing crop residues like corn stover; and (iii) by converting current agricultural land to production of second-generation biofeedstock crops like perennial grasses.

Expansion of First-Generation Biofeedstock Production

Expansion of first-generation biofeedstock production could happen both extensively and intensively. The extensive expansion will mainly happen through deforestation or converting grassland. This will be constrained by either the land supply for specific regions or the potential negative impacts on global warming as mentioned above. The intensive expansion will depend on genetically modified varieties of crops or better management practices. The "Billion-Ton Update" projected that corn yield will be increased by 1% annually (Demissie et al., 2012). The authors estimated the expansion of first-generation biofeedstock, mainly corn with higher yield potential, could be increased by 41% through corn yield increase and 59% through corn production area increase from 2006 to 2022 in the Upper Mississippi River Basin. However, the development of first-generation biofuel is capped at 15 billion L by 2015 in the EISA (Schnepf and Yacobucci, 2010)

Another consideration of expanding first-generation biofeedstock production is the environmental consequences, especially those related to the impacts on water. Considerable losses of sediment, nutrients, and pesticides take place from intensively managed row-crop agricultural fields. Expanding grain-based biofeedstock production in the geographic areas with limited availability of water may exacerbate water scarcity problems in the region. For economic and

Table 3. Reported impacts on hydrology and water quality variables with different targeting techniques.

Targeted area	Feedstock	Hydrology	Sediment	N	P	References
Cropland	Corn	ET: +0.1%; SM: −0.2%; Flow: −0.1%	−7.7%	−6.0%	+13.7%	Demissie et al., 2012
	Corn stover	Flow: −1.4 to 3.2%	+4.8 to +268.0%	−2.0 to −52%	−3 to 30% under CT; +24 to 157% under NT	Cibin et al., 2012; Thomas et al., 2011; Wu and Liu, 2012
	Switchgrass	ET: +36%	−4.5 to −99%	−34%		Le et al., 2011; Nelson et al., 2006; Parajuli, 2012; Wu and Liu, 2012
	Miscanthus	ET: +58%, Flow: −11%	−4.5%			
Grassland	Switchgrass	Flow: +2.1%		+1.2%		Christian and Riche, 1998; Wu and Liu, 2012
	Miscanthus	Flow: +4.6%	No change	+5.1%		
Marginal land						
Less productive crop and pasture land	Switchgrass	Flow: −0.4 to −4.3%	−2.3 to −50%	−3.1 to −46.6%	+171.4 to −15.4%	Cibin, 2013; Feng, 2013; Thomas et al., 2014
	Miscanthus	Flow: −0.3 to −12.2%	−2.6 to −87.9%	−3.1 to −95.3%	−0.58 to −21%	
Environmental risk land	Switchgrass	−1.40	−22.20	−14.10	1.5	Cibin, 2013
	Miscanthus	−1.00	−21.40	−14.10	−2	
Filter strip	Switchgrass	Runoff: −34 to −41%	−63% to 99%	−46% to 81%	−34% to 78%	Blanco-Canqui et al., 2004, 2006a; Dabney et al., 2012; Lee et al., 1998, 2003; Sanderson et al., 2001; Tomer et al., 2007
	Miscanthus		−50%			

environmental reasons and regulatory constraints, the considerable expansion of first-generation biofeedstock production is less likely to be feasible.

Crop Residue Removal

Agricultural residue is currently the most widely available source of second-generation biofeedstock in the United States (Cruse and Herndl, 2009; Gramig et al., 2013; Kadam and McMillan, 2003), with 82 to 153 Tg yr^{-1} from corn stover, 58 Tg yr^{-1} from other agricultural residues and 72 Tg yr^{-1} from wood products (Kadam and McMillan, 2003). Starting September 2014, the first commercial-scale ethanol plant (Project LIBERTY) producing 95 million L yr^{-1} of cellulosic ethanol from corn residue was opened in Emmetsburg, IA. Another facility (DuPont Nevada Site Cellulosic Ethanol Facility) is also under construction in Nevada, IA, with projected capacity of producing 114 million L of cellulosic ethanol per year (http://biofuels.dupont.com/).

At present, most of the agricultural residue is left in the field after harvest to protect soil erosion and maintain land productivity. For example, corn stover staying on the soil surface could help protect soil surface from direct raindrop, which reduces soil detachment and crust formation (Or and Ghezzehei, 2002). Soils covered by corn stover also tend to have lower temperature and high soil moisture content than bare soil. These effects, by having residue, could affect planting and following crop development, which reduce the uniformity of crop spacing (Klocke et al., 2009). When corn stover is harvested as biofeedstock, the risk of soil erosion increases with concurrent effects on soil properties related to physical, biological, and thermal processes (Blanco-Canqui et al., 2006b). Karlen et al. (1994) reported removal of corn stover could reduce soil organic C and total N, increasing the requirement of additional fertilizer application and subsequent impacts on water quality.

For conservation purpose and reduction of negative impacts on soil properties and pollutant losses, an optimal crops residue removal rate needs to be determined. The optimal rate depends on biophysical characteristic of the field, management practice, and environmental goals (Gramig et al., 2013). Some researchers have suggested that the amount of erosion should not exceed the tolerable limit (T-value) after the crop residue is removed. The common recommendation of residue removal rate based on this criterion is 30 to 50% (Blanco-Canqui et al., 2006b), while this method and rate need to be used with caution because the T-value does not consider the impacts on land degradation and water quality (Newman et al., 2010). Another way to determine removal rate was based on harvest technology (Thomas et al., 2011). Brechbill and Tyner (2008) recommend three harvest efficiencies based on harvest technologies including baling a windrow (38%); raking and baling (52.5%); and shredding, raking, and baling (70%). Wu and Liu (2012) evaluated impacts of 0, 40, 80, and 100% stover removal; however, it is unlikely that the large-scale residue removal rate would exceed 50%.

The impacts of corn stover removal on hydrologic processes and water quality generally have a close relationship with the rate of removal but could vary a lot because of differences in soil, climatic conditions, and management practices. For example, Cibin et al. (2012) reported reduction of streamflow by 1.4% with 38% removal and by 2.7% with 70% removal rate in the Wildcat Creek watershed located in Indiana. Thomas et al. (2011) reported that removal of corn stover by 38, 52, and 70% increased soil loss by 6, 10, and 17%, respectively, on a Blount silt loam soil (fine, illitic, mesic Aeric Epiaqualfs). But on a Hoytville clay soil (fine, illitic, mesic Mollic Epiaqualfs), the increase rate of soil loss was 50, 75, and 125% for the three removal rates, respectively. The impacts on soil loss reported in literature ranged from <5 to >250% (Table 1). As for nutrient loss, nitrate-N is generally reduced with corn stover removal and the reduction ranged from 2 to 52%. The reported response of total P to corn stover removal is found to be more affected by management than those of soil erosion and N. Under conventional tillage, total P is generally reduced and the reduction ranges from 3 to 30% for different soil types with different removal rates. While under no-till, total P is generally increased by from 24 to 157%. These higher variations of responses from hydrologic processes and water quality with different removal rate indicate that a site-specific evaluation is required for specific locations.

Conversion to Perennial Grass Production

Converting productive agricultural land to perennial grass for biofeedstock production is generally not considered as a viable choice for biofuel development mainly because of the competition of land for food and fuel. As pointed out by Love and Nejadhashemi (2011), the use of cropland for biofeedstock is most likely to happen only when the price of biofeedstock is higher enough to be competitive with existing crops on agricultural land. General suggestion of land for perennial grass production is on grassland (Demissie et al., 2012). Even though, positive environmental impacts by producing perennial grasses on cropland was reported such as reduction of surface runoff (by 11%), sediment yield (by 99%), and nitrate loss (by 34%) (Le et al., 2011; Nelson et al., 2006; Parajuli, 2012); environmental benefits alone may not be a sufficient driver of large-scale conversion of cropland to perennial grasses.

Targeting at Productive Pasture Land

Existing pasturelands that have not been engaged in active forage production may be potential sites for biofeedstock production. Gu et al. (2012) estimated that about 30.4% of the Greater Pallet River Basin area, located in heartland of the United States, was a possible region of grassland for cellulosic biofeedstock production. Further analysis of these identified grasslands indicated that biofeedstock could be sustainably produced in these areas under future climate change conditions. Even though the availability of grassland is large, the conversion to biofeedstock still requires consideration of the impacts on different ecosystem services including policy constraints for natural reserve, wildlife habitat removing, and negative environmental consequences.

Wu and Liu, (2012) evaluated the impacts on hydrologic processes and water quality in the Iowa River basin. They reported that 100% of pastureland conversion to switchgrass and *Miscanthus* will reduce water yield by 2.1 and 4.6%, respectively. Sediment yield might not be impacted considerably; however, N losses could increase by 1.2 and 5.1%, respectively. In addition, nitrate leaching from drainage water could be high, ranging from 4 to 87 kg ha^{-1} depending on fertilizer application rate and the growth stage of switchgrass (Christian and Riche, 1998). For conservation planning, the potentially impacts on water quality should be carefully evaluated when existing pasturelands are converted to produced biofeedstock.

Targeting Marginal Land

To avoid the competition of land between food and fuel production, marginal land is proposed as alternative land resource for biofeedstock production, especially for producing perennial grasses (Cai et al., 2011; Kang et al., 2013; Liu et al., 2011; Zhang et al., 2012). Generally, marginal land is considered as set-aside land that has low productivity and not suitable for regular agricultural crop but suitable for perennial grasses. Land could be considered as marginal land for various reasons such as low productivity, poor accessibility, certain landscape position including land area along roads or rivers and streams, high environmental risk, and degraded or abandoned from agricultural use. Large areas of marginal land have been estimated available for biofeedstock production worldwide with about 100 million to 1 billion ha (Kang et al., 2013). However, their actual availability and flexibility to be used for biofeedstock production depends on their suitability for growth of biofeedstock crops, economic trade-offs, policy guidance, environmental consequence, and farmers' willingness to produce biofeedstock in those areas (Bryngelsson and Lindgren, 2013; Nalepa and Bauer, 2012; Schweier and Becker, 2013; Shortall, 2013; Skevas et al., 2014; Swinton et al., 2011).

Less Productive Crop and Pasture Land

Even within a generally productive agricultural landscape, there may be areas with relatively lower productivity. Maintaining crop production in these areas may require greater inputs of fertilizers and may not be economically viable. Traditional row-crop agricultural production in these areas may also result in creation of environmental hot spots because of greater losses of sediment, nutrients, and pesticides. It has been pointed out that land with low productivity tend to be more vulnerable to soil erosion (Lubowski et al., 2006). When perennial biofeedstock grasses are planted on these lands, they could help to prevent these potential environmental risks. In addition, soil productivity is also expected to be improved by having more organic matter and better soil structure for holding nutrient and water (Bonin et al., 2012; Speight and Singh, 2014). Similarly, growing switchgrass on land where corn grain production is economically negative could increase both sustainability and profitability of corn.

Generally, conversion of these low productivity lands to perennial grass could bring long-term benefits to water quality but may alter hydrologic cycle. Evaluation of impacts from producing switchgrass and *Miscanthus* on hydrologic and water quality with models indicates that water yield could be reduced by from 1 to 12.2% (Cibin, 2013; Cibin et al., 2016; Feng, 2013; Thomas et al., 2014). Sediment load is expected to be reduced significantly at both field and watershed scales, ranging from 2.3 to 87.9% (Cibin, 2013; Cibin et al., 2016; Feng, 2013; Thomas et al., 2014). Similarly, total N losses could be reduced by 3.1 to 95.3% and total P losses reduced from 0.6 to 21.2%. Field observation also found that N leaching from switchgrass grown on cropland could be reduced by 96.5% (McIsaac et al., 2010).

Environmental Risk Land

As mentioned in the last section, low productivity lands generally have higher environmental risk such as higher soil erosion potential. Other marginal lands with environmental risk include land with high slope, land enrolled within the Conservation Reserve Program (CRP), frequently flooded land, land located in areas with high nitrate concentration in ground water, and brownfields (Cibin,

2013; Gopalakrishnan et al., 2009, 2011). The major intention of using these types of land is to meet the energy goals as well as environmental requirements for the purpose of building multifunctional landscapes (Gopalakrishnan et al., 2009). For example, CRP lands are generally highly erodible land and require effective harvesting techniques to make sure land degradation are minimized while maximizing yield (Hartman et al., 2011).

Because of the already-fragile environmental condition, a careful evaluation of the current status and conservation plans can bring numerous environmental benefits. However, current understanding of the environmental consequences of converting these types of land into biofeedstock production is limited. Cibin, (2013) evaluated the impacts from conversion of current corn and soybean lands that have slopes >2% to switchgrass and *Miscanthus* in the Wildcat Creek watershed (2080 km²) located in the US Midwest. He reported that flow in the watershed outlet could be reduced by 1 to 1.4%, respectively, when converted to switchgrass and *Miscanthus*, while, the reduction of sediment load could be 22.2 and 21.4%, respectively, for the two crops. This effect is much higher than the effects on soil erosion when low productive land (<5% corn and soybean land) and pasture land is converted. Soil erosion could be reduced by 2.6 to 9.4% when low-productive land or pastureland was converted to switchgrass and *Miscanthus*, respectively. The same trend was also observed for total N losses. Total N could be reduced by 14.1% for conversion to both switchgrass and *Miscanthus* when high-slope area is converted. When low-productive land and pastureland is converted, total N reduction ranged from 3.1 to 7.5%. The impact on total P loss was less obvious than other water quality variables in this evaluation. These results confirm that more benefits might be expected when the environmental risk area are engaged in perennial grass production.

Filter Strips

Use of filter strips is a traditional BMP implemented to reduce runoff rate; increase infiltration; trap sediments, nutrients, and pesticides; and reduce non-point-source pollution at the edge of the field (Blanco-Canqui et al., 2004; Dosskey et al., 2008). Filter strips along roadsides are also recommended as suitable locations for biofeedstock production (Gopalakrishnan et al., 2009; Lu et al., 2009). Many studies have evaluated the effectiveness of using perennial grasses as filter strips (Blanco-Canqui et al., 2004; Dabney et al., 1995, 2009, 2012; Eghball et al., 2000; Lee et al., 1998; Meyer et al., 1995; Tomer et al., 2007). These field experiments indicate that perennial grasses could help reduce sediment losses from field and reduce non-point-source pollution. With the potential value of perennial grasses as biofeedstock species, managing filter strips as potential locations for producing perennial grass could be an attractive conservation practice and can serve dual purpose of biomass production and water quality protection.

Conservation plans for involving biofeedstock production in filter strip areas needs evaluation of many factors including width of the filter strip, management of the filter area, and biomass production potential. The width of a filter strip needs to be determined based on required level of treatment for non-point-source pollution, rainfall characteristics, land availability, and plant species (Dosskey et al., 2008; Gopalakrishnan et al., 2009). Even though minimum recommendations of width are provided with only loose review of experimental results (Dosskey et al., 2008), a careful evaluation based on the site-specific conditions and non-point-source pollution levels should be conducted with either field experiment or mathematical models. Similarly, the filter area will need to be managed actively in terms of harvesting time, rate, and frequency. Because of nutrient translocation by perennial grasses from aboveground to roots, harvesting time will affect the amount of nutrient being recycled to the rhizome of the plants (Giannoulis and Danalatos, 2014; Heaton et al., 2009). A late harvest would allow these perennial grasses to complete senesce and recycle nutrients, which could potentially reduce the fertilizer requirements thus nutrient loss from the field (Heaton et al., 2009). Harvest frequency might vary depending on the growth performance and functionality of the grasses as a filter strip. Total area of filter strips and biomass productivity should also be evaluated. Feng (2013) evaluated the production of switchgrass and *Miscanthus* on filter strips of four different widths (10, 25, 50, and 100 m) in the St. Joseph River watershed (2810 km^2) located in the US Midwest. The total areas of filter strips in the watershed were 1.4, 6.2, 12.4, and 31.9 km^2 for the four widths, respectively. Even with the widest strip, total biomass production was reported to be relatively small and may not be justifiable for the investment to produce perennial grass for biofeedstock harvest purposes indicating that filter strips could serve as one of the possible locations among a portfolio of candidate sites within a watershed.

Filter strips areas may need to be managed actively to maximize their potential for biofeedstock production. However, choice and intensity of the management operations can make the filter strip area become both sink and source of sediment and nutrient losses. For example, during the first year of establishment, planting operation may cause disturbance to soil surface and increase risk of soil erosion. Harvest operation might also increase erosion risk with the combined effect from surface disturbance and surface cover removal. Fertilizer application may not be recommended in the first year for weed control and to ensure successful establishment. However, after the filter strip areas are successfully established, nutrient needs for plant growth may or may not be met depending on choice of the bioenergy crop in the filter strip area and the amount of nutrients trapped from upstream runoff. If the nutrient trapped from run-on is not sufficient to meet crop nutrient needs, additional fertilizer may need to be applied. However, additional fertilizer input may jeopardize the very basic function of the filter strips to improve water quality and may need to be managed carefully. When the nutrient stress is experienced, the biomass produced will be less than biomass yields under ideal conditions. Conversion of marginal cropland to production of switchgrass and *Miscanthus* with 28 kg N ha^{-1} application in the St. Joseph River watershed showed significant reduction of N loss at the edge of field (Feng, 2013).

Optimization Techniques for Sustainable Biofuel Production

Selection of suitable biofeedstock sources and their strategic placement can minimize potential negative impacts and maximize benefits of biofuel production (Parish et al., 2012; Robertson et al., 2008). However, identifying suitable energy crops and locations is a complex problem considering the multiple choices of biofeedstock sources and the interests of different stakeholders to have priorities toward food and fuel targets and environmental and socioeconomic sustainability. Use of optimization tools may be effective to identify optimum suitable energy crops in best locations satisfying various stakeholder priorities considering multiple dimensions of environmental sustainability, bioenergy production, and food security.

Spatial optimization (optimization methodology to identify spatial relationships between land areas to maximize or minimize objective functions subject to defined constraints) has been widely used in determining selection and placement of agricultural BMPs to reduce non-point-source pollutants from agricultural and mixed-land-use watersheds (e.g., Arabi et al., 2006; Bekele and Nicklow, 2005; Gramig et al., 2013; Kalcic et al., 2014; Maringanti et al., 2009; Rodriguez et al., 2011; Srivastava et al., 2002). Application of spatial optimization in land-use selection and placement is very limited (Cibin, 2013). The current emphasis on biofuel production gives an opportunity to optimally select and place biofuel crops to satisfy food and fuel targets and environmental and socioeconomic sustainability. The precision conservation planning for biofuel production challenging as many decision variables required are highly uncertain, for example, the location and size of ethanol production plant, the type of feedstock requirements, etc.

Spatial Optimization Framework
for Precision Conservation Planning of Biofuel Production

Figure 5 shows a general spatial optimization framework for biofuel production system planning. An efficient optimization algorithm is the main driver in the spatial optimization framework, where the optimization algorithm creates sample populations of feasible biofuel production plans for the study area based on the strength of the objective functions defined. Two important key attributes are (i) designing of optimization problem with appropriate objective function and the production system boundaries (the type of bioenergy crops, system constraints,

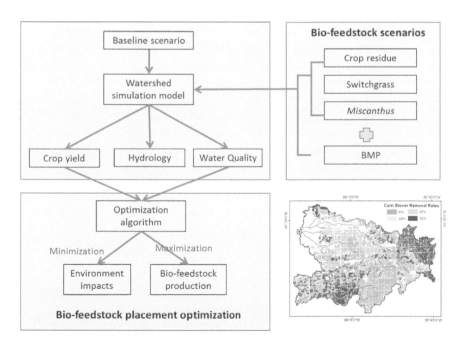

Fig. 5. Spatial optimization framework for precision planning of biofuel production

etc.) and (ii) the objective function evaluation. Generally, a watershed simulation model and an economic model are linked to an optimization algorithm for evaluating the defined objective functions such as minimize environmental impacts and maximize economic benefits of the sample biofuel production plan created by the optimization algorithm.

The various components described in a precision conservation planning framework (Fig. 2) are critical in designing a spatial optimization problem. The selection of objective function and constraints is the most important aspect of spatial optimization framework. There is no criteria or rules of thumb that can be used in all spatial optimization problems as the objective function and constraint definition vary between problems and are subjective to specific targets in the study region, area characteristics, resources availability, and socioeconomic constraints, among other factors. The conservation planning criteria defined above in the Conservation Planning Criteria for Biofuel Production section can be used to identify the suitable energy crops in the region and to frame the objective functions and constraints of the spatial optimization framework.

Challenges of Spatial-Optimization-Based Conservation Planning

Although spatial optimization of conservation practices and conservation planning for biofuel provide optimal solutions that satisfy decision criteria and constraints, there are several challenges:

1. The spatial-optimization-based conservation planning is complex and requires a high level of technical expertise to formulate and conduct spatial optimization. Spatial optimization is also computationally intensive and requires heavy computational infrastructure. Generally, spatial optimization requires thousands of objective function evaluations and, in case of complex simulation models linked to optimization algorithms, it could take several days or even weeks to complete (Arabi et al., 2006).

2. The biofuel production system is futuristic and is challenging to design the objective functions and system boundaries.

3. Conservation plans developed using spatial optimization techniques are often difficult to be generalized. Conservation plans are highly biased with the objective functions and constraints used to find optimal solutions.

4. Spatial-optimization-based conservation plans are difficult to communicate with stakeholders and to get implemented. Optimal solutions obtained assume that all farms identified for adoption of conservation practices will participate. In reality, such plans are difficult to implement, as all farmers may not be willing to adopt those conservation practices at the locations identified by the optimization algorithm.

Suggestions to reduce computational time include (i) application of parallel computing methods (Joseph and Guillaume, 2013; Rouholahnejad et al., 2012; Wu et al., 2013; Yalew et al., 2013; Zhang et al., 2014; Her et al., 2015), (ii) application of simplified versions of complex simulation models by creating surrogate models (Sreekanth and Datta, 2011) or lookup tables (Gitau et al., 2005; Maringanti et al., 2009, 2011), and (iii) use of efficient optimization methods such as multialgorithm, genetically adaptive multiobjective (AMALGAM) (Vrugt and Robinson,

2007) or efficient optimization frameworks such as multilevel spatial optimization (MLSOPT) (Cibin and Chaubey, 2015).

Biofuel production is still in its early stages of development in the United States. Many decision variables are still highly uncertain such as the location of ethanol plant, demand of feedstock, feedstock price to farmers, adaptability or acceptability of biofeedstock crops to commercial farming system, etc. The precision conservation plans needs to be designed with clearly defined objectives, targets, and constraints. The objective function and constraints definition are very critical in spatial optimization. Cibin (2013) demonstrated the significance of objective function by comparing results from two biofuel-production-based spatial optimization scenarios with a biofeedstock target of 800,000 Mg and objective function (i) to minimize sediment erosion at field scale in a watershed and (ii) to minimize sediment load at the watershed outlet (Fig. 6). For the erosion (at field scale)-based scenario, the crop residue (corn stover) removal rates were relatively evenly distributed across the watershed (Fig. 6, top). For sediment load at an outlet-based scenario, the residue removal rates at fields were sensitive to the distance to watershed outlet with minimum residue harvested from fields near to outlet and more residue removed from farther fields (Fig. 6, bottom). Scatter plots between fraction of stover removed from subbasin and distance to outlet indicates linear increase in stover harvest with distance till ~40 to 50 km and no relationship thereafter for sediment load at outlet-based optimization (Fig. 7).

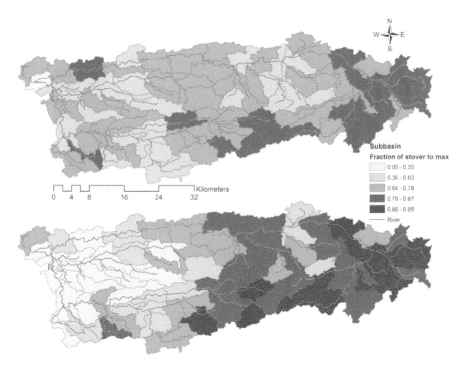

Fig. 6. Spatial distribution of corn stover removed at subbasin level to the maximum possible for optimization to minimize field-scale erosion (top) and to minimize sediment loading at watershed outlet (bottom) with watershed target of 800,000 Mg biomass.

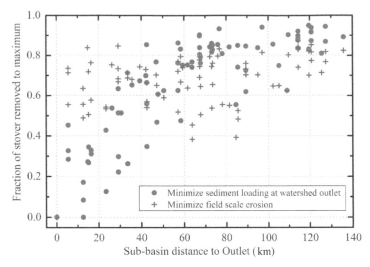

Fig. 7. Relationship between fractions of stover removed in subbasin with distance to outlet for optimization to minimize sediment loading at watershed outlet and to minimize field-scale erosion with watershed target of 800,000 Mg biomass.

Subbasins in the reach of 40–50 km from watershed outlet were critical areas for sediment load at the outlet based optimization and minimum stover removal was estimated from these areas. But for the erosion (at field scale)-based optimization, all subbasins have equal significance toward the objective function and field characteristics such as slope and soil affect the spatial distribution of stover removal. As illustrated in Fig. 6 and 7, conservation plans derived from spatial optimization could be specific to the defined objective function and constraints. Stakeholders carefully define objective functions that meet the watershed management goals. Nonetheless, spatial optimization presents a powerful tool to developing precision conservation plans to achieve watershed management and biomass production goals.

There are many spatial-optimization-based studies related to BMP selection and placement; however, most of these studies are limited to research problems and have not been implemented in a watershed. One major reason is that the optimization does not have enough stakeholder consultations in formulating objective functions, defining candidate conservation practices that could be implemented in the watershed, and evaluating optimization results. For example, most of the spatial optimization studies assume that all farmers are ready to accept and implement the results from the optimization. In practice, many farmers have skepticism about optimization results and may not be willing to implement. Kalcic et al. (2014) developed an adaptive optimization framework that engages farmers and landowners in the optimization process to encourage adoption of conservation practices proposed for the watershed using optimization methods. The approach included initial interview with farmers to understand their willingness to adopt. The authors conducted optimization that included the information from initial interviews, and the optimization results were subsequently discussed with farmers in follow-up interviews. The optimizations were further refined based on farmers' and landowners' feedback. The results obtained from these adaptive optimization may be suboptimum; however, they will likely be more adoptable by farmers if their input are solicited during the optimization process.

Simplified Targeting Rules from Spatial Optimization

As discussed above, spatial optimization requires considerable technical expertise, and the results obtained are often considered definitive and are not analyzed in detail to derive general rules in optimization trends. These challenges make it difficult for stakeholders, watershed managers, and county soil and water conservation districts to not only perform spatial optimization but also to communicate results with the farmers and land owners to get them implemented. There is a need to perform analysis of optimization results with a goal to develop simple and easy to implement procedures for biofuel production. This can be accomplished by relating optimization results with biophysical characteristics of the field. The biophysical characteristics could be slope of the field, soil properties, existing land use, and land management, etc. The simpler method could be in the form of regression equations or a set of rules of thumb . Developing such procedures will allow farmers, landowners, conservation specialists, and the biofuel industry to make decisions without having to perform complex spatial optimization.

Geospatial Techniques for Targeting and Optimization

Geospatial techniques are critical for biofuel production system evaluations. Field-scale, watershed-scale, and regional-scale analysis of biofuel production systems with environmental, economic, and ecological perceptive are required for developing sustainable conservation plans. Engel et al. (2010) discussed the applicability of Groundwater Loading Effects of Agricultural Management Systems and National Agricultural Pesticide Risk Analysis (GLEAMS-NAPRA) (Leonard et al., 1987; Lim and Engel, 2003), Environmental Policy Integrated Climate (EPIC) (Izaurralde et al., 2006), Agricultural Policy/Environmental Extender (APEX) (Gassman et al., 2010), and Soil and Water Assessment Tool (SWAT) (Arnold et al., 1998). All these models were used in biofuel production system evaluation at various spatial and temporal scales (Table 4). The SWAT model is, so far, the most widely used with more than 30 peer-reviewed publications based on studies conducted at various watershed scales. The model has been improved for perennial bioenergy crop simulation with simulation from other crop models (Ng et al., 2010), field-measured data (Trybula et al., 2014), and bioenergy tree simulation (Guo et al., 2015).

Landscape Environmental Assessment Framework (LEAF) (Bonner et al., 2014a,b; Muth and Bryden, 2013; Muth et al., 2013) is another emerging framework for evaluating biofuel production systems. LEAF is an integrated decision support platform for evaluating impacts of agricultural practices on the environmental and economic performance of a landscape (Bonner et al., 2014b). The modeling framework brings together diverse data sources and simulation models such as WEPS (Wind Erosion Prediction System), RUSLE2 (Revised Universal Soil Loss Equation, Version 2), and DAYCENT (daily time-step version of the CENTURY model).

Table 4. List of mathematical simulation models used for biofuel production system evaluation.

Simulation model	Sustainability attribute	Model spatial scale	Model temporal scale	Bioenergy production system evaluated (references)
SWAT	Hydrology, water quality, and crop productivity	Watershed Scale	Long-term daily scale	Crop expansion (Secchi et al., 2011), crop residues (Cibin et al., 2012; Wu and Liu, 2012; Wu et al., 2012), dedicated bioenergy crops such as Miscanthus and switchgrass (Ng et al., 2010; Trybula et al., 2014; Wu and Liu, 2012; Kim and Parajuli, 2012; Cibin, 2013), and woody biomass (Guo et al., 2015)
GLEAMS-NAPRA	Hydrology and water quality	Field scale	Long-term annual	Crop expansion (Thomas et al., 2009), crop residues (Thomas et al., 2011), and dedicated bioenergy crops such as Miscanthus and switchgrass (Thomas et al., 2014)
EPIC	Hydrology, water quality, crop productivity, and economic budgets	Field scale	Long-term daily scale	Dedicated bioenergy crops such as Miscanthus and switchgrass (Kang et al., 2013)
APEX	Hydrology, water quality, crop productivity, and economic budgets	Multiple fields	Long-term daily scale	Dedicated bioenergy crops such as Miscanthus and switchgrass (Feng, 2013)
LEAF	Hydrology, water quality, crop productivity, greenhouse gas and economic budgets	Multiple fields	Long-term daily scale	Crop expansion, crop residues (Muth and Bryden., 2013; Muth et al., 2013; Bonner et al. 2014a,b)
DAYCENT	Greenhouse gas	Field scale	Long-term daily scale	Crop expansion and crop residues (Gramig et al., 2013; Liska et al., 2014), dedicated bioenergy crops such as Miscanthus and switchgrass (Dwivedi et al., 2015)
Agro-IBIS	Hydrology, crop productivity,	Grid cell, Regional scale	Various temporal scales	Miscanthus (Vanloocke et al., 2010)
MLCan	Hydrology	Grid cell, Regional scale	Long-term daily scale	Miscanthus and switchgrass (Le et al., 2011)
RUSLE	Soil erosion	Field scale	Daily scale	Dedicated bioenergy crops such as switchgrass (Khanal et al., 2013)

There are other effective models used in biofuel production system evaluations such as DAYCENT model for greenhouse gas emission evaluation (Dwivedi et al., 2015; Gramig et al., 2013; Liska et al., 2014), Agro-IBIS (Vanloocke et al., 2010) and MLCan (multilayer canopy model) (Le et al., 2011) for hydrology evaluation, and RUSLE2 for soil erosion evaluation (Khanal et al., 2013). Development and use of these and additional models for biofuel system analysis and conservation planning can be expected to increase in future.

Summary

Landscape design to coproduce food, feed, and bioenergy crops presents an unprecedented opportunity to develop precision conservation plans to meet biomass yield goals in an environmentally sustainable, socially acceptable, and economically feasible manner. Various sustainability indicators have been developed to evaluate landscape design plans and their long-term sustainability. However, there is a need to conduct studies that evaluate these sustainability indicators developed for bioenergy crop productions. Various targeting- and optimization-based methods are available or currently being developed to help evaluate selection and placement of crops and conservation practices in agricultural landscape and their associated environmental impacts. Targeting methods present simple-to-use recommendations (e.g., <50% crop residue removal from field with slope <2%) that can be easily implemented in the field. However, such recommendations may not yield optimal results. Spatial optimization methods can overcome such limitations but suffer from many simplified assumptions such as farmers' willingness to adopt optimal results. There is a need to develop targeting methods that are derived from complex spatial optimization so that results are easy to implement and meet crop production and environmental and economic decision criteria. Using such methods, agricultural landscape can be designed so that our society is able to meet food, feed, and biofeedstock production goals in an environmentally sustainable manner.

References

Arabi, M., R.S. Govindaraju, and M.M. Hantush. 2006. Cost-effective allocation of watershed management practices using a genetic algorithm. Water Resour. Res. 42:W10429. doi:10.1029/2006WR004931

Arnold, J.G., R. Srinivasan, R.S. Muttiah, and J.R. Williams. 1998. Large area hydrologic modeling and assessment part I: Model development1. J. Am. Water Resour. Assoc. 34:73–89. doi:10.1111/j.1752-1688.1998.tb05961.x

Bekele, E.G., and J.W. Nicklow. 2005. Multiobjective management of ecosystem services by integrative watershed modeling and evolutionary algorithms. Water Resour. Res. 41:W10406. doi:10.1029/2005WR004090

Berndes, G. 2002. Bioenergy and water: The implications of large-scale bioenergy production for water use and supply. Glob. Environ. Change 12:253–271. doi:10.1016/S0959-3780(02)00040-7

Blanco-Canqui, H., C.J. Gantzer, S.H. Anderson, and E.E. Alberts. 2004. Grass barriers for reduced concentrated flow induced soil and nutrient loss. Soil Sci. Soc. Am. J. 68:1963–1972. doi:10.2136/sssaj2004.1963

Blanco-Canqui, H., C.J. Gantzer, and S.H. Anderson. 2006a. Performance of grass barriers and filter strips under interrill and concentrated flow. J. Environ. Qual. 35:1969–1974. doi:10.2134/jeq2006.0073

Blanco-Canqui, H., R. Lal, W.M. Post, and L.B. Owens. 2006b. Changes in long-term no-till corn growth and yield under different rates of stover mulch. Agron. J. 98:1128–1136. doi:10.2134/agronj2006.0005

Bonin, C., R. Lal, M. Schmitz, and S. Wullschleger. 2012. Soil physical and hydrological properties under three biofuel crops in Ohio. Acta Agric. Scand., Sect. B 62:595–603.

Bonner, I.J., K.G. Cafferty, D.J. Muth, M.D. Tomer, D.E. James, S.A. Porter, and D.L. Karlen. 2014a. Opportunities for energy crop production based on subfield scale distribution of profitability. Energies 7:6509–6526. doi:10.3390/en7106509

Bonner, I.J., D.J. Muth, Jr., J.B. Koch, and D.L. Karlen. 2014b. Modeled impacts of cover crops and vegetative barriers on corn stover availability and soil quality. BioEnergy Res. 7:576–589. doi:10.1007/s12155-014-9423-y

Brechbill, S.C., and W.E. Tyner. 2008. The economics of renewable energy: Corn stover and switchgrass. ID-404-W. Purdue University Cooperative Extension Service, West Lafayette, IN.

Bryngelsson, D.K., and K. Lindgren. 2013. Why large-scale bioenergy production on marginal land is unfeasible: A conceptual partial equilibrium analysis. Energy Policy 55:454–466. doi:10.1016/j.enpol.2012.12.036

Cai, X., X. Zhang, and D. Wang. 2011. Land availability for biofuel production. Environ. Sci. Technol. 45:334–339. doi:10.1021/es103338e

Christian, D.G., and A.B. Riche. 1998. Nitrate leaching losses under *Miscanthus* grass planted on a silty clay loam soil. Soil Use Manage. 14:131–135. doi:10.1111/j.1475-2743.1998.tb00136.x

Cibin, R. 2013. Optimal land use planning on selection and placement of energy crops for sustainable biofuel production. Ph.D. diss. Purdue University, West Lafayette, IN.

Cibin, R., and I. Chaubey. 2015. A computationally efficient approach for watershed scale spatial optimization. Environ. Model. Softw. 66:1–11. doi:10.1016/j.envsoft.2014.12.014

Cibin, R., I. Chaubey, and B. Engel. 2012. Simulated watershed scale impacts of corn stover removal for biofuel on hydrology and water quality. Hydrol. Processes 26:1629–1641. doi:10.1002/hyp.8280

Cibin, R., R. Logsdon, I. Chaubey, and K.A. Cherkauer. 2015. Ecosystem services evaluation of futuristic bioenergy based land use change and their uncertainty from climate change and variability. Paper no. 152121620. ASABE 1st Climate Change Symposium: Adaptation and Mitigation Conference Proceedings, Chicago, IL. 3–5 May 2015. doi:10.13031/cc.20152121620

Cibin, R., E. Trybula, I. Chaubey, S.M. Brouder, and J.J. Volenec. 2016. Watershed-scale impacts of bioenergy crops on hydrology and water quality using improved SWAT model. GCB Bioenergy 8:837–848. doi: 10.1111/gcbb.12307

Cruse, R.M., and C.G. Herndl. 2009. Balancing corn stover harvest for biofuels with soil and water conservation. J. Soil Water Conserv. 64:286–291. doi:10.2489/jswc.64.4.286

Dabney, S.M., K.C. McGregor, G.V. Wilson, and R.F. Cullum. 2009. How management of grass hedges affects their erosion reduction potential. Soil Sci. Soc. Am. J. 73:241–254. doi:10.2136/sssaj2007.0434

Dabney, S.M., L.D. Meyer, W.C. Harmon, C.V. Alonso, and G.R. Foster. 1995. Depositional patterns of sediment trapped by grass hedges. Trans. ASAE 38:1719–1729. doi:10.13031/2013.27999

Dabney, S.M., G.V. Wilson, K.C. McGregor, and D.A.N. Vieira. 2012. Runoff through and upslope of contour switchgrass hedges. Soil Sci. Soc. Am. J. 76:210–219. doi:10.2136/sssaj2011.0019

Dale, V.H., R.A. Efroymson, K.L. Kline, and M.S. Davitt. 2015. A framework for selecting indicators of bioenergy sustainability. Biofuels Bioprod. Biorefining 9:435–446. doi:10.1002/bbb.1562

Dale, V.H., R.A. Efroymson, K.L. Kline, M.H. Langholtz, P.N. Leiby, G.A. Oladosu, M.R. Davis, M.E. Downing, and M.R. Hilliard. 2013. Indicators for assessing socioeconomic sustainability of bioenergy systems: A short list of practical measures. Ecol. Indic. 26:87–102. doi:10.1016/j.ecolind.2012.10.014

Demissie, Y., E. Yan, and M. Wu. 2012. Assessing regional hydrology and water quality implications of large-scale biofuel feedstock production in the Upper Mississippi River Basin. Environ. Sci. Technol. 46:9174–9182. doi:10.1021/es300769k

Dosskey, M.G., M.J. Helmers, and D.E. Eisenhauer. 2008. A design aid for determining width of filter strips. J. Soil Water Conserv. 63:232–241. doi:10.2489/jswc.63.4.232

Dwivedi, P., W. Wang, T. Hudiburg, D. Jaiswal, W. Parton, S. Long, E. DeLucia, and M. Khanna. 2015. Cost of abating greenhouse gas emissions with cellulosic ethanol. Environ. Sci. Technol. 49:2512–2522. doi:10.1021/es5052588

Eghball, B., J.E. Gilley, L.A. Kramer, and T.B. Moorman. 2000. Narrow grass hedge effects on phosphorus and nitrogen in runoff following manure and fertilizer application. J. Soil Water Conserv. 55:172–176.

Engel, B., I. Chaubey, M. Thomas, D. Saraswat, P. Murphy, and B. Bhaduri. 2010. Biofuels and water quality: Challenges and opportunities for simulation modeling. Biofuels 1:463–477. doi:10.4155/bfs.10.17

Feng, Q. 2013. Biomass production and hydrological/water quality impacts of perennial crop production on marginal land. M.S. thesis. Purdue University, West Lafayette, IN.

Fieldsend, A., and H.P. Singh. 2013. Biofuel crops, ecosystem services, and biodiversity. In: B.P. Singh, editor, Biofuel crop sustainability. John Wiley & Sons, Ltd, Oxford, UK. p. 357–382.

Gassman, P.W., J.R. Williams, X. Wang, A. Saleh, E. Osei, L.M. Hauck, R.C. Izaurralde, and J.D. Flowers. 2010. The agricultural policy/environmental eXtender (APEX) model: An emerging tool for landscape and watershed environmental analyses. Trans. ASABE 53:711–740. doi:10.13031/2013.30078

Giannoulis, K.D., and N.G. Danalatos. 2014. Switchgrass (*Panicum virgatum* L.) nutrients use efficiency and uptake characteristics, and biomass yield for solid biofuel production under Mediterranean conditions. Biomass Bioenergy 68:24–31. doi:10.1016/j.biombioe.2014.05.030

Gitau, M.W., W.J. Gburek, and A.R. Jarrett. 2005. A tool for estimating best management practice effectiveness for phosphorus pollution control. J. Soil Water Conserv. 60:1–10.

Gopalakrishnan, G., M.C. Negri, and S.W. Snyder. 2011. A novel framework to classify marginal land for sustainable biomass feedstock production. J. Environ. Qual. 40:1593–1600. doi:10.2134/jeq2010.0539

Gopalakrishnan, G., M.C. Negri, M. Wang, M. Wu, S.W. Snyder, and L. LaFreniere. 2009. Biofuels, land, and water: A systems approach to sustainability. Environ. Sci. Technol. 43:6094–6100. doi:10.1021/es900801u

Gramig, B.M., C.J. Reeling, R. Cibin, and I. Chaubey. 2013. Environmental and Economic trade-offs in a watershed when using corn stover for bioenergy. Environ. Sci. Technol. 47:1784–1791. doi:10.1021/es303459h

Groom, M.J., E.M. Gray, and P.A. Townsend. 2008. Biofuels and biodiversity: Principles for creating better policies for biofuel production. Conserv. Biol. 22:602–609. doi:10.1111/j.1523-1739.2007.00879.x

Gu, Y., S.P. Boyte, B.K. Wylie, and L.L. Tieszen. 2012. Identifying grasslands suitable for cellulosic feedstock crops in the Greater Platte River Basin: Dynamic modeling of ecosystem performance with 250 m eMODIS. GCB Bioenergy 4:96–106. doi:10.1111/j.1757-1707.2011.01113.x

Guo, T., B.A. Engel, G. Shao, J.G. Arnold, R. Srinivasan, and J.R. Kiniry. 2015. Functional approach to simulating short-rotation woody crops in process-based models. BioEnergy Res. 4:1598–1613. doi:10.1007/s12155-015-9615-0

Hartman, J.C., J.B. Nippert, R.A. Orozco, and C.J. Springer. 2011. Potential ecological impacts of switchgrass (*Panicum virgatum* L.) biofuel cultivation in the Central Great Plains, USA. Biomass Bioenergy 35:3415–3421. doi:10.1016/j.biombioe.2011.04.055

Heaton, E.A., F.G. Dohleman, and S.P. Long. 2009. Seasonal nitrogen dynamics of *Miscanthus 'giganteus* and *Panicum virgatum*. GCB Bioenergy 1:297–307. doi:10.1111/j.1757-1707.2009.01022.x

Her, Y., R. Cibin, and I. Chaubey. 2015. Application of parallel computing methods for improving efficiency of optimization in hydrologic and water quality modeling. Appl. Eng. Agric. 31:455–468. doi:10.13031/aea.31.10905

Izaurralde, R.C., J.R. Williams, W.B. McGill, N.J. Rosenberg, and M.C. Quiroga Jakas. 2006. Simulating soil C dynamics with EPIC: Model description and testing against long-term data. Ecological Modelling 192:362–384.

Joseph, J.F., and J.H.A. Guillaume. 2013. Using a parallelized MCMC algorithm in R to identify appropriate likelihood functions for SWAT. Environ. Model. Softw. 46:292–298. doi:10.1016/j.envsoft.2013.03.012

Kadam, K.L., and J.D. McMillan. 2003. Availability of corn stover as a sustainable feedstock for bioethanol production. Bioresour. Technol. 88:17–25. doi:10.1016/S0960-8524(02)00269-9

Kalcic, M., L. Prokopy, J. Frankenberger, and I. Chaubey. 2014. An in-depth examination of farmers'. perceptions of targeting conservation practices. Environ. Manage. 54:795–813.

Kang, S.J., W.M. Post, J.A. Nichols, D. Wang, T.O. West, V. Bandaru, and R.C. Izaurralde. 2013. Marginal lands: Concept, assessment and management. J. Agric. Sci. 5:129–139. doi:10.5539/jas.v5n5p129

Karlen, D.L., N.C. Wollenhaupt, D.C. Erbach, E.C. Berry, J.B. Swan, N.S. Eash, and J.L. Jordahl. 1994. Long-term tillage effects on soil quality. Soil Tillage Res. 32:313–327. doi:10.1016/0167-1987(94)00427-G

Khanal, S., R.P. Anex, C.J. Anderson, D.E. Herzmann, and M.K. Jha. 2013. Implications of biofuel policy-driven land cover change for rainfall erosivity and soil erosion in the United States. GCB Bioenergy 5:713–722. doi:10.1111/gcbb.12050

Kim, H., and P.B. Parajuli. 2012. Economic analysis using SWAT-simulated potential switchgrass and *Miscanthus* yields in the Yazoo River Basin. Trans. ASABE 55:2123–2134.

Klocke, N.L., R.S. Currie, and R.M. Aiken. 2009. Soil water evaporation and crop residues. Trans. ASABE 52:103–110. doi:10.13031/2013.25951

Le, P.V.V., P. Kumar, and D.T. Drewry. 2011. Implications for the hydrologic cycle under climate change due to the expansion of bioenergy crops in the Midwestern United States. Proc. Natl. Acad. Sci. USA 108:15085–15090. doi:10.1073/pnas.1107177108

Lee, K.H., T.M. Isenhart, and R.C. Schultz. 2003. Sediment and nutrient removal in an established multi-species riparian buffer. J. Soil Water Conserv. 58:1–8.

Lee, K.H., T.M. Isenhart, R.C. Schultz, and S.K. Mickelson. 1998. Nutrient and sediment removal by switchgrass and cool-season grass filter strips in Central Iowa, USA. Agrofor. Syst. 44:121–132. doi:10.1023/A:1006201302242

Leonard, R.A., W.G. Knisel, and D.A. Still. 1987. GLEAMS: Groundwater loading effects of agricultural management systems. Trans. ASAE 30:1403–1418. doi:10.13031/2013.30578

Lim, K.J., and B.A. Engel. 2003. Extension and enhancement of national agricultural pesticide risk analysis (NAPRA) WWW decision support system to include nutrients. Comput. Electron. Agric. 38:227–236. doi:10.1016/S0168-1699(03)00002-4

Liska, A.J., H. Yang, M. Milner, S. Goddard, H. Blanco-Canqui, M.P. Pelton, X.X. Fang, H. Zhu, and A.E. Suyker. 2014. Biofuels from crop residue can reduce soil carbon and increase CO_2 emissions. Nat. Clim. Change 4:398–401. doi:10.1038/nclimate2187

Liu, T.T., B.G. McConkey, Z.Y. Ma, Z.G. Liu, X. Li, and L.L. Cheng. 2011. Strengths, weaknesses, opportunities and threats analysis of bioenergy production on marginal land. Energy Procedia 5:2378–2386. doi:10.1016/j.egypro.2011.03.409

Logsdon, R.A., and I. Chaubey. 2013. A quantitative approach to evaluating ecosystem services. Ecol. Modell. 257:57–65. doi:10.1016/j.ecolmodel.2013.02.009

Love, B.J., and A.P. Nejadhashemi. 2011. Water quality impact assessment of large-scale biofuel crops expansion in agricultural regions of Michigan. Biomass Bioenergy 35:2200–2216. doi:10.1016/j.biombioe.2011.02.041

Lu, L., Y. Tang, J. Xie, and Y. Yuan. 2009. The role of marginal agricultural land-based mulberry planting in biomass energy production. Renew. Energy 34:1789–1794. doi:10.1016/j.renene.2008.12.017

Lubowski, R.N., S. Bucholtz, R. Claassen, M.J. Roberts, J.C. Cooper, A. Gueorguieva, and R.C. Johansson. 2006. Environmental effects of agricultural land-use change: The role of economics and policy. USDA, Economic Res. Serv., Washington, DC.

Maringanti, C., I. Chaubey, M. Arabi, and B. Engel. 2011. Application of a multi-objective optimization method to provide least cost alternatives for NPS pollution control. Environ. Manage. 48:448–461. doi:10.1007/s00267-011-9696-2

Maringanti, C., I. Chaubey, and J. Popp. 2009. Development of a multiobjective optimization tool for the selection and placement of best management practices for nonpoint source pollution control. Water Resour. Res. 45:W06406. doi:10.1029/2008WR007094

McBride, A.C., V.H. Dale, L.M. Baskaran, M.E. Downing, L.M. Eaton, R.A. Efroymson, et al. 2011. Indicators to support environmental sustainability of bioenergy systems. Ecol. Indic. 11:1277–1289. doi:10.1016/j.ecolind.2011.01.010

McIsaac, G.F., M.B. David, and C.A. Mitchell. 2010. *Miscanthus* and switchgrass production in central Illinois: Impacts on hydrology and inorganic nitrogen leaching. J. Environ. Qual. 39:1790–1799. doi:10.2134/jeq2009.0497

Meyer, L.D., S.M. Dabney, and W.C. Harmon. 1995. Sediment-trapping effectiveness of stiff-grass hedges. Trans. ASAE 38:809–815. doi:10.13031/2013.27895

Millennium Ecosystem Assessment. 2005. Ecosystem services and human well-being: Synthesis. Island Press, Washington, DC.

Muth, D.J., and K.M. Bryden. 2013. An integrated model for assessment of sustainable agricultural residue removal limits for bioenergy systems. Environ. Model. Softw. 39:50–69. doi:10.1016/j.envsoft.2012.04.006

Muth, D.J., K.M. Bryden, and R.G. Nelson. 2013. Sustainable agricultural residue removal for bioenergy: A spatially comprehensive US national assessment. Appl. Energy 102:403–417. doi:10.1016/j.apenergy.2012.07.028

Nalepa, R.A., and D.M. Bauer. 2012. Marginal lands: The role of remote sensing in constructing landscapes for agrofuel development. J. Peasant Stud. 39:403–422. doi:10.1080/03066150.2012.665890

Nelson, R.G., J.C. Ascough, II, and M.R. Langemeier. 2006. Environmental and economic analysis of switchgrass production for water quality improvement in northeast Kansas. J. Environ. Manage. 79:336–347. doi:10.1016/j.jenvman.2005.07.013

Newman, J.K., A.L. Kaleita, and J.M. Laflen. 2010. Soil erosion hazard maps for corn stover management using National Resources Inventory data and the Water Erosion Prediction Project. J. Soil Water Conserv. 65:211–222. doi:10.2489/jswc.65.4.211

Ng, T.L., J.W. Eheart, X. Cai, and F. Miguez. 2010. Modeling *Miscanthus* in the Soil and Water Assessment Tool (SWAT) to simulate its water quality effects as a bioenergy crop. Environ. Sci. Technol. 44:7138–7144. doi:10.1021/es9039677

Nonhebel, S. 2005. Renewable energy and food supply: Will there be enough land? Renew. Sustain. Energy Rev. 9:191–201. doi:10.1016/j.rser.2004.02.003

Or, D., and T.A. Ghezzehei. 2002. Modeling post-tillage soil structural dynamics: A review. Soil Tillage Res. 64:41–59. doi:10.1016/S0167-1987(01)00256-2

Parajuli, P.B. 2012. Comparison of potential bio-energy feedstock production and water quality impacts using a modeling approach. J. Water Resour. Prot. 4:763–771. doi:10.4236/jwarp.2012.49087

Parish, E.S., M.R. Hilliard, L.M. Baskaran, V.H. Dale, N.A. Griffiths, P.J. Mulholland, A. Sorokine, N.A. Thomas, M.E. Downing, and R.S. Middleton. 2012. Multimetric spatial optimization of switchgrass plantings across a watershed. Biofuels Bioprod. Biorefining 6:58–72. doi:10.1002/bbb.342

Renewable Fuels Association. 2011. Industry statistics: World fuel ethanol production. http://ethanolrfa.org/resources/industry/statistics/world/ (accessed 31 Aug. 2016)

Robertson, G.P., V.H. Dale, O.C. Doering, S.P. Hamburg, J.M. Melillo, M.M. Wander, et al. 2008. Sustainable biofuels redux. Science 322:49–50. doi:10.1126/science.1161525

Rodriguez, H.G., J. Popp, C. Maringanti, and I. Chaubey. 2011. Selection and placement of best management practices used to reduce water quality degradation in Lincoln Lake watershed. Water Resour. Res. 47:W01507. doi:10.1029/2009WR008549

Rouholahnejad, E., K.C. Abbaspour, M. Vejdani, R. Srinivasan, R. Schulin, and A. Lehmann. 2012. A parallelization framework for calibration of hydrological models. Environ. Model. Softw. 31:28–36. doi:10.1016/j.envsoft.2011.12.001

Sahu, M., and R.R. Gu. 2009. Modeling the effects of riparian buffer zone and contour strips on stream water quality. Ecol. Eng. 35:1167–1177. doi:10.1016/j.ecoleng.2009.03.015

Sanderson, M.A., R.M. Jones, M.J. McFarland, J. Stroup, R.L. Reed, and J.P. Muir. 2001. Nutrient movement and removal in a switchgrass biomass–filter strip system treated with dairy manure. J. Environ. Qual. 30:210. doi:10.2134/jeq2001.301210x

Schnepf, R., and B.D. Yacobucci. 2010. Renewable fuel standard (RFS): Overview and issues. CRS Report no. R40155. Congressional Research Service, Washington, DC.

Schweier, J., and G. Becker. 2013. Economics of poplar short rotation coppice plantations on marginal land in Germany. Biomass Bioenergy 59:494–502. doi:10.1016/j.biombioe.2013.10.020

Secchi, S., L. Kurkalova, P.W. Gassman, and C. Hart. 2011. Land use change in a biofuels hotspot: The case of Iowa, USA. Biomass Bioenergy 35:2391–2400. doi:10.1016/j.biombioe.2010.08.047

Shortall, O.K. 2013. "Marginal land" for energy crops: Exploring definitions and embedded assumptions. Energy Policy 62:19–27. doi:10.1016/j.enpol.2013.07.048

Skevas, T., S.M. Swinton, and N.J. Hayden. 2014. What type of landowner would supply marginal land for energy crops? Biomass Bioenergy 67:252–259. doi:10.1016/j.biombioe.2014.05.011

Solomon, B.D. 2010. Biofuels and sustainability. Ann. N.Y. Acad. Sci. 1185:119–134. doi:10.1111/j.1749-6632.2009.05279.x

Speight, J.G., and K. Singh. 2014. Environmental management of energy from biofuels and biofeedstocks John Wiley & Sons, Oxford, UK.

Sreekanth, J., and B. Datta. 2011. Coupled simulation-optimization model for coastal aquifer management using genetic programming-based ensemble surrogate models and multiple-realization optimization. Water Resour. Res. 47:W04516. doi:10.1029/2010WR009683

Srivastava, P., J.M. Hamlett, P.D. Robillard, and R.L. Day. 2002. Watershed optimization of best management practices using AnnAGNPS and a genetic algorithm. Water Resour. Res. 38:3-1–3-14. doi:10.1029/2001WR000365

Swinton, S.M., B.A. Babcock, L.K. James, and V. Bandaru. 2011. Higher US crop prices trigger little area expansion so marginal land for biofuel crops is limited. Energy Policy 39:5254–5258. doi:10.1016/j.enpol.2011.05.039

Thomas, M.A., L.M. Ahiablame, B.A. Engel, and I. Chaubey. 2014. Modeling water quality impacts of growing corn, switchgrass, and Miscanthus on marginal soils. J. Water Resour. Prot. 6:1352–1368. doi:10.4236/jwarp.2014.614125

Thomas, M.A., B.A. Engel, and I. Chaubey. 2009. Water quality impacts of corn production to meet biofuel demands. J. Environ. Eng. 135:1123–1135. doi:10.1061/(ASCE)EE.1943-7870.0000095

Thomas, M.A., B.A. Engel, and I. Chaubey. 2011. Multiple corn stover removal rates for cellulosic biofuels and long-term water quality impacts. J. Soil Water Conserv. 66:431–444. doi:10.2489/jswc.66.6.431

Tomer, M.D., T.B. Moorman, J.L. Kovar, D.E. James, and M.R. Burkart. 2007. Spatial patterns of sediment and phosphorus in a riparian buffer in western Iowa. J. Soil Water Conserv. 62:329–338.

Trostle, R. 2011. Why have food commodity prices risen again? DIANE Publishing, Collingdale, PA

Trybula, E.M., R. Cibin, J.L. Burks, I. Chaubey, S.M. Brouder, and J.J. Volenec. 2014. Perennial rhizomatous grasses as bioenergy feedstock in SWAT: Parameter development and model improvement. GCB Bioenergy 7:1185–1202. doi:10.1111/gcbb.12210

USDA. 2010. Biofuels strategic production report: A USDA regional roadmap to meeting the biofuels goals of the renewable fuels standard by 2022. Technical Report 201024. USDA, Washington, DC

US Department of Energy. 2011. U.S. billion-ton update: Biomass supply for a bioenergy and bioproducts industry, R.D. Perlack and B.J. Stokes, leads, ORBL/TM-2011/224. Oak Ridge National Laboratory, Oak Ridge, TN. http://www1.eere.energy.gov/bioenergy/pdfs/billion_ton_update.pdf

Vanloocke, A., C.J. Bernacchi, and T.E. Twine. 2010. The impacts of *Miscanthus 'giganteus* production on the Midwest US hydrologic cycle. GCB Bioenergy 2:180–191.

Vrugt, J.A., and B.A. Robinson. 2007. Improved evolutionary optimization from genetically adaptive multimethod search. Proc. Natl. Acad. Sci. USA 104:708–711. doi:10.1073/pnas.0610471104

Wu, M., Y. Chiu, and Y. Demissie. 2012. Quantifying the regional water footprint of biofuel production by incorporating hydrologic modeling. Water Resour. Res. 48:W10518.

Wu, Y., T. Li, L. Sun, and J. Chen. 2013. Parallelization of a hydrological model using the message passing interface. Environ. Model. Softw. 43:124–132. doi:10.1016/j.envsoft.2013.02.002

Wu, Y., and S. Liu. 2012. Impacts of biofuels production alternatives on water quantity and quality in the Iowa River Basin. Biomass Bioenergy 36:182–191. doi:10.1016/j.biombioe.2011.10.030

Yalew, S., A. van Griensven, N. Ray, L. Kokoszkiewicz, and G.D. Betrie. 2013. Distributed computation of large scale SWAT models on the Grid. Environ. Model. Softw. 41:223–230. doi:10.1016/j.envsoft.2012.08.002

Zhang, Q., J. Ma, G. Qiu, L. Li, S. Geng, E. Hasi, C. Li, G. Wang, and X. Li. 2012. Potential energy production from algae on marginal land in China. Bioresour. Technol. 109:252–260. doi:10.1016/j.biortech.2011.08.084

Zhang, Y., Y. Sun, J. Zhang, J. Xu, and Y. Wu. 2014. An efficient framework for parallel and continuous frequent item monitoring. Concurrency Computat.: Pract. Exper. 26:2856–2879. doi:10.1002/cpe.3182

Precision Conservation to Enhance Wildlife Benefits in Agricultural Landscapes

Mark D. McConnell* and L. Wes Burger, Jr.

Abstract

Agriculture is the world's largest industry, continues to dominate human land use, and will become more intensive to meet global food demands associated with population growth. Sustainability of global agricultural systems will require strategic integration of conservation practices to protect ecosystems services, health, and productivity. Natural communities as a component of agricultural landscapes support wildlife populations that provide essential ecosystem services with broad societal value. However, allocation of land to noncrop uses entails economic opportunity costs to producers. Effective conservation delivery is dependent on being able to quantify and visualize both the expected costs and benefits. We argue that by identifying economic opportunities for conservation enrollment, increased adoption by landowners is achievable. Our primary goal was to illustrate the necessity, technology, and application of precision conservation in a wildlife management framework. The tools, technologies, and processes associated with precision agriculture can be adapted to inform conservation practice adoption when wildlife objectives are explicitly incorporated into farm- and landscape-level decision framework. We illustrate strategic, objective-driven conservation planning and delivery with case studies from an intensive agricultural landscape in the Lower Mississippi Alluvial Valley.

Agriculture is the world's largest industry and continues to dominate human land use (Robertson and Swinton, 2005). With the human population expected to reach 9.4 billion and per capita arable land expected to be reduced by nearly 40% by 2050 (Lal, 2000), intensification of agricultural production is expected. The mechanism of increase will involve either allocation of additional land to production or maximization of the potential (i.e., increase yield) of land already in use. Considering most of the world's arable land is already in agricultural production (Baligar et al., 2001), future production demands will likely come from land currently in use. Precision agriculture provides a method for implementing the latter

---

Abbreviations: CRP, Conservation Reserve Program; DST, decision support tool; WRP, Wetland Reserve Program.

M.D. McConnell, Warnell School of Forestry & Natural Resources, Univ. of Georgia, 180 E. Green St., Athens, GA 30602-2152. L.W. Burger, Jr., Mississippi Agricultural and Forestry Experiment Station, 211F Bost Center, 190 Bost Extension Drive, Mississippi State, MS 39762-9740 (lwb6@msstate.edu). *Corresponding author (mdm@uga.edu).

doi:10.2134/agronmonogr59.2013.0031

Precision Conservation: Geospatial Techniques for Agricultural and Natural Resources Conservation
J.A. Delgado, G.F. Sassenrath, and T. Mueller, editors. Agronomy Monograph 59.

of these options by allowing producers to maximize yield and profitability in a spatially explicit and economically advantageous manner (Stull et al., 2004).

Precision agriculture is "the application of technologies and principles to manage spatial and temporal variability associated with all aspects of agricultural production" (Pierce and Nowak, 1999). Whelan and McBratney (2000) describe precision agriculture as "a philosophical shift in the management of variability within agricultural industries aimed at improving profitability and/or environmental impact (both short and long term)." The precision agriculture concept is based on reorganization of the agricultural system to low-input, high-efficiency, sustainable agriculture (Shibusawa, 1998). The principal goal of precision agriculture is to maximize yield (Mg ha^{-1}) and profitability (\$ ha^{-1}). When yield is maximized, amount of land needed to meet food demands and financial obligations is reduced. If financial obligations can be met with less cropped acreage, the opportunity for land reallocation is created. Less-productive agricultural lands (i.e., those with reduced yields) are logical candidates for conservation implementation (Hyberg and Riley, 2009). Conservation and food production goals can be linked through increasing yield on cultivated land, thereby freeing up land for conservation use (Green et al., 2005). Precision agriculture can increase profitability for producers and concomitantly provide ecological benefits to the public (Zhang et al., 2002).

The field of precision conservation uses precision agriculture technology to achieve conservation objectives. Precision conservation is "a set of spatial technologies and procedures linked to mapped variables directed to implement conservation management practices that take into account spatial and temporal variability across natural and agricultural systems" (Berry et al., 2003, 2005; Delgado and Berry, 2008). Precision conservation, much like precision agriculture, depends on geospatial tools such as global positioning systems, geographic information systems, digital landscape information, spatially explicit mathematical models, and intensive computer analysis across natural and agricultural ecosystems (Berry et al., 2003; Delgado and Berry, 2008). Numerous studies on precision agriculture's application in conservation planning have been conducted (Berry et al., 2003; Dosskey et al., 2005; Kitchen et al., 2005; Delgado et al., 2005; Delgado and Bausch, 2005) but generally focus on nutrient loading and erosion control. Precision agriculture has also been used in strategic establishment of conservation buffers to reduce nutrient runoff and topsoil erosion (Stull et al., 2004; Dosskey et al., 2005) and has been shown to increase buffer effectiveness. However incorporating precision agriculture's or precision conservation's use in wildlife conservation planning is an underutilized field of research.

To achieve strategic conservation of multiple wildlife species while concomitantly enhancing farming efficiency and productivity, we propose a systematic change in how we use the tools and interpret the outputs of precision conservation. We illustrate a novel approach for using precision conservation tools in a wildlife management framework. This approach can be used for any wildlife species and production system. Because of the recent development, very little research has been conducted on its application. We will present a case study from Mississippi that illustrates the conservation and economic effectiveness of this approach.

A Broader Definition of Conservation

Precision conservation provides a suite of tools and processes for optimizing economic and environmental outcomes of conservation design and delivery within the context of site- and production system–specific constraints. In agricultural systems, precision conservation has most successfully been applied to physical and edaphic processes such as sediment and nutrient transport with few applications to other environmental services such as wildlife habitat, pollination, or biological diversity. Although the broad goal of precision conservation is to use site-specific information to inform design and delivery of conservation management practices in agricultural systems (Berry et al., 2003), the conservation targets or outcomes to which precision conservation principles have been applied are relatively few. This is in part because of the rather restrictive connotation of conservation historically incorporated under USDA–NRCS conservation planning. In agricultural contexts, the term conservation is often used synonymously with minimizing nutrient loading, maximizing erosion control, and increasing water quality (Berry et al., 2003; Dosskey et al., 2005; Kitchen et al., 2005). Precision conservation using site-specific information to inform amount, width, and placement of conservation practices has been used in strategic establishment of conservation buffers to reduce nutrient runoff and topsoil erosion (Stull et al., 2004; Dosskey et al., 2005) and has been shown to increase profitability, return on investment, and buffer effectiveness. While we agree these practices certainly fall under the umbrella of conservation and improve the ecological integrity of the system, we contend that ecological function involves many other environmental services, and therefore, conservation planning and delivery must also address broader targets including biological diversity, wildlife, and their interaction with the landscape. Although landscape-scale wildlife conservation design, as illustrated in "Adaptive Management for a Turbulent Future" (Allen et al., 2011) and "Strategic Habitat Conservation: Final Report of the National Ecological Assessment Team" (US Department of the Interior, 2006), is a maturing concept, the use of precision conservation principles to improve wildlife conservation design and delivery at the field and farm scale is a largely underdeveloped field of science and application (although, see McConnell and Burger [2011]). Our goal in this chapter is to illustrate the necessity, technology, and application of precision conservation in a wildlife management framework.

Conservation Concerns in Agricultural Landscapes

Essential to implementing precision conservation for wildlife is an understanding of the species-specific habitat requirements and the multitude and magnitude of threats facing wildlife species in agricultural landscapes. The increasing demand for food and fiber to feed an exponentially growing human population has resulted in continued agricultural expansion resulting in conversion of natural communities to crop production and large-scale habitat loss and degradation. Multiple species of birds, mammals, insects, amphibians, and reptiles show large-scale population declines attributed to agricultural intensification. Although some species have benefited from modern agriculture (Askins, 1999), many species that historically inhabited this landscape have experienced population declines, range reductions, and local extinctions associated with agricultural intensification (Medan et al., 2011). Wilcove et al. (1998) suggested that agricultural practices affected 38% of endangered species in the United States. While the ramifications of increasing needs for food and fiber have measurable impacts on native fauna, there exists a theoretical and empirical balance between meeting goals of production and maintaining ecosystem quality and function. Unfortunately, the perception exists that management practices that benefit agricultural and wildlife species are often not in sync with production goals. However, when objectively delivered, conservation goals can promote and possibly enhance production goals (McConnell and Burger, 2011).

Effects of Intensive Agriculture on Native Wildlife

Birds

Numerous grassland songbirds have experienced steep declines associated with conversion of grasslands to agriculture and introduction of exotic forages grasses (Herkert, 1994; Chamberlain et al., 2000; Murphy, 2003; Brennan and Kuvlesky, 2005; Sauer et al., 2008). Although large-scale agricultural expansion has benefited some grassland bird species (Askins, 1999), farming (conversion and intensification) is considered the single greatest danger to threatened bird species (Green et al., 2005) and the leading cause of grassland songbird decline (Vickery and Herkert, 1999; Blackwell and Dolbeer, 2001; Murphy 2003). However, agricultural landscapes can support populations of many grassland species given sufficient quantity, quality, and connectivity of native grassland patches within the agricultural matrix.

Mammals

Agriculture has been identified as a leading cause of global mammalian decline (Hoffmann et al., 2011). Ceballos et al. (2005) estimated that 80% of the land area needed to maintain just 10% of the geographic ranges of global mammal species has been affected by agriculture. The primary effect of agriculture on mammals is habitat fragmentation (Mackenzie et al., 1998). Small mammals are directly affected by loss of ecological integrity associated with agriculture intensification via reductions in diversity, abundance, and distribution (Todd et al., 2000; Jacob 2003; de la Peña et al., 2003). Declines in mammalian prey (Calvete et al., 2004) and, consequently, predators (Palma et al., 1999) have been attributed to agriculture intensification. Certain insectivorous mammal declines have also been attributed to agriculture intensification (Pocock and Jennings, 2008). Habitat loss and degradation have effectively endangered mammalian diversity and abundance in agricultural landscapes, thus illustrating the need for effective conservation strategies.

Reptiles and Amphibians

Global herpetofauna populations are declining (Gibbons et al., 2000; Gallant et al., 2007) as a result of multiple land-use practices including agricultural conversion (Hutchens and DePerno, 2009). Agricultural land use has fragmented herpetofauna habitat and will likely continue to at an increasing rate as a result of global demands for agricultural productivity (Lehtinen et al., 1999). A multitude of herpetofaunal species inhabit agricultural landscapes and their persistence can be used as an indicator of ecosystem health (Gibbons et al., 2000) and integrity

(Davic and Welsh, 2004). Furthermore, reptiles and amphibians significantly contribute to estimates of biomass in agricultural wetland habitats (Gibbons et al., 2006). Many species use terrestrial habitats surrounding wetlands at different stages of their life (Semlitsch and Bodie, 2003). However, the abundance and distribution of wetlands influences the distribution of herpetofauna across the landscape (Rittenhouse and Semlitsch, 2007). Therefore, effective management of agricultural wetlands and their surrounding terrestrial habitats is essential for agricultural ecosystem health, function, and biodiversity.

Ecological Services of Wildlife in Agricultural Systems

Agricultural landscapes produce a multitude of goods and services that benefit the producer, the owner, the local community, and society at large. Each of these stakeholders may have different and sometimes competing production goals and priorities. From a global food systems perspective, the primary goal for agricultural land is to produce an abundant, safe, secure, and affordable supply of food and fiber to meet the needs of a growing human population with ever increasing per capita consumption. The goals of a landowner operator may differ somewhat from a renter operator, but both focus primarily on profitability and, to varying degrees, sustainability and stewardship. However, these landscapes provide a plethora of ecological services (Swinton et al., 2007), which are largely unrecognized by the general populace. Many of these ecological services are rendered by and reliant on healthy wildlife populations. Viable wildlife populations require habitat in sufficient quality, quantity, and configuration to survive and reproduce. Therefore, maintaining ecological service functionality requires providing quality wildlife habitat. We contend a new paradigm of agricultural ecosystem management is necessary for the future viability and productivity of agricultural systems, one in which a philosophical marriage between agricultural production and conservation is applied to optimize the ecological integrity and productivity of the whole system while still achieving production and economic objectives.

Pollination

Wildlife species (mostly insects) contribute to more than 90% of the pollination of flowering plants (Kearns et al., 1998). Many agricultural crops depend on pollinators to varying degrees (Free, 1993). Estimates from McGregor (1976) and Klein et al. (2007) conclude that roughly one-third of our food is derived from animal-pollinated crops. As stated by Ingram et al. (1996), "for one in every three bites you eat, you should thank a bee, butterfly, bat, bird or other pollinator." Insect pollinators affect more than 75% of major global crops (Klein et al., 2007). Consequently, agricultural production is largely dependent on biotic pollination and therefore the sustainability of pollinator populations (Aizen et al., 2009). Crop pollinators include a myriad of vertebrate and invertebrate species that are essential to global food production. Honey bees (*Apis mellifera*) represent the single most important pollinator species. Other native wild bee species also represent a significant proportion of pollination services (Aizen et al., 2009). However, global honeybee declines illustrate a significant threat to the ecological services pollinators provide to global food production. Declines of honeybees have been attributed to disease, parasites, and other unverified factors (Cameron et al., 2011), while decline in other native bee species has been attributed to habitat destruction

(Biesmeijer et al., 2006; Fitzpatrick et al., 2007). As the number of honeybee stocks decline (Aizen et al., 2009), more reliance will fall on native pollinators that are often better adapted to local conditions and collectively more effective at pollinating a wider range of crops (Goulson, 2003). Wild pollinators become less diverse and abundant when seminatural habitats are absent from the agricultural landscape (Richards, 2001). Other insect pollinators, such as butterflies, moths, and beetles, require seminatural habitats in agricultural landscapes. This source of refugia has considerable impacts on the abundance, distribution, and effectiveness of pollinators. It is this limiting factor of habitat availability that presents conservation opportunities in these landscapes. Using precision conservation to spatially optimize pollinator habitat effectiveness, while minimizing economic opportunity costs, is just one example of how this technology can be used in a wildlife conservation framework.

Pest Control

Natural pest control has been identified as one of many ecological services at risk as a result of global agricultural intensification (Wilby and Thomas, 2002). Maintenance of habitat that supports beneficial insects has been shown to control crop pests (Myers et al., 1989), increase crop yield (Ostman et al., 2003), and reduce the need for chemical pesticide use (Naylor and Ehrlich, 1997) thereby providing environmental and economic benefits (Bianchi et al., 2013). However, landscape diversity affects the abundance and diversity of natural enemies and their ability to regulate herbivore populations (Cardinale et al., 2003; Bianchi et al., 2005). Noncrop habitats such as field borders and hedge rows account for much of the biodiversity in agricultural systems (Bianchi et al., 2013) and are more suitable to natural enemies of herbivorous insects (Meek et al., 2002). The distribution of noncrop habitats affects the ability and efficiency of natural enemies to control pest populations (Wissinger, 1997). Therefore, creation and maintenance of diverse, natural, noncrop habitats is essential to sustain effective beneficial insect populations (Bianchi et al., 2013).

Bianchi et al. (2013) discusses several research needs regarding use of natural enemies for pest control and crop production. Improving the cost-benefit of natural enemy management was listed as one of the least evaluated aspects. Optimizing natural enemy habitat through creation of noncrop habitats can be addressed with precision conservation tools. Bianchi and van der Werf (2003) noted that the spatial arrangement and shape of noncrop habitats influenced their effectiveness. Spatial tools for field margin creation can be coupled with spatially explicit species suitability models to optimize pest suppression, farm efficiency, and economic gain using precision conservation tools in a novel integrated pest management framework.

Conservation Programs

Multiple governments around the world provide various forms of voluntary, incentive-based conservation enrollment on working agricultural fields. Such conservation programs and their associated conservation practices vary considerably in their design, objectives, and incentives. For example, in the United States, the Conservation Reserve Program (CRP) and the Wetland Reserve Program (WRP) have become the primary programmatic vehicles for delivering conservation in agricultural landscapes. The CRP consists of more than 40 conservation practices that are designed to achieve practice-specific natural resource objectives on working agricultural lands. The WRP implements wetland restoration on a massive scale and is designed to increase wetland function and improve wetland wildlife habitat on marginal farmlands. The economic incentives differ among programs with CRP providing soil and county-specific rental payments over a fixed-length contract (e.g., 10–15 yr), whereas WRP provides a per-acre easement payment for a long-term or perpetual abandonment of certain developmental and use rights. Both programs provide cost-share assistance to offset the costs of practice establishment. Other conservation programs, such as Environmental Qualities Incentive Program and Wildlife Habitat Incentive Program, provide only cost share for practice establishment. Practices that remove entire fields from production and plant them to permanent groundcover are referred to as whole-field enrollments. These practices are usually enrolled on environmentally sensitive (i.e., highly erodible soils) fields and have been credited with numerous, large-scale environmental benefits. For an exhaustive review, see Heard et al. (2000), Haufler (2005), and Waddle et al. (2013). However, whole field practices, while effectively removing marginal land, have the potential to remove productive, nonsensitive portions of fields that could produce more revenue to landowners if retained in production. To more specifically target environmentally sensitive field regions, conservation buffers are commonly used for optimizing natural resource concerns while maintaining sustainable crop production.

Wildlife Benefits of Conservation Buffers

Targeted delivery of resource concern–specific practices has become the primary philosophy for increasing conservation value and ecological function of agricultural landscapes. Conservation buffers are one such targeted conservation practices, predicated on the assumption that a relatively small change in primary land use can produce a disproportionate change in environmental outcomes when strategically delivered (Evans et al., 2013). Conservation buffers are noncrop habitats left idle or planted to a specific vegetative type to achieve a specific conservation objective (e.g., erosion control, sedimentation retention, stream bank stabilization, filtration, nutrient retention, water quality enhancement, or wildlife habitat). Their size, distribution, composition, and management regime depend on the specific conservation goals for which they are designed. Implementation of conservation buffers has been widely successful in North American and European agricultural systems where cropland conversion has eliminated large amounts of native habitat. The wildlife and ecosystem services provided by buffers can easily be monitored and evaluated to ensure strategic implementation and management to maximize their effectiveness. Objective-driven design and strategic implementation informed by systematic monitoring in an adaptive resource management framework empower conservation planners, natural resource biologists, and agricultural landowners to simultaneously achieve production goals and enhance ecological integrity of agricultural landscapes.

Conservation buffers have the potential to increase biodiversity and improve the ecological health of agricultural landscapes. Numerous studies document the conservation benefits of buffers to soil retention, water quality, and nutrient filtering, but buffers also provide other benefits such as wildlife and pollinator habitat (Lovell and Sullivan, 2006). Buffers are often the only habitat in a hostile landscape and, therefore, are crucial to persistence of many populations. Herbaceous conservation buffers have been shown to increase abundance and diversity of numerous grassland birds (Cederbaum et al., 2004; Smith et al., 2005; Conover et al., 2007, 2009; Evans et al., 2013) and provide suitable nesting habitat (Conover et al., 2011a,b; Adams et al., 2013). Buffers also provide corridors to facilitate movement between fragmented habitat patches characteristic of agricultural matrices (Fahrig and Merriam 1985; Henry et al., 1999; Shueller et al., 2000). Agriculture intensification has reduced the abundance and distribution of forests resulting in small isolated woodlots with riparian buffers as the only remaining linkage between patches, thus increasing pressure on noncrop patches to provide wildlife habitat (Maisonneuve and Rioux, 2001). Riparian conservation buffers have been shown to increase mammalian and herpetofaunal abundance and diversity in agricultural landscapes (Maisonneuve and Rioux, 2001). Buffers have also been shown to sustain beneficial arthropod species (Landis et al., 2000; Marshall and Moonen 2002; Gurr et al., 2005), and numerous studies have documented that buffers increase the reproductive success of natural enemies of crop pests (Gurr et al., 2005; Heimpel and Jervis, 2005).

Understanding Conservation Implementation

Allocation of land from crop production to conservation results in direct and opportunity costs to landowners that forgo the economic returns associated with the commodities that otherwise would have been produced. The agricultural producer incurs a private cost to produce a public good (ecological services). Understanding the site-specific economic tradeoffs of conservation and the degree to which conservation program incentives might offset these costs should be central to conservation planning and delivery and guide producer adoption decisions. Precision agriculture technologies can help to inform both sides of this equation. But such data-driven conservation delivery requires an understanding of factors that motivate agricultural producers, a conceptual framework of conservation delivery (Burger, 2006), decision support tools to quantify costs and benefits, and geospatial tools to help visualize alternative conservation landscapes. We demonstrate how geospatial technologies can be used collectively in a precision conservation context for wildlife conservation planning. We use the USDA's Farm Bill conservation programs framework from here on because it is a model for global conservation delivery.

Factors That Influence Adoption

Agricultural producers operate under uncertainty created by environmental and market stochasticity. Consequently, financial concerns strongly influence producer decisions (Kitchen et al., 2005). Variations in global economies, federal policies (e.g., Farm Bill), commodity prices, subsidy payments, weather and climatic events, input costs, farm ownership, and equipment expenses together provide numerous financial obstacles for producers. Removing land from production for conservation imposes an opportunity cost associated with loss in revenue from commodities that otherwise would have been produced (USDA,

2003). "Conservation must be compatible with profitability" (Kitchen et al., 2005), and to make conservation implementation economically attractive to agricultural landowners, conservation programs must address economic concerns of producers (USDA, 2003). Conservation and profitability can coexist if ecological and economic demands are taken into account (Holzkamper and Seppelt, 2007). Because farm policy in the United States (implemented through the Farm Bill) has evolved to recognize the importance of financial concerns and profitability in adoption of conservation practices, numerous conservation programs provide financial incentives to compensate for opportunity costs of land retirement. Conservation buffer practices address producers' financial and environmental concerns by providing substantial financial incentives for enrollment of environmentally sensitive lands.

Strategic Conservation Enrollment

Currently, a combination of land eligibility and landowner objectives are the decision making components of conservation program adoption. Under a program-driven approach, landowners choose a program and are restricted to the management practices available under that program, which may or may not be conducive to desired objectives (Burger, 2006). Furthermore, implementation of such programs may not fully optimize the landowner's economic and conservation goals or potential (Burger, 2006). Under the general-signup CRP, eligible fields must meet a highly erodible land criterion. Continuous-signup CRP practices are not limited to highly erodible land, which creates the opportunity of removing moderate to highly productive land from cultivation. Although overall environmental benefits may be produced, profitability for a landowner may be reduced by enrollment. Removing highly profitable land from agricultural production is not necessarily an effective strategy for maximizing overall benefits of conservation programs if opportunity costs create an impediment to adoption. Efficacy of conservation implementation depends on maximizing whole-field profitability and concomitantly providing the greatest environmental and wildlife benefits. Agricultural landowners will enroll in conservation programs that address environmental and wildlife concerns provided financial incentives are adequate (USDA, 2003). To maximize societal, environmental, and economic benefits through conservation programs, strategic implementation is crucial. The vehicle for strategic conservation will be precision agriculture technology.

Conservation buffers represent a suite of best management practices that are conducive to a precision conservation approach because they take the most environmentally sensitive lands out of production and address specific resource concerns (e.g., soil erosion, water quality, wildlife conservation) in a manner that is compatible with row-crop production systems. These targeted conservation practices often carry extra economic incentives (i.e., signup incentive payments, increased cost-share, elevated rental rates) to induce adoption. To increase the degree of targeting, eligibility of cropland for conservation buffer practices is constrained based on spatial relationships such as location within a conservation priority area, hill slope position, proximity to water bodies and wetlands, proximity to field margins, or other ecologically sensitive features. Buffer width, configuration, and plant materials are constrained so as to achieve desired resource outcomes. However, enrollment of all eligible land might not necessarily maximize financial returns and thus might not be the best land use from a profitability standpoint. A strategic enrollment that maximizes conservation benefits, subject to the constraint that economic benefits equal or exceed that under agricultural production, might be considered optimal from a producer standpoint and might increase adoption.

Tools for Identifying Conservation Eligibility

Considering the multiple conservation programs and practices offered under the US Farm Bill, and their practice-specific eligibility criteria, agricultural landowners are faced with the daunting task of understanding which programs and practices fit and where they fit on their farm. Such a challenge can hinder adoption and limit the conservation potential. Therefore, a precision conservation tool that spatially illustrates conservation practice eligibility could aid in conservation adoption and implementation. Quantifying conservation eligibility is paramount because most producers and natural resource planners cannot visualize where and how conservation programs fit into their production systems. Illustrating eligible land for multiple conservation practices provides options to producers to optimize not only their economic interests but also their specific natural resource concerns (i.e., water quality, soil loss, wildlife habitat). McConnell and Burger (2011) describe a spatially explicit decision support tool (DST) to illustrate the spatial eligibility of multiple conservation buffer practices (each of which benefits multiple wildlife species). Use of geospatial technology is essential to this process and the DST produces simple, spatially explicit maps that producers can use to make informed land-use decisions. Their tool operates from the toolbox in ArcMap (Environmental Systems Research Institute, 2009) and outputs editable shapefiles that can be depicted, on screen, over an aerial photograph. This approach allows landowners to visualize conservation eligibility and make informed decisions about conservation implementation.

Results from McConnell and Burger (2011) were limited to one production farm (~1200 ha) in Mississippi. However, on that one farm, their tool identified more than 400 ha of eligible conservation buffers for two conservation practices (Fig. 1 and 2). Their research demonstrates the utility and effectiveness of precision conservation technologies coupled with a geospatial DST to identify conservation opportunities in agricultural landscapes. While the McConnell and Burger (2011) DST was specifically designed to identify eligibility for a specific suite of practices, it provides a model that could be adopted by the precision conservation community to build on and further develop such approaches to increase the speed and efficiency of conservation planning and delivery for a variety of natural resource concerns. Considering the abundance and depth of literature documenting the benefits of conservation buffers to birds, mammals, amphibians, insects, and ecosystem health and function, it is essential for global biodiversity to develop and implement novel approaches to conservation delivery. We contend that precision conservation technology is the best method for delivering wildlife-friendly conservation in agricultural landscapes.

Fig. 1. Total eligible area for Conservation Practice 33, habitat buffers for upland birds on a 1200-ha grain farm in Tallahatchie, MS, USA, 2007.

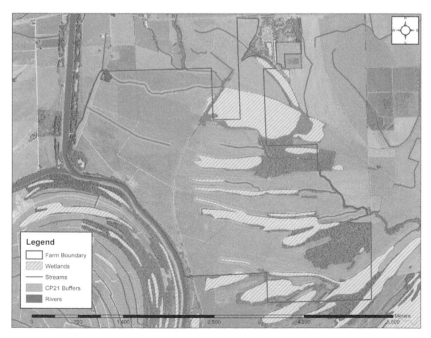

Fig. 2. Total eligible area for Conservation Practice 21, filter strips on a 1200-ha grain farm in Tallahatchie, MS, USA 2007.

Tools for Identifying Economic Opportunities for Conservation

Precision conservation technology provides a wealth of data to inform the decision-making process on agricultural land management. Specifically, yield monitors provide spatially explicit information about field productivity, which provides managers with an opportunity to adjust management strategies. Yield monitors accurately illustrate spatial variability of yield (Mg ha⁻¹) but provide no economic information on how yield affects revenue ($ ha⁻¹). Connecting yield to profit is paramount to adoption of conservation programs. Because traditional yield maps provide no financial information, profit maps are a more efficient tool for identifying conservation opportunities. Given that financial considerations generally have the greatest influence on producer decisions (Kitchen et al., 2005), profit maps are a logical tool for identifying conservation and economic opportunities and quantifying conservation tradeoffs of adoption. Profit maps illustrate regions of decreased revenue that managers can use to make informed decisions (Fig. 3 and 4). Calculating whole-field profitability under agricultural production alone identifies field regions where revenue is lost (i.e., negative net revenue) or minimal, whereas calculating whole-field profitability under alternative conservation buffer enrollments identifies field regions where profitability under conservation enrollment is greater than that of production alone (Fig. 5 and 6). Running this analysis independently for multiple conservation practices and alternative enrollments within a practice provides a multitude of land-use options for agricultural producers.

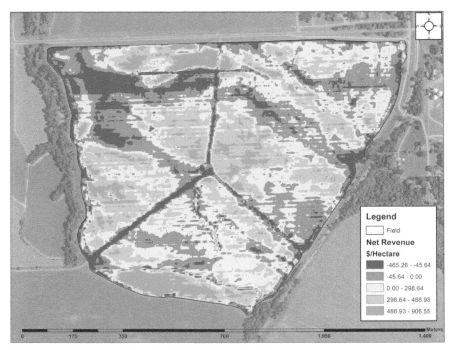

Fig. 3. Profit surface for center-pivot irrigated soybean field assuming a $331 Mg⁻¹ commodity price and $597.87 ha⁻¹ production cost in Mississippi, USA.

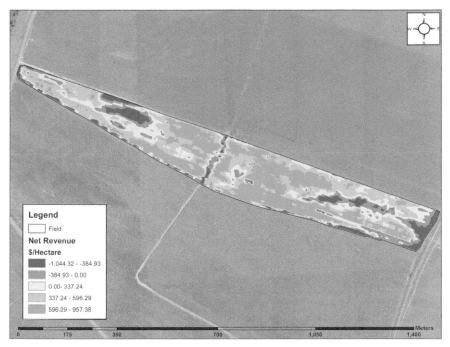

Fig. 4. Profit surface for center-pivot irrigated corn field assuming a $138 Mg⁻¹ commodity price and $1237.53 ha⁻¹ production cost in Mississippi, USA.

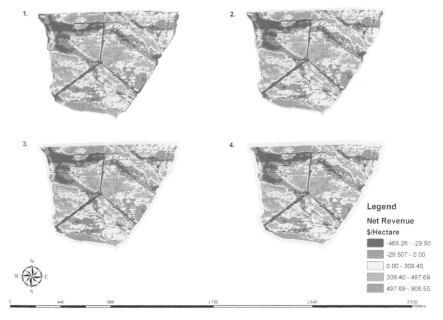

Fig. 5. Profit surface for alternative CP-33 buffer widths on center-pivot irrigated soybean field in Mississippi, USA. (1) 9.1-m CP-33 buffer; (2) 18.2-m CP-33 buffer; (3) 27.4-m CP-33 buffer; and (4) 36.5-m CP-33 buffer.

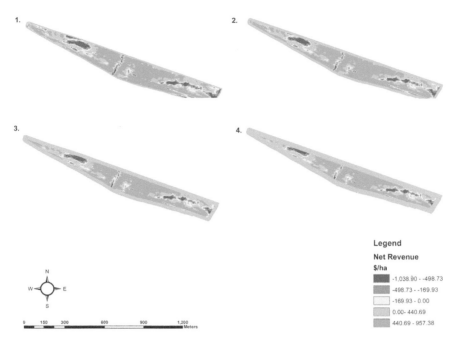

Fig. 6. Profit surface for alternative CP-33 buffer widths on center-pivot irrigated corn field in Mississippi, USA. (1) 9.1-m CP-33 buffer; (2) 18.2-m CP-33 buffer; (3) 27.4-m CP-33 buffer; and (4) 36.5-m CP-33 buffer.

McConnell and Burger (2011) outlined an approach for using profit maps to identify conservation and economic opportunities. They illustrate how low-yielding field regions represent the best candidate locations for strategic conservation enrollment (buffers or whole-field). However, theirs was not the first use of yield monitors for conservation buffer enrollment. Stull et al. (2004) and Barbour (2006) used precision agriculture technology (i.e., global positioning system yield monitors) to identify field regions where monetary benefits of conservation enrollment outweighed agricultural production. Stull et al. (2004) strategically optimized conservation buffer enrollment using historic yield data to identify field margins where revenue from conservation payments exceeded production. Historic yield data was useful for identifying field regions where conservation buffer enrollment could increase field revenue more so than enrolling the whole field in conservation or not enrolling at all (Stull et al., 2004). Specifically, use of precision agriculture to enroll only those areas where current management was below a break-even economic point increased average whole field net revenue most (Stull et al., 2004). Barbour (2006) quantified effects of adjacent plant communities on crop yield near field margins and showed that some adjacent plant communities reduced yield ≤60% relative to field interior. Thus, replacing low-yielding field edges with conservation buffers could be more profitable than cropping (Barbour, 2006). Conservation buffers were economically advantageous up to two combine swaths (14.64 m wide) from the field edge for corn (*Zea mays* L.) fields but not economically advantageous for soybean [*Glycine max* (L.) Merr.] fields in the Gulf

Coast Plain of Mississippi (Barbour, 2006). These two studies represent the pioneering forefront to using precision technology for wildlife conservation.

McConnell and Burger (2011) illustrated the economic outcomes of alternative conservation buffer enrollments on individual production fields. Their results showed varying economic gains from alternative buffer enrollments, thus elucidating the utility of this approach for making informed management decisions. On one field in their study, a minimal conservation buffer width (9.1 m) was the most economically advantageous option (Fig. 7), while on another field, a wider buffer width (27.4 m) maximized mean net revenue of the whole field (Fig. 8). Therefore, precision conservation technology allows producers to evaluate their conservation and economic options at the field level. This type of strategic enrollment will increase producer revenue and conservation adoption.

For a broader investigation across multiple fields, McConnell (2011) used precision conservation technology to simulate the economic outcome of strategic

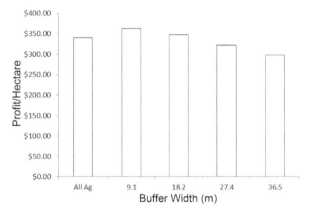

Fig. 7. Whole field net revenue of alternative CP-33 buffer widths on center-pivot irrigated corn field in Mississippi, USA (Mean yield = 11.19 Mg ha^{-1}; commodity price = $138 Mg ha^{-1}).

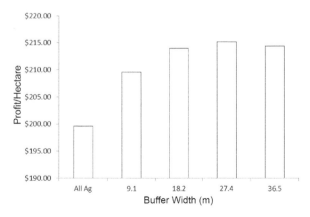

Fig. 8. Whole field net revenue of alternative CP-33 buffer widths on center-pivot irrigated corn field in Mississippi, USA (Mean yield = 11.19 Mg ha^{-1}; commodity price = $138 Mg ha^{-1}).

conservation buffer enrollment on 34 row-crop fields (corn and soybean) across Mississippi. He compared the mean economic tradeoffs of four conservation buffer widths and production without conservation buffers to illustrate the potential economic benefits of strategic conservation enrollment. Results indicated that for corn and soybean fields in Mississippi, conservation buffers increased mean net revenue at differing levels across a range of commodity prices (Fig. 9 and 10). Buffers increased mean net revenue on a percentage of fields for all buffer width and commodity price

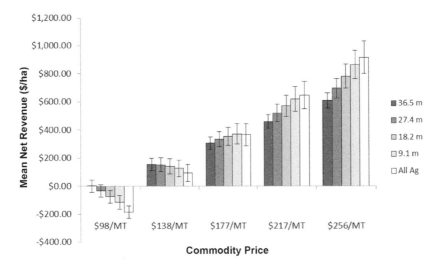

Fig. 9. Per hectare net revenue (±SE) for production only and alternative CP-33 buffer widths averaged for corn fields (*N* = 8) in Monroe County, MS, USA 2007 across multiple commodity prices.

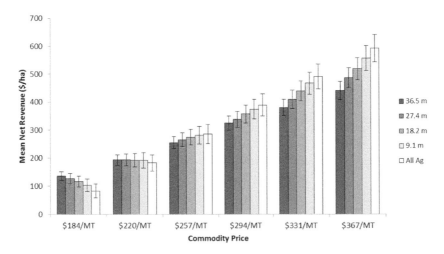

Fig. 10. Per hectare net revenue (±SE) for production only and alternative CP-33 buffer widths averaged for soybean fields (*N* = 26) in Mississippi, USA across multiple commodity prices.

simulations (Fig. 11 and 12). As commodity prices increased, revenue derived from low-yielding land became increasingly competitive with conservation payments. Consequently, increasing commodity prices increased mean net revenue, even at low grain yields, which eventually exceed incentive payments for buffer enrollment. However, even at greater commodity prices, there were locations within a high proportion of fields where buffers offered a competitive economic advantage

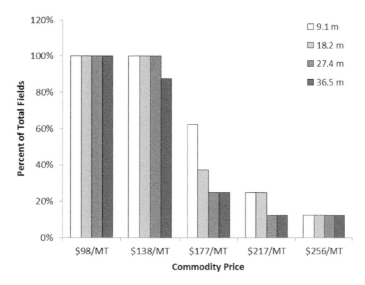

Fig. 11. Percentage of total fields (*N* = 8) where alternative CP-33 buffer widths increase mean net revenue across a range of commodity prices on corn fields in Mississippi, USA.

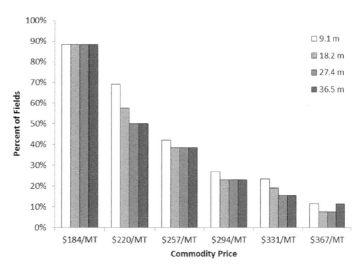

Fig. 12. Percentage of total fields (*N* = 26) where alternative CP-33 buffer widths increase mean net revenue across a range of commodity prices on soybean fields in Mississippi, USA.

to cropping. Although the economic advantage of buffer enrollment decreased at greater commodity price simulations, the DST identified multiple fields that would have increased revenue with buffer implementation. From an economic perspective, applying conservation buffers to all fields within a farm or management area would be illogical if conservation enrollment did not maximize economic returns. However, using precision conservation technology to identify fields and field regions where buffer revenue exceeds that of cropping is a viable management strategy.

For fields where fixed-width buffers decreased revenue, it is important to evaluate the proportion of eligible buffer area where revenue was increased by conservation enrollment. Conservation buffers are not constrained to fixed widths for the whole field (i.e., buffer widths can vary for each field margin). Spatial distribution of yield and profitability is often nonuniform among field margins. Therefore, nonuniform distribution of reduced profitability would warrant nonuniform design of conservation buffers. Evaluating the proportion of eligible buffer area where conservation increases revenue provides information about how spatial arrangement of buffers should be implemented (Fig. 13). Identifying those eligible conservation buffer areas that can generate more revenue enrolled in a conservation practice than cropped across a range of commodity prices provides spatially explicit data to inform the decision making process of buffer placement (Fig. 14 and 15).

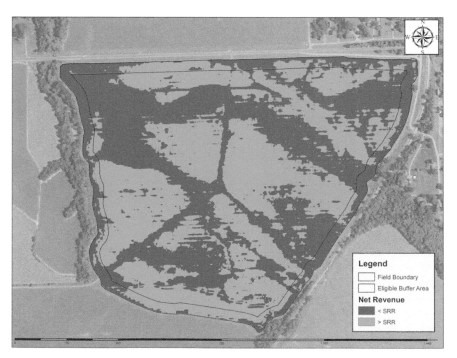

Fig. 13. Adjusted profit surface for a center-pivot irrigated soybean field in Mississippi, USA.

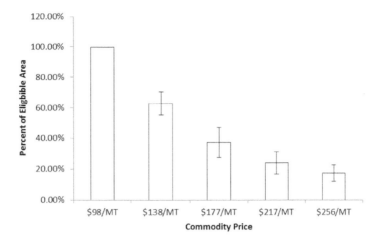

Fig. 14. Percentage of eligible buffer area (±SE) where mean net revenue under CP-33 enrollment exceeds revenue of crop production across a range of commodity prices on corn fields (N = 8) in Mississippi, USA.

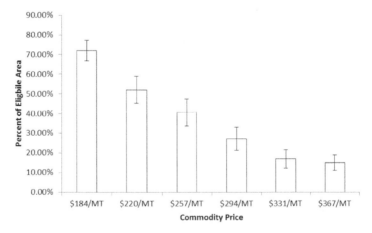

Fig. 15. Percentage of eligible buffer area (±SE) where mean net revenue under CP-33 enrollment exceeds revenue of crop production across a range of commodity prices on soybean fields (N = 26) in Mississippi, USA.

Evaluating the Wildlife Response
to Precision Conservation Enrollment

Linking Monitoring to Conservation Planning

Precision conservation approaches can also be used to predict environmental benefits or services produced by alternative enrollment scenarios. Under 2002 and 2008 Farm Bills, Congress charged USDA with more effectively quantifying environmental outcomes to justify societal investments in agricultural conservation. Blanketing the landscape with a myriad of conservation practices may yield multiple environmental benefits, but those outcomes must be quantifiable. Nontargeted approaches to conservation implementation not only potentially limit environmental benefits but also fail to optimize limited resources available for agrienvironmental conservation (Batary et al., 2010). Similarly, Schonhart et al. (2011) indicated that spatial targeting of agrienvironmental programs is more cost-effective. Effective conservation requires monitoring and evaluation of practices that target specific natural resource goals. Effective monitoring will provide a plethora of information regarding how, when, and where conservation programs and practices work in the landscape, thus improving efficacy of agrienvironmental management schemes (Davey et al., 2010). Monitoring will also provide information needed to build predictive models that can be used to optimize future enrollments. Models that assess which landscape variables, conservation programs, and management practices influence species occurrence, abundance, and life history characteristics will provide a new innovative foundation on which to base future precision conservation enrollment.

Species-specific conservation practices like CP-33 (Habitat Buffers for Upland Birds) are designed to meet a specific conservation objective (e.g., increase Northern Bobwhite [*Colinus Virginianus*] abundance; hereafter, bobwhite). The CP-33 practice was established to address the population recovery goals set by the Northern Bobwhite Conservation Initiative (USDA, 2004a). Upland habitat buffers are native herbaceous communities maintained along cropped field edges. Under CP-33, agricultural landowners can enroll 9.1 to 36.5 m of upland habitat buffers along crop field edges by planting native warm-season grasses, forbs, legumes, and shrubs or by allowing natural succession to occur and maintain them in an early-seral stage. The premise of CP-33 is that relatively small changes in a working agricultural landscape can significantly affect bobwhite and grassland bird abundance. To evaluate this assumption, the USDA Farm Service Agency mandated that bobwhite and priority songbird response to CP-33 implementation be monitored (USDA, 2004b). Results of monitoring have shown greater bobwhite and select grassland bird densities on fields enrolled in CP-33 than fields with no CP-33 (Evans et al., 2009; Evans et al., 2013). These data can also be used to inform conservation planning.

Case Study: Northern Bobwhite

McConnell (2011) used data from the CP-33 national monitoring program to model landscape–population response relationships for bobwhite in northern Mississippi. He then used these relationships to simulate the predicted response to strategic conservation buffer enrollment on 34 row-crop fields in Mississippi. Using variable buffer widths from the previously described economic analysis coupled with grassland bird habitat models, he simulated population response of bobwhite to the amount of CP-33 in the immediate landscape

Results of that study indicated that the predicted bobwhite abundance increased with increasing amount of CP-33 in the landscape. As CP-33 buffer width increased, the amount of CP-33 in the landscape also increased (Fig. 16). On average, every 9 m of increase in buffer width yielded a ~3.72% increase in the amount of CP-33 in the landscape. Similarly, for every incremental increase in CP-33, bobwhite abundance increased 7.66% on average. Predicted bobwhite

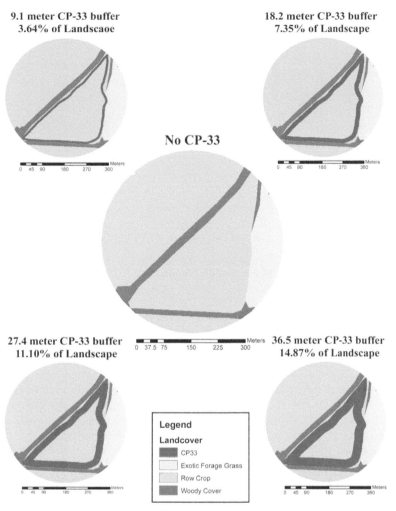

Fig. 16. Landcover simulations of alternative CP-33 buffer widths on agricultural fields in Mississippi, USA.

abundance increased from 0.55 males detected with no CP-33 to 0.85 males detected with 36.5 m of CP-33. Thus, there is a 30.63% increase in predicted abundance from 0 to 14.87% CP-33 in the landscape (Fig. 17). Analysis indicated modest changes in predicted bobwhite abundance with an increase in CP-33 buffers; however, addition of CP-33 (0–3.36%) alone increased abundance ~23.22%. Further incremental increases in CP-33 area yielded a smaller, on average, increase in abundance (i.e., 2.47%). Most noteworthy is the increase in abundance from 0 to 3.64% of the landscape in CP-33, which was equivalent to a 9.1-m buffer around the center field. The presence of a minimum CP-33 enrollment (i.e., 9.1 m) can have a measurable effect on bobwhite abundance. These estimates for bobwhite abundance for landscapes with no CP-33 are likely over estimated as a result of

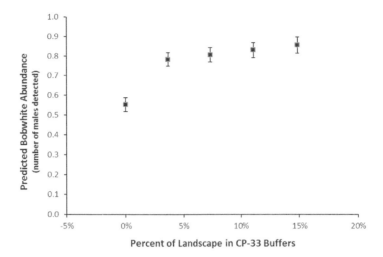

Fig. 17. Predicted northern bobwhite abundance (±SE) in response to percentage of CP-33 in the landscape for 34 250-m radius landscapes in Mississippi, USA.

sampling design and modeling limitations. Therefore estimates of the magnitude of change in bobwhite abundance were likely conservative. The direct outcomes from the national monitoring program, coupled with simulations described here, support the underlying assumption that relatively small changes in primary land use can produce disproportional responses in environmental services. Moreover, characterization of predicted environmental outcomes compliment economic analyses, better informing conservation planning and producer decisions.

The McConnell (2011) results were not dissimilar from previous research investigating bobwhite response to grassland field borders (Puckett et al., 2000; Palmer et al., 2005), which saw measurable increases in bobwhite abundance between buffered and nonbuffered landscapes. Puckett et al. (2000) reported a 59.1% increase in breeding abundance on sites with herbaceous filter strips than those without. Field borders in that study represented 4.9 to 9.4% of the landscape. Similarly, Palmer et al. (2005) reported a 40% increase in breeding abundance on sites with field borders than those without. Smith and Burger (2009) also observed a 23.3% increase in breeding abundance on bordered vs. nonbordered sites with field buffers comprising 0.8 to 1.3% of the landscape. Our modeled response similarly predicted a 23.22% increase in breeding abundance with CP-33 comprising only 3.64% of the landscape. These results represent simulations based on empirical data to predict bobwhite abundance relative to changes in percentage of CP-33 in the landscape and not a measure of difference between controls and treatments. Results are consistent with previous research assessing effects of herbaceous field borders on bobwhite abundance and indicate that a relatively small change in primary land use can produce a disproportionate population response. McConnell (2011) represents the first investigation of both economic and wildlife benefits to strategic conservation enrollment. The approach taken provides insight into the optimal buffer width to maximize field revenue along with the predicted bobwhite population response to that enrollment. When applied at the farm and field level, this precision conservation framework provides landowners with the information they need to make responsible, informed land-use decisions. Although we illustrate this method with a specific conservation practice (CP-33) and a specific wildlife species (northern bobwhite), this approach can be applied to any wildlife species or ecosystem service where tangible data is available. Be it pest control, pollinator habitat, or amphibian corridor creation, the precision conservation framework outlined here is applicable and testable. We encourage the scientific community to build on our framework to improve and elaborate on the multitude of conservation objectives across the landscape.

Conclusions

Effective conservation delivery is dependent on being able to quantify and visualize both the expected costs and benefits. We argue that by identifying economic opportunities for conservation enrollment, increased adoption by landowners is achievable. Our primary goal was to illustrate the necessity, technology, and application of precision conservation in a wildlife management framework. This requires a new and innovative evaluation of how we use the tools of precision farming (i.e., yield monitors). The next step in the evolution of precision conservation is to incorporate wildlife conservation objectives into the framework with which farm-level conservation decisions are made. To do this, soil conservationists, agriculture economists, and wildlife biologists will have to achieve new levels of synergy to face the multitude of global natural resource challenges associated with agricultural intensification. The frontier of precision conservation is the unification of soil, water, air, and wildlife conservation into a common conservation framework. The technological tools are available to inform conservation planning and delivery to a far greater degree than has been previously achieved. The responsibility now lies in the hands of researchers, conservation planners, agricultural consultants, and landowners who, together, must continue to develop new, innovative ways to protect and enhance the ecological integrity of agricultural landscapes.

References

Adams, H.L., L.W. Burger, Jr., and S. Riffell. 2013. Disturbance and landscape effects on avian nests in agricultural conservation buffers. J. Wildlife Manag. 77:1213–1220. doi:10.1002/jwmg.568

Aizen, M.A., L.A. Garibaldi, S.A. Cunnigham, and A.M. Klein. 2009. How much does agriculture depend on pollinators? Lessons from long-term trends in crop production. Ann. Bot. (Lond.) 103:1579–1588. doi:10.1093/aob/mcp076

Allen, C.R., J.J. Fontaine, K.L. Pope, and A.S. Garmestani. 2011. Adaptive management for a turbulent future. J. Environ. Manage. 92:1339–1345. doi:10.1016/j.jenvman.2010.11.019

Askins, R.A. 1999. History of grassland birds in eastern North America. Stud. Avian Biology 19:60–71.

Baligar, V.C., N.K. Fageria, and D.I. He. 2001. Nutrient use efficiency in plants. Commun. Soil Sci. Plant Anal. 32:921–950. doi:10.1081/CSS-100104098

Barbour, P.J. 2006. Ecological and economic effects of field borders in row crop agriculture production systems in Mississippi. Ph.D. diss., Mississippi State University, Starkville.

Batary, P., A. Baldi, D. Kleijn, and T. Tscharntke. 2010. Landscape-moderated biodiversity effects of agri-environmental management: A meta-analysis. Proc. R. Soc. Edinburgh, Sect. B: Biol. 278:1894–1902. doi:10.1098/rspb.2010.1923

Berry, J.K., J.A. Delgado, R. Khosla, and F.J. Pierce. 2003. Precision conservation for environmental sustainability. J. Soil Water Conserv. 58:332–339.

Berry, J.K., J.A. Delgado, F.J. Pierce, and R. Khosla. 2005. Applying spatial analysis for precision conservation across the landscape. J. Soil Water Conserv. 60:363–370.

Bianchi, F.J.J.A., C.J.H. Booij, and T. Tscharntke. 2013. Sustainable pest regulation in agricultural landscapes: A review on landscape composition, biodiversity and natural pest control. Proc. R. Soc. Edinburgh, Sect. B: Biol. 273:1712–1727.

Bianchi, F.J.J.A., W.K.R.E. van Wingerden, A.J. Griffoen, M. van der Veen, M.J.J. van der Straten, R.M.A. Wegman, H.A.M. Meeuwsen. 2005. Landscape factors affecting the control of *Mamestra brassicae* by natural enemies in Brussells sprout. Agric., Ecosyst. Environ. 107:145–150. doi:10.1016/j.agee.2004.11.007

Bianchi F.J.J.A. and W. van der Werf. 2003. The effect of the area and configuration of hibernation sites on the control of aphids by *Coccinella septempunctata* (Coleoptera: Coccinellidae) in agricultural landscapes: A simulation study. Environ. Entomol. 32:1290–1304. doi:10.1603/0046-225X-32.6.1290

Biesmeijer, J.C., S.P.M. Roberts, M. Reemer, R. Ohlemuller, M. Edwards, T. Peeters, A.P. Schaffers, S.G. Potts, R. Kleukers, C.D. Thomas, J. Settele, and W.E. Kunin. 2006. Parallel declines in pollinators and insect-pollinated plants in Britain and the Netherlands. Science 313:351–354. doi:10.1126/science.1127863

Blackwell, B.F., and R.A. Dolbeer. 2001. Decline of Red-winged Blackbird population in Ohio correlated to changes in agriculture (1965-1996). J. Wildl. Manage. 65:661–667. doi:10.2307/3803017

Brennan, L.A., and W.P. Kuvlesky, Jr. 2005. North American grassland birds: An unfolding conservation crisis? J. Wildl. Manage. 69:1–13. doi:10.2193/0022-541X(2005)069<0001:NAGBAU>2.0.CO;2

Burger, L.W., Jr. 2006. Creating wildlife habitat through federal farm programs: An objective-driven approach. Wildl. Soc. Bull. 34:994–999. doi:10.2193/0091-7648(2006)34[994:CWHTFF]2.0.CO;2

Calvete, C., R. Estrada, E. Angulo, and S. Cabezas-Ruiz. 2004. Habitat factors related to wild-rabbit conservation in an agriculture landscape. Landscape Ecol. 19:531–542. doi:10.1023/B:LAND.0000036139.04466.06

Cameron, S.A., J.D. Lozier, J.P. Strange, J.B. Koch, N. Cordes, L.F. Solter, and T.L. Griswold. 2011. Patterns of widespread decline in North American bumble bees. Proc. Natl. Acad. Sci. 108:662–667. doi:10.1073/pnas.1014743108.

Cardinale, B.J., C.T. Harvey, K. Gross, and A.R. Ives. 2003. Biodiversity and control: Emergent impacts of a multi-enemy assemblage on pest suppression and crop yield in an agroecosystem. Ecol. Lett. 6:857–865. doi:10.1046/j.1461-0248.2003.00508.x

Ceballos, G., P.R. Ehrlich, J. Soberón, I. Salazar, and J.P. Fay. 2005. Global mammal conservation: What must we manage? Science 39:603–607. doi:10.1126/science.1114015

Cederbaum, S.B., J.P. Carroll, and R.J. Cooper. 2004. Effects of alternative cotton agriculture on avian and arthropod populations. Conserv. Biol. 18:1272–1282. doi:10.1111/j.1523-1739.2004.00385.x

Chamberlain, D.E., R.J. Fuller, R.G.H. Bunce, J.C. Duckworth, and M. Shrubb. 2000. Changes in the abundance of farmland birds in relation to the timing of agricultural intensification in England and Wales. J. Appl. Ecol. 37:771–788. doi:10.1046/j.1365-2664.2000.00548.x

Conover, R.R., L.W. Burger, Jr., and E.T. Linder. 2007. Winter avian community and sparrow response to field border width. J. Wildl. Manage. 71:1917–1923. doi:10.2193/2006-119

Conover, R.R., L.W. Burger, Jr., and E.T. Linder. 2009. Breeding bird response to field border presence and width. Wilson J. Ornithol. 121:548–555. doi:10.1676/08-082.1

Conover, R.R., L.W. Burger, Jr., and E.T. Linder. 2011a. Grassland Bird Nest Ecology and Survival in Upland Habitat Buffers near Wooded Edges. Wildl. Soc. Bull. 35:353–361. doi:10.1002/wsb.87

Conover, R.R., S.J. Dinsmore, and L.W. Burger, Jr. 2011b. Effects of conservation practices on bird nest density and survival in intensive agriculture. Agric. Ecosyst. Environ. 141:126–132. doi:10.1016/j.agee.2011.02.022

Davey, C., J. Vickery, N. Boatman, D. Chamberlain, H. Parry, and G. Siriwardena. 2010. Regional variation in the efficacy of entry level stewardship in England. Agric. Ecosyst. Environ. 139:121–128. doi:10.1016/j.agee.2010.07.008

Davic, R.D., and H.H. Welsh, Jr. 2004. On the ecological roles of salamanders. Annu. Rev. Ecol. Evol. Syst. 35:405–434. doi:10.1146/annurev.ecolsys.35.112202.130116

de la Peña, N.M., A. Butet, and Y. Delettre. 2003. Response of small mammal community to changes in western French agricultural landscapes. Landscape Ecol. 18:265–278. doi:10.1023/A:1024452930326

Delgado, J.A., and W.C. Bausch. 2005. Potential use of precision conservation techniques to reduce nitrate leaching in irrigated crops. J. Soil Water Conserv. 60:379–387.

Delgado, J.A., and J.K. Berry. 2008. Advances in precision conservation. Adv. Agron. 98:1–44. doi:10.1016/S0065-2113(08)00201-0

Delgado, J.A., R. Khosla, W.C. Bausch, D.G. Westfall, and D. Inman. 2005. Nitrogen fertilizer management based on site specific management zones reduce potential for nitrate leaching. J. Soil Water Conserv. 60:402–410.

Dosskey, M.G., D.E. Eisenhauer, and M.J. Helmers. 2005. Establishing conservation buffers using precision information. J. Soil Water Conserv. 60:349–354.

Environmental Systems Research Institute. 2009. ArcGIS Desktop and Spatial Analyst. ESRI, Inc. Redlands, CA.

Evans, K.O., L.W. Burger, Jr., C. Oedekoven, M.D. Smith, S.K. Riffell, J.A. Martin, and S.T. Buckland. 2013. Multi-region response to conservation buffers targeted for northern bobwhite. J. Wildl. Manage. 77:716–726. doi:10.1002/jwmg.502

Evans, K.O., L.W. Burger, M.D. Smith, and S. Riffell. 2009. Conservation reserve program. CP33–Habitat buffers for upland birds. Bird monitoring and evaluation plan. 2006–2011 Final Report. Mississippi State Univ., Forest and Wildlife Res. Center.

Fahrig, L., and G. Merriam. 1985. Habitat patch connectivity and population survival. Ecology 66:1762–1768. doi:10.2307/2937372

Fitzpatrick, U., T.E. Murray, R.J. Paxton, J. Breen, D. Cotton, V. Santorum, and M.J.F. Brown. 2007. Rarity and decline in bumblebees: A test of causes and correlates in the Irish fauna. Biol. Conserv. 136:185–194. doi:10.1016/j.biocon.2006.11.012

Free, J.B. 1993. Insect pollination of crops. Academic Press, London.

Gallant, A.L., R.W.W. Klaver, G.S. Casper, and J.J. Lannoo. 2007. Global rates of habitat loss and implications for amphibian conservation. Copeia 2007:967–979. doi:10.1643/0045-8511(2007)7[967:GROHLA]2.0.CO;2

Gibbons, J.W., D.E. Scott, T.J. Ryan, K.A. Buhlmann, T.D. Tubervill, B.S. Metts, J.L. Greene, T. Mills, Y. Leiden, S. Poppy, and C.T. Winne. 2000. The global decline of reptiles, déjà vu amphibians. Bioscience 50:655–666.

Gibbons, J.W., C.T. Winne, D.E. Scott, J.D. Willson, X.A. Glaudas, K.M. Andrews, B.D. Todd, L.A. Fedewa, L. Wilkinson, R.N. Tsaliagos, S.J. Harper, J.L. Greene, T.D. Tuberville, B.S. Metts, M.E. Dorcas, J.P. Nestor, C.A. Young, T. Akre, R.N. Reed, K.A. Buhlmann, J. Norman, D.A. Croshaw, C. Hagen, and B.B. Rothermel. 2006. Remarkable amphibian biomass and abundance in an isolated wetland: Implications for wetland conservation. Conserv. Biol. 20:1457–1465. doi:10.1111/j.1523-1739.2006.00443.x

Goulson, D. 2003. Conserving wild bees for crop pollination. J. Food Agric. Environ. 1:142–144.

Green, R.E., S.J. Cornell, J.P.W. Scharlemann, and A. Balmford. 2005. Farming and the fate of wild nature. Science 307:550–555. doi:10.1126/science.1106049

Gurr, G.M., S.D. Wratten, J. Tylianakis, J. Kean, and M. Keller. 2005. Providing plant foods for natural enemies in farming systems: Balancing practicalities and theory. In: F.L. Wackers, P.C.J. van Rijn, and J. Bruin, editors, Plant provided food for carnivorous insects: A protective mutualism and its applications. Cambridge Univ. Press, New York, NY. p. 341–347.

Haufler, J.B., editor. 2005. Fish and wildlife benefits of Farm Bill conservation programs: 2000–2005 update. The Wildlife Society Technical Review 05-02, Bethesda, Maryland, USA.

Heard, L.P., A.W. Allen, L.B. Best, S.J. Brady, L.W. Burger, Jr., A.J. Esser, E. Hackett, R.R. Helinksi, W.L. Hohman, D.H. Johnson, R.L. Pederson, R.E. Reynolds, C. Rewa, M.R. Ryan, R.T. Molleur, and P. Buck. 2000. A comprehensive review of Farm Bill contributions to wildlife conservation, 1985–2000. USDA–NRCS, Wildlife Habitat Management Institute, Technical Report, USDA/NRCS/WHMI-2000, Madison, MS.

Heimpel, G.E., and M.A. Jervis. 2005. Does nectar improve biological control by parasitoids. In: F.L. Wackers, P.C.J. van Rijn, and J. Bruin, editors, Plant provided food for carnivorous insects: A protective mutualism and its applications. Cambridge Univ. Press, New York, NY. p. 267–304.

Henry, A.C., D.A. Hosack, C.W. Johnson, D. Rol, and G. Bentrup. 1999. Conservation corridors in the United States: Benefits and planning guidelines. J. Soil Water Conserv. 54:645–650.

Herkert, J.R. 1994. Breeding bird communities of mid-western prairie fragments: The effect of prescribed burning and habitat area. Nat. Areas J. 14:128–135.

Hoffmann, M., J.L. Belant, J.S. Chanson, N.A. Cox, J. Lamoreaux, A.S.L. Rodrigues, J. Schipper, and S.N. Stuart. 2011. The changing fates of the world's mammals. Philos. Trans. R. Soc., B 366:2598–2610. doi:10.1098/rstb.2011.0116

Holzkamper, A., and R. Seppelt. 2007. Evaluating cost-effectiveness of conservation management actions in an agricultural landscape on a regional scale. Biol. Conserv. 136:117–127. doi:10.1016/j.biocon.2006.11.011

Hutchens, S., and C. DePerno. 2009. Measuring species diversity to determine land-use effects on reptile and amphibian assemblages. Amphibia–Reptilia 30:81–88. doi:10.1163/156853809787392739

Hyberg, B.T., and R. Riley. 2009. Floodplain ecosystem restoration: Commodity markets, environmental services, and the Farm Bill. Wetlands 29:527–534. doi:10.1672/08-132.1

Ingram, M., G. Nabhan, and S. Buchmann. 1996. Our forgotten pollinators: Protecting the birds and bees. Global Pesticide Campaigner 6.1–8.

Jacob, J. 2003. Short-term effects of farming practices on populations of common voles. Agric. Ecosyst. Environ. 95:321–325. doi:10.1016/S0167-8809(02)00084-1

Kearns, C.A., D.W. Inouye, and N.M. Waser. 1998. Endangered mutualisms: The conservation of plant-pollinator interactions. Annu. Rev. Ecol. Syst. 29:83–112. doi:10.1146/annurev.ecolsys.29.1.83

Kitchen, N.R., K.A. Sudduth, D.B. Myers, R.E. Massey, E.J. Sadler, R.N. Lerch, J.W. Hummel, and H.L. Palm. 2005. Development of a conservation-oriented precision agriculture system: Crop production assessment and plan implementation. J. Soil Water Conserv. 60:421–430.

Klein, A.M., B.E. Vaissiere, J.H. Cane, I. Steffan-Dewenter, S.A. Cummingham, C. Kremen, and T. Tscharntke. 2007. Importance of pollinators in changing landscapes for world crop. Proc. Biol. Sci. 247:1226–1228.

Lal, R. 2000. A modest proposal for the year 2001: We can control greenhouse gases and feed the world…with proper soil management. J. Soil Water Conserv. 55:429–433.

Landis, D.A., S.D. Wratten, and G.M. Gurr. 2000. Habitat management to conserve natural enemies of arthropod pests in agriculture. Annu. Rev. Entomol. 45:175–201. doi:10.1146/annurev.ento.45.1.175

Lehtinen, R.M., S.M. Galatowitsch, and J.R. Testet. 1999. Consequences of habitat loss and fragmentation for wetland amphibian assemblages. Wetlands 19:1–12. doi:10.1007/BF03161728

Lovell, S.T. and W.C. Sullivan. 2006. Environmental benefits of conservation buffers in the United States: Evidence, promise, and open questions. Agric., Ecosyst. Environ. 112:249–260. doi:10.1016/j.agee.2005.08.002

Mackenzie, A., A.S. Ball, and S.R. Virdee. 1998. Section V1, Rare species, habitat loss, and extinction. In: A. Mackenzie, A.S. Ball, and S.R. Virdee, editors, Instant notes in ecology. BIOS Scientific Pub. Ltd., New York. p. 265–270.

Maisonneuve, C., and S. Rioux. 2001. Importance of riparian habitats for small mammal and herpetofaunal communities in agricultural landscapes of southern Québec. Agric. Ecosyst. Environ. 83:165–175. doi:10.1016/S0167-8809(00)00259-0

Marshall, E.J.P., and A.C. Moonen. 2002. Field margins in northern Europe: Their functions and interactions with agriculture. Agric. Ecosyst. Environ. 89:5–21. doi:10.1016/S0167-8809(01)00315-2

McConnell, M.D. 2011. Using precision agriculture technology to evaluate environmental and economic tradeoffs of alternative CP-33 enrollments. M.S. thesis, Mississippi State University, Starkville.

McConnell, M.D., and L.W. Burger. 2011. Precision conservation: A geospatial decision support tool for optimizing conservation and profitability in agricultural landscapes. J. Soil Water Conserv. 66:347–354. doi:10.2489/jswc.66.6.347

McGregor, S.E. 1976. Insect pollination of cultivated plants. Agric. Handbook 496. USDA–ARS, Washington, DC.

Medan, D., J.P. Torreta, K. Hodara, E.B. de la Fuente, and N.H. Montaldo. 2011. Effects of agriculture expansion and intensification on the vertebrate and invertebrate diversity in the Pampas of Argentina. Biodivers. Conserv. 20:3077–3100. doi:10.1007/s10531-011-0118-9

Meek, B., D. Loxton, T.H. Sparks, R.F. Pywell, H. Pickett, and M. Nowakowski. 2002. The effect of arable field margin composition on invertebrate biodiversity. Biol. Conserv. 106:259–271. doi:10.1016/S0006-3207(01)00252-X

Murphy, M.T. 2003. Avian population trends with the evolving agricultural landscape of eastern and central United States. Auk 120:20–34. doi:10.1642/0004-8038(2003)120[0020:APTWTE]2.0.CO;2

Myers, J.H., C. Higgins, and E. Kovacs. 1989. How many insect species are necessary for the biological control of insects? Environ. Entomol. 18:541–547. doi:10.1093/ee/18.4.541

Naylor, R.L., and P.R. Ehrlich. 1997. Natural pest control services and agriculture. In: G.C. Daily, editor, Nature's services: Societal dependence on natural ecosystems. Island Press, Washington, DC. p. 151–174.

Ostman, O., B. Ekbom, and J. Bengtsson. 2003. Yield increase attributable to aphid predation by ground-living polyphagous natural enemies in spring barely in Sweden. Ecol. Econ. 45:149–158. doi:10.1016/S0921-8009(03)00007-7

Palma, L., P. Beja, and M. Rodrigues. 1999. The use of sighting data to analyze Iberian lynx habitat distribution. J. Appl. Ecol. 36:812–824. doi:10.1046/j.1365-2664.1999.00436.x

Palmer, W.E., S.D. Wellendorf, J.R. Gills, and P.T. Bromley. 2005. Effects of field borders and nest-predator reduction on abundance of northern bobwhites. Wildl. Soc. Bull. 33:1398–1405. doi:10.2193/0091-7648(2005)33[1398:EOFBAN]2.0.CO;2

Pierce, F.J., and P. Nowak. 1999. Aspects of precision agriculture. Adv. Agron. 67:1–85. doi:10.1016/S0065-2113(08)60513-1

Pocock, M.J., and N. Jennings. 2008. Testing biotic indicator taxa: The sensitivity of insectivorous mammals and their prey to intensification of lowland agriculture. J. Appl. Ecol. 45:151–160. doi:10.1111/j.1365-2664.2007.01361.x

Puckett, K.M., W.E. Palmer, P.T. Bromely, J.R. Anderson, and T.L. Sharpe. 2000. Effects of filter strips on habitat use and home range of northern bobwhites on Alligator River National Wildlife Refuge. In: L.A. Brennan, W.E. Palmer, L.W. Burger, Jr., and T.L. Pruden, editors, Quail IV: Proc. of the Fourth National Quail Symp. Tall Timbers Research Station, Tallahassee, FL. p. 26–31

Richards, A.J. 2001. Does low biodiversity resulting from modern agricultural practice affect crop pollination and yield? Ann. Bot. (Lond.) 88:165–172. doi:10.1006/anbo.2001.1463

Rittenhouse, T.A.G., and R.D. Semlitsch. 2007. Distribution of amphibians in terrestrial habitat surrounding wetlands. Wetlands 27:153–161. doi:10.1672/0277-5212(2007)27[153:DOAITH]2.0.CO;2

Robertson, G.P., and S.M. Swinton. 2005. Reconciling agricultural productivity and environmental integrity: A grand challenge for agriculture. Front. Ecol. Environ. 3:38–46.

Sauer, J.R., J.E. Hines, and J. Fallon. 2008. The North American Breeding Bird Survey, results and analysis 1966–2007. Version 5.15.2008. USGS Patuxent Wildlife Research Center, Laurel, MD.

Schonhart, M., T. Schauppenlehner, E. Schmid, and A. Muhar. 2011. Integration of biophysical and economic models to analyze management intensity and landscape structure effects at farm and landscape level. Agric. Syst. 104:122–134. doi:10.1016/j.agsy.2010.03.014

Semlitsch, R.D., and J.R. Bodie. 2003. Biological criteria for buffer zones around wetlands and riparian habitats for amphibians and reptiles. Conserv. Biol. 17:1219–1228. doi:10.1046/j.1523-1739.2003.02177.x

Shibusawa, S. 1998. Precision farming and terra-mechanics. Fifth ISTVS Asia-Pacific Regional Conference in Korea, 20–22 October.

Shueller, D., H. Brunken-Winkler, P. Busch, M. Forster, P. Janiesch, R. Lemm, R. Niedringhaus, and H. Stasser. 2000. Sustainable land use in an agriculturally misused landscape in northwest Germany through ecotechnical restoration by a 'Patch-Network-Concept'. Ecol. Eng. 16:99–117. doi:10.1016/S0925-8574(00)00094-X

Smith, M.D., P.J. Barbour, L.W. Burger, and S.J. Dinsmore. 2005. Density and diversity of overwintering birds in managed field borders in Mississippi. Wilson Bull. 117:258–269. doi:10.1676/04-097.1

Smith, M.D., and L.W. Burger. 2009. Population response of northern bobwhite to field border management practices in Mississippi. In: S.B. Cedarbaum, B.C. Faircloth, T.M. Terhune, J.J. Thompson, and J.P. Carroll, editors, Gamebird 2006: Quail VI and Perdix XII. 31 May-4 June 2006. Warnell School of Forestry and Natural Resources, Athens, GA. p. 220–231.

Stull, J., C. Dillon, S. Shearer, and S. Isaacs. 2004. Using precision agriculture technology for economically optimal strategic decisions: The case of CRP filter strip enrollment. J. Sustain. Agric. 24:79–96. doi:10.1300/J064v24n04_07

Swinton, S.M., F. Lupi, G.P. Robertson, and S.K. Hamilton. 2007. Ecosystem services and agriculture: Cultivating agricultural ecosystem services for diverse benefits. Ecol. Econ. 64:245–252. doi:10.1016/j.ecolecon.2007.09.020

Todd, I.A., T.E. Tew, and D.W. McDonald. 2000. Arable habitat use by wood mice (*Apodemus sylvaticus*) 1. Macrohabitat. J. Zool. 250:299–303. doi:10.1111/j.1469-7998.2000.tb00773.x

USDA. 2003. Natural resource inventory. USDA–NRCS, Resource Inventory Division. http://www.nrcs.usda.gov/wps/portal/nrcs/main/national/technical/nra/nri/ (accessed 31 July 2008).

USDA. 2004a. Notice CRP-479 practice CP-33: Habitat buffers for upland birds. USDA, Farm Service Agency, Washington, DC.

USDA. 2004b. Practice CP33 habitat buffers for upland wildlife. Farm Service Agency, Notice CRP-479, Washington, DC.

US Department of the Interior. 2006. Strategic habitat conservation: Final report of the national ecological assessment team. US Department of the Interior. Washington, DC. http://www.fws.gov/landscape-conservation/pdf/SHCReport.pdf (accessed 26 June 2013)

Vickery, P.C., and J.R. Herkert. 1999. Ecology and conservation of grassland birds of the Western Hemisphere. Studies in Avian Biology No. 19. Cooper Ornithological Society, Camarillo, CA.

Waddle, J.H., B.M. Gloriso, and S.P. Faulkner. 2013. A quantitative assessment of the conservation benefits of the Wetlands Reserve Program to amphibians. Restor. Ecol. 21:200–206. doi:10.1111/j.1526-100X.2012.00881.x

Whelan, B.M., and A.B. McBratney. 2000. The "Null Hypothesis" of precision agriculture management. Precis. Agric. 2:265–279. doi:10.1023/A:1011838806489

Wilby, A., and M.B. Thomas. 2002. Natural enemy diversity and pest control: Patterns emergence with agricultural intensification. Ecol. Lett. 5:353–360. doi:10.1046/j.1461-0248.2002.00331.x

Wilcove, D.S., D. Rosthein, J. Dubow, A. Phillips, and E. Losos. 1998. Quantifying threats to imperiled species in the United States. Bioscience 48:607–615. doi:10.2307/1313420

Wissinger, S.A. 1997. Cyclic colonization in predictably ephemeral habitat: A template for biological control in annual crop systems. Biol. Control 10:4–15. doi:10.1006/bcon.1997.0543

Zhang, N., M. Wang, and N. Wang. 2002. Precision agriculture: A worldwide overview. Comput. Electron. Agric. 36:113–132. doi:10.1016/S0168-1699(02)00096-0

Precision Conservation and Water Quality Markets

Ali Saleh* and Edward Osei

Abstract

Water quality trading affords an alternative to regulatory policies for reducing pollutant emissions to water bodies from nonpoint sources such as agriculture. The estimation process of the water quality credits generated by farmers play a significant role in trading. In general, the number of credits generated by the farmer is expressed as a function of the portfolio of conservation practices (CPs) implemented on the farm in lieu of more expensive field monitoring. The generated water quality credits are highly related to spatial attributes, which include factors such as soil type, topography, and soil chemical and physical properties as well as land management practices. By making use of the information on spatial distributions of these field attributes, precision conservation can play a key role in improving the effectiveness of CPs and reducing the uncertainty associated with those practices. In this chapter, the major precision CPs that provide opportunity for generating credits in water quality markets are described. Tools that are readily available, such as the Comprehensive Economic Environmental Optimization Tool (CEEOT) at the watershed-scale level and the more user-friendly Nutrient Tracking Tool (NTT) at the farm and small-watershed level, can be used by users to determine the optimal practice implementation for their farms that embodies precision conservation approaches to maximize the benefit from each dollar of CP investment.

Water quality trading affords an opportunity for pollutant emissions in a watershed to be reduced at lower cost than by regulatory policies that require specific emissions reductions at each pollution source. For many decades, economists touted the rationale of emissions trading. However, while trading between point sources has flourished in the United States since the 1970s, trading involving nonpoint sources, such as agriculture, have only gained momentum in the past two decades. A recent USEPA–USDA partnership for trading provided much

Abbreviations: APEX, Agricultural Policy eXtender; ASAE, American Society of Agricultural Engineers; CEEOT, Comprehensive Economic Environmental Optimization Tool; CP, conservation practice; FEM, Farm-Level Economic Model; NTT, Nutrient Tracking Tool; SWAT, Soil and Water Assessment Tool; TIAER, Texas Institute for Applied Environmental Research.

Ali Saleh, Associate Director/Professor, and Edward Osei (osei@tiaer.tarleton.edu), Senior Research Economist, Texas Institute for Applied Environmental Research, Tarleton State Univ., Stephenville, TX. *Corresponding author (saleh@tiaer.tarleton.edu).

doi:10.2134/agronmonogr59.2013.0032

© ASA, CSSA, and SSSA, 5585 Guilford Road, Madison, WI 53711, USA.
Precision Conservation: Geospatial Techniques for Agricultural and Natural Resources Conservation
J.A. Delgado, G.F. Sassenrath, and T. Mueller, editors. Agronomy Monograph 59.

needed impetus to water and air quality trades between agricultural land uses and point sources of pollutant emission (USDA–USEPA, 2013).

In typical natural resource trading markets, water quality trading between point source polluters and agricultural landowners allows point-source polluters to avoid costly pollution abatement investments by buying emissions credits from farmers who install CPs to offset the pollutant emissions of the point sources. The economic rationale is that trades will occur if it is more cost-effective for farmers to reduce pollutant emissions from their fields beyond required levels than for the point source to bring levels of the same pollutant emitted from their sources down to required levels. In reality, the difference in cost-effectiveness between the point and nonpoint sources must be large enough to more than offset the transactions costs entailed during the trade.

The practical implementation of water quality markets between point and nonpoint sources has been hampered primarily by uncertainty about how to estimate the water quality credits the farmer has available to legally engage in the trade. In other words, while point-source emission levels can readily be estimated, the pollutant offsets generated by the nonpoint source are much more difficult to quantify. To address this issue, water quality markets often associate a prespecified degree of pollution reduction with the type of CP implemented. Thus the amount of credits generated by the farmer is expressed as a function of the portfolio of CPs implemented on the farm in lieu of more expensive field monitoring.

Assigning exogenous pollution reduction rates to CPs reduces the burden of proof associated with quantifying emissions credits. However, the effectiveness of CPs themselves is subject to a wide degree of variability. Such variability introduces concerns in water quality markets that are often addressed by special adjustments in credits known as margins of safety. Among others, the margin of safety often accounts for the uncertainty inherent in the effectiveness of the CPs. The margin of safety essentially increases the amount of credits a farmer must generate to offset a unit of pollutant generated by the point source. Consequently, a higher margin of safety generally means fewer opportunities for trading parties to benefit from trading.

Variability in CP effectiveness is a largely spatial phenomenon. Across a watershed, CPs implemented in different locations may have widely different impacts as a result of weather patterns. Within a field, differences can also be seen in the effectiveness of the same practice as a result of differences in soil type, topography, and soil chemical and physical properties as well as land management practices. By making use of the information on spatial distributions of these field and watershed attributes, precision conservation can play a key role in improving the effectiveness of CPs and reducing the uncertainty associated with those practices. This improvement, in turn, leads to lower margins of safety and greater trading opportunities.

The remainder of this chapter is organized as follows. First, a discussion of the variability of CP effectiveness is presented along with tools that can be used to quantify this variability. Second, precision conservation is presented as an approach that achieves two primary objectives: (i) it improves the overall per-unit-area effectiveness of CPs and (ii) reduces the variability and uncertainty inherent in the effectiveness of these practices. The implications of these improvements are discussed followed by closing comments.

Variability of Conservation Practice Effectiveness

A wide variety of CPs have been proposed for reducing sediment and nutrient losses in runoff from agricultural lands. In many instances, field experiments suggest much higher levels of effectiveness than what is realized in practice particularly at the watershed outlet. To enhance appreciation of the variability of these practices, the following sections provide a brief outline of the most prevalent CPs and the ranges of effectiveness associated with them.

Nonstructural Practices

Alternative Tillage Practices

Tillage practices, as well as nutrient applications, are part of the major nonstructural practices on cropland or pasture. Tillage operations are done mainly as part of the preparation of land for planting, weed control, and harvesting operations. These operations also impact the amount, mode, and form of nutrient (fertilizer or manure) and herbicides applied.

Tillage operations including the depth, timing, frequency, and type also affect water quality and quantity and crop production. Conservation tillage practices can reduce soil erosion from wind and water. Conservation tillage can also positively impact water quality by reducing nutrient losses (both soluble and insoluble).

Planting systems, such as no-till, that leave the soil surface undisturbed until the time of planting consistently leave the highest levels of crop residue and involve planting or drilling seed into a narrow seedbed prepared by coulters (no-till), ridge-scrapers (ridge-till), disk openers, and other attachments. These systems result in lower loss of sediment and sediment-bound nutrients, higher water uptake by the plant, and higher soil organic matter from residues remaining on the soil surface. Conservation tillage systems change the soil's physical properties such as an increased infiltration rate caused by the development of macropores in the soil created by earthworm activity, soil cracking, and root growth.

Similar to N and P, most herbicides attach to soil particles or dissolve in surface runoff. Thus, conservation tillage also effectively reduces herbicide runoff. Most studies show that no-till systems usually reduce herbicide runoff by up to 70% than conventional systems.

Nutrient Management

Nutrient management is specified by date, frequency within a year, rate, form (i.e., commercial fertilizer or manure application), and mode (i.e., surface or injected) of application. The method, amount, timing, and type of applied nutrients play an important role in the fate and transport of nutrients and the resulting impacts on crop yields and sediment and nutrient losses from the area of interest. The following is a brief discussion of the most important impacts of nutrient management on water quality, quantity, and crop production:

Amount of Applied Nutrient: Insufficient nutrient application results in lower crop yield and consequently less surface residue and ultimately higher soil erosion. On the other hand, excessive application results in loss of nutrients to water bodies and higher cost of farming for farmers. Therefore, the right amount of nutrient is needed to produce the highest crop yield while sustaining the water quality standards. Overapplication of nutrients can occur by not crediting the existing nutrient and organic matter levels in the soil profile. This could easily be prevented by conducting annual soil nutrient (e.g., soil P) tests and adjusting the nutrient application rate accordingly.

Nutrient Application Timing: Timing of nutrient application has a significant effect on the loss of nutrients and sediment and crop production. Various studies have shown that split application of N before planting and during the high-crop-demand period (e.g., June) results in over 30% reduction in required N for maximum crop yield as compared to fall application. The correct nutrient application timing will also result in reduction of N loss to water bodies and higher income potential for producers.

Nutrient Type: The type of nutrient applied in the field affects the loss of nutrient and crop production level in two ways: The first way is through excessive application of P and organic N by high rates of manure application. Furthermore, manure application is often done without accounting for manure nutrients, which consequently results in over-application of commercial fertilizer. The second way is by incorrect calculation of the nutrient applied to the field. For instance, applying the correct amount of P in the form of P_2O_5 requires the correct calculation of P using the molecular weight of P in the solution.

Nutrient Placement: Nutrient placement has a significant effect on the efficiency of nutrient uptake by plants. Broadcast and surface application of nutrients during rainy seasons could cause significant nutrient loss by runoff events. This is more pronounced under no-till conditions, since nutrients are often left on the soil surface when no-till is practiced. To prevent this loss, one needs to incorporate the nutrients by knifing. Also, incorporating some nutrients, such as urea, immediately after application will reduce the risk of volatilization.

Cover Crops

Cover crops are crops grown to protect and enrich the soil. In agricultural conservation efforts, cover crops are grown primarily to protect the soil. Traditionally, cover crops are winter annuals—legumes or grasses—that are planted after harvest of the main cash crop and removed by tillage or herbicide application just before planting the subsequent year's cash crop. Cover crops also vary widely in effectiveness of reducing sediment and nutrient losses. The variability in cover crop effectiveness is normally based on cover crop type, planting date, and management in addition to weather and the physical properties of the field.

Grazing

Grazing operations also have significant effects on nutrient and sediment losses (Fig. 1). Appropriate grazing management will serve to reduce denuded areas, minimize soil loss, and enhance pasture establishment. Key herd management characteristics that impact nutrient losses include forage intake rate (kg head^{-1} d^{-1}), grazing efficiency (accounts for waste by trampling, etc.), manure production rate (kg head^{-1} d^{-1}), urine production (L head^{-1} d^{-1}), and soluble and organic N, P, and C fractions in the manure.

Fig. 1. An example of a dairy cow grazing operation (Source: USDA-NRCS, http://www. nrcs.usda.gov).

Irrigation and Fertigation

Irrigation types, including sprinkler (Fig. 2), drip, furrow and flood, and furrow diking, could play an important role in crop production in many regions of the world. Irrigation can also cause runoff and consequently loss of nutrient and sediment from the field. Irrigation also plays an important role in protecting delicate crops (e.g., vineyards) from freezing events during winter.

Structural Practices

A number of the farm structural practices, as described in the USDA-NRCS's "National Conservation Practice Standards" (NRCS, 2016), impact the crop yield and nutrient and sediment losses from an area of interest. The following are among the most commonly adopted management practices by producers to sustain and, in some cases, improve water and soil quality while increasing farm

Fig. 2. A center-pivot (sprinkler) irrigation; an example of one of the different types of irrigation simulated in the Nutrient Tracking Tool (Source: USDA-NRCS, http://www. al.nrcs.usda.gov).

production. These structural practices address soil management through stabilization and control of water and soil on and off agricultural fields.

Tile (Controlled) Drainage

Tile drainage is a practice that removes excess water from the subsurface layers of soils intended for agriculture (Fig. 3). Drainage brings excessive soil moisture levels down for optimal crop growth. Tile drainage is often the best recourse for reducing high subsurface water levels to improve crop yields (Table 1). Too much subsurface water can be counterproductive to agriculture by preventing root development and thereby inhibiting the growth of crops. Excessive water can also limit access to the land particularly by farm machinery.

Wetlands

Wetlands are those areas that are inundated or saturated by surface or groundwater at a frequency and duration sufficient to support—under normal circumstances— a prevalence of vegetation typically adapted for life in saturated soil conditions (Fig. 4). Wetlands include swamps, marshes, bogs, and similar areas. In addition, wetlands can be constructed on the farmland by producers (NRCS, 2010) to treat wastewater and contaminated runoff from agricultural processing, livestock, and

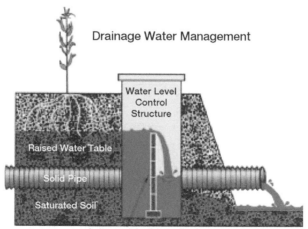

Fig. 3. A typical agricultural tile drainage system (Source: USDA-NRCS, http://www.ny.nrcs.usda.gov).

Table 1. Tile (controlled) drainage efficiency from experimental studies. Adapted from Kelly et al. (2014).

Study area and scale	Pollutant	Removal efficiency	Reference
		%	
Ontario, farm level	Nitrate	65	Agriculture and Agri-Food
	Phosphorus	65	Canada (2013)
Ontario, watershed scale	Nitrogen	50–100	

Fig. 4. A type of wetland simulated in the Nutrient Tracking Tool as a shallow reservoir with growing vegetation (Source: Chisago Soil & Water Conservation District, www. chisagoswcd.org).

aquaculture facilities. A well-constructed wetland could improve the quality of storm water runoff or other contaminated water by withholding runoff water (Table 2). Wetlands, in most cases, trap sediment, reduce runoff, increase infiltration, and also increase uptake of water and nutrients by the vegetation in the area covered by the wetland.

Filter Strips

A filter strip is an area of vegetation, generally narrow in width and long across the downslope edge of a field, that slows the rate of runoff, allowing sediments, organic matter, and other pollutants that are being conveyed by the water to be removed by settling out (Fig. 5). Filter strips reduce erosion and the accompanying sediment-bound pollution (Dosskey et al., 2016, this volume). Filter strips function by providing for better infiltration of soluble nutrients, trapping of sediment, and uptake of water and nutrients by the filter strip vegetation. One side effect of the filter strip is the reduction of crop yield because of removal of land

Table 2. Constructed wetland efficiency from experimental studies.

Study area	Pollutant	Removal efficiency	Reference
		%	
Western Australia	Litter	>95%	Fletcher et al., 2003
	Sediment	65–95	
	Total N	40–80	
	Total P	60–85	
	Coarse sediment	>95	
	Heavy metals	55–95	
Urban stormwater runoff,	Sediment	65	USEPA, 1993
United States†	Total P	25	
	Total N	20	
	Chemical oxygen demand	50	
	Lead	65	
	Zinc	35	

† Adapted from Spaulding (1996).

Fig. 5. An agricultural field with a grass filter strip at the edge of the field (Source: USDA-NRCS, http://www.oh.nrcs.usda.gov).

for the filter strip. Another side effect is the high concentration of sediment in the filter strip area under excessive upland soil erosion when the proper land management is not practiced in the farmland. Grismer et al. (2006) summarizes the results of various field studies on the effectiveness of filter strips in reducing non-point pollutant runoff (Table 3). Variability in effectiveness is largely predicated on design considerations and the geophysical attributes of the landscape.

Riparian Forest Buffer

A riparian forest buffer is a narrow, permanent grassed strip along with a forested (predominantly trees and shrubs) area of land adjacent to a body of water such as a river, stream, pond, lake, marshland, estuary, canal, playa, or reservoir (Fig. 6). The grass and forested area traps sediment and increases infiltration thereby reducing sediment and nutrient losses. Plant uptake of nutrients in the grass and forested area also reduces nutrient losses downstream. Management for most of the forest zone includes planting pine and poplar trees as well as a perennial grass.

Grassed Waterways

Grassed waterways are natural or constructed vegetated channels that convey and dispose of overland flow from upstream areas (Fig. 7; Feiner et al., 2016, this volume). Simulated grassed waterways reduce flow, trap sediment, and increase infiltration within the area of the grassed waterway. Grassed waterways typically work by increasing surface roughness, which reduces the runoff velocity. The effectiveness of grassed waterways is a function of its width and the type of vegetation planted. However, similar to filter strips and forest riparian buffer treatments, crop production area is reduced by the grassed waterways area, resulting in lower total crop production.

Contour Buffer Strip

A contour buffer strip is an area of land maintained in permanent vegetation that helps to improve air, soil, and water quality and other environmental problems associated with agriculture land (Fig. 8). Buffer strips trap sediment and enhance

Table 3. Filter strip efficiency from experimental studies. Adapted from Grismer et al. (2006).

Filter type	Filter width	Pollutant	Removal efficiency	Reference
	m		%	
Bermudagrass	4.8	Chlorpyrifos	62–99	Cole et al., 1997
		Dicamba	90–100	
		2,4-D	89–98	
		Mecroprop	89–95	
Bermudagrass–crabgrass mixture	4.3–5.3	Total P	26	Parsons et al., 1990
		Total N	50	
Bluegrass and fescue sod (9% slope)	4.6	NH4-N	92	Barfield et al., 1992
		Atrazine	93	
	9.1	NH4-N	100	
		Atrazine	100	
	13.7	NH4-N	97	
		Atrazine	98	
Corn–oat or orchardgrass mixture (4% slope)	13.7	Total P	88	Young et al., 1980
		Total N	87	
Fescue (10% slope)	1.5	Dissolved P	8	Doyle et al., 1977
		NO_3	57	
	4.0	Dissolved P	62	
		NO_3	68	
Orchardgrass (5–16% slope)	4.6	Total P	39	Dillaha et al., 1989
		Total N	43	
	9.1	Total P	52	
		Total N	52	
Sorghum-Sudan grass mix (4%)	13.7	Total P	81	Young et al., 1980
		Total N	84	

Fig. 6. An image of a riparian forest buffer (Source: USDA-NRCS, http://www.oh.nrcs. usda.gov).

filtration of nutrients and pesticides by slowing down runoff that could enter the local surface waters.

The buffer strip practice is usually costly because of the land area taken out of production and is adopted only for the situations when soil loss and nutrient losses are jeopardizing the land health and sustainable crop production. The area of upland field is reduced in the contour buffer strip area, which ultimately

Fig. 7. An example of agricultural land with grass waterways system (Source: USDA-NRCS, http://www.wi.nrcs.usda.gov).

Fig. 8. An image of a contour buffer farming system (Source: USDA-NRCS, http://www. wi.nrcs.usda.gov).

results in lower crop production. However, this type of costly practice could yield benefits in terms of the significant reduction in nutrient and sediment losses leaving the farmland area.

Stream Fencing

Stream fencing, to keep cattle out of streams, is seen as a way to improve water quality (Fig. 9). Fencing will prevent animal defecation in streams, which constitutes a direct contribution of manure, manure nutrients, and pathogens to the surface water resource. In addition, stream fencing prevents bank erosion and protects the aquatic habitat.

The presence of stream fencing also serves as a filter strip based on the distance of fence to the edge of the stream. This will help to improve the water quality as a secondary CP. Defecated manure from animals in the stream is considered as a direct point source of nutrient in the water. The amount of nutrient (N and P), in solid and liquid forms, are released in the water based on the type of

Fig. 9. A stream in a pasture land protected from animal access by a stream fence (Source: USDA-NRCS, http://www.pa.nrcs.usda.gov).

manure produced from the specific animal and the period of the year they spend their time in the stream.

Terrace System

A terrace system is a leveled section of a hill formed for a cultivated area designed as a method of soil conservation to slow or prevent the rapid surface runoff of water (Thompson, 2016, this volume). Terraces decrease hill slope length, reduce formation of gullies, and intercept and conduct runoff to a safe outlet, thereby reducing sediment content in runoff water. Often, in application, the landscape is formed into multiple terraces, giving a stepped appearance (Fig. 10).

Fig. 10. A steep landscape with a terrace system to prevent or minimize soil erosion (Source: USDA-NRCS, http://www.ga.nrcs.usda.gov).

Pond

A pond is a water impoundment made by constructing an embankment or by excavating a pit or dugout (Fig. 11). Ponds are usually constructed to provide water for livestock, fish and wildlife, recreation, fire control, renewable energy systems and other related uses, and to maintain or improve water quality.

Stream-Bank Stabilization

Erosive stream banks (Fig. 12) are reshaped and seeded and sometimes protected with rock rip-rap or seeded with bioengineering materials under this management practice. Stabilizing the bank of the streams protects water quality and improves fish habitat, and the vegetation provides habitat for birds and small animals.

Fig. 11. An example of an on-farm holding pond (Source: USDA-NRCS, http://www. il.nrcs.usda.gov).

Fig. 12. A type of stream-bank stabilization practice with rock protection (Source: USDA-NRCS, http://www.nrcs.usda.gov).

Land Leveling

Land leveling (Fig. 13) will facilitate the efficient use of water on irrigated land and is accomplished by reshaping the surface of land to be irrigated to planned lines and grades.

Furrow Dikes

Furrow dikes are small earthen dams formed periodically between the ridges of a ridge–furrow tillage system or, alternatively, small basins created in the loosened soil behind a ripper shank or chisel. The furrow diking practice is known by many names including tied ridges, furrow damming, basin tillage, basin listing, and microbasin tillage. The dikes or basins store potential runoff on the soil surface, allowing the water to infiltrate (Fig. 14), thus decreasing storm or irrigation runoff and increasing storage of plant available water in the soil.

Fig. 13. Land leveling as a conservation practice to reduce sediment and nutrient losses (Source: USDA–NRCS, http://www.la.nrcs.usda.gov).

Fig. 14. A farm with furrow dike irrigation system (www.ars.usda.gov).

Tools for Measuring Spatial
Variability of Conservation Practices

To assess the degree of uncertainty of CP effectiveness and the role of precision conservation in reducing this uncertainty, reliable tools are needed that can reflect the spatial variability at reasonable spatial scales. Field-scale models that simulate the effectiveness of practices as functions of spatial attributes of fields play a useful role in quantifying the role of precision CPs. The following sections outline readily available tools that can be used to track practice effectiveness at subfield, field, and higher scales of resolution.

Watershed-Level Tool

The CEEOT is an integrated suite of economic and environmental models designed to simulate economic and environmental impacts of policy alternatives and individual practices or combinations of management practices (Osei et al., 2000b). The CEEOT was initially developed in 1995 at the Texas Institute for Applied Environmental Research (TIAER) at Tarleton State University. The CEEOT currently incorporates the following environmental and economic models (Fig. 15).

1. The SWAPP program (Saleh and Gallego, 2007): the fully linked Soil and Water Assessment Tool (SWAT) and Agricultural Policy eXtender (APEX) models. The SWAPP allows comprehensive simulation of management practices at the watershed level in a way that is not possible with either APEX or SWAT alone. For instance, SWAT is currently not very suitable for simulating multiple cropping systems; however, APEX is designed for such systems. Similarly,

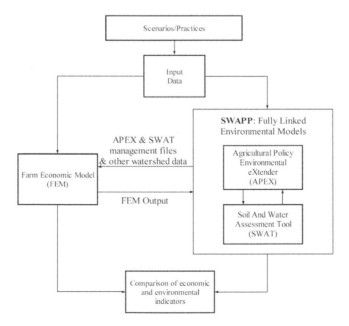

Fig. 15. Schematic of the Comprehensive Economic Environmental Optimization Tool (CEEOT modeling system.

APEX by itself is designed for farm or small watershed-scale simulations and not for large watershed-level assessments, which are the strengths of SWAT.

2. The Farm-Level Economic Model (FEM) is a comprehensive model, also developed at TIAER, that combines simulation, optimization, and accounting features to estimate the economic impacts of scenarios on representative agricultural operations.

A brief discussion of APEX, SWAT, and FEM is given here; a more detailed description of CEEOT is provided in Osei et al. (2000b).

APEX (Williams et al., 2000) is a modified version of the Environmental Policy Integrated Climate (Williams, 1990), a farm-level model that was developed in the early 1980s to assess the effects of management strategies on water quality. APEX extends the functionality of the Environmental Policy Integrated Climate model by allowing simultaneous simulation of multiple contiguous subareas (fields) for a wide range of soil, landscape, climate, crop rotation, and management practice combinations. It is designed for whole-farm or small-watershed analyses and can be used for applications such as filter strip impacts on nutrient losses from manure application fields that require the configuration of at least two subareas. The ability to simulate liquid waste applications from animal feeding operation waste storage ponds and waste treatment lagoons is a key component in the model for application to the North Bosque River watershed. Other components include weather, hydrology, soil temperature, erosion–sedimentation, nutrient and carbon cycling, tillage, dairy management practices, crop management and growth, pesticide and nutrient movement, and costs and returns of various management practices.

SWAT (Arnold et al., 1998) is a daily-time-step watershed model and was developed to overcome the limitations associated with nonpoint source modeling at the watershed scale by allowing continuous-time simulations with a high level of spatial detail through the division of a watershed or river basin into hundreds or thousands of grid cells or subwatersheds. SWAT operates on a command structure for routing runoff and chemicals through a watershed. These commands allow the user to route flows through streams and reservoirs, combine flows, and input measured data (e.g., weather) and point-source loading. The major components of SWAT include hydrology, weather, sedimentation, soil temperature, crop growth, nutrients, pesticides, and agricultural management. SWAT has a history of successful applications in addressing problems related to water, sediment, nutrients, and pesticides ranging from watershed scale to national scale across the United States and in other countries.

FEM (Osei et al., 2000b), the economic model used in CEEOT, is a whole-farm annual-time-step model that simulates the economic impacts of various scenarios on agricultural operations (Osei et al., 2000b). Routines within FEM are used to estimate costs and returns of a representative farm based primarily on livestock and crop operations, ownership and characteristics of structures, facilities and equipment, financing terms, land areas and uses, livestock nutrition, and manure production and handling. Feed ration optimizations for livestock farms are handled within the General Algebraic Modeling System (Brooke et al., 1992) submodule that is linked to special model routines for the transfer of relevant decision and exogenous variables. The model calculates fixed and variable costs of field operations using agricultural machinery management specifications tabulated in American Society of Agricultural Engineers (ASAE) engineering practice EP496.1 (ASAE, 1995a) and ASAE data D497.1 (ASAE, 1995b). The economic life of machinery is also dynamically calculated based on annual hours of use and ASAE agricultural machinery management specifications.

The CEEOT has been applied to the dairy industry in the upper North Bosque River watershed (Osei et al., 2003a, 2000a; Pratt et al., 1997; Saleh et al., 2000) and the Lake Fork Reservoir watershed in eastern Texas (Osei et al., 2003b; McNitt et al., 1999). In subsequent years, the CEEOT modeling system was applied to the Upper Maquoketa River watershed in northeast Iowa, a region with dairy operations but also including beef cattle, swine, and mixed operations (Keith et al., 2000), and to the Duck Creek watershed, a central Texas region experiencing rapid growth of broiler operations (Keplinger and Abraham, 2002). In the upper North Bosque River watershed study, the CEEOT modeling system was also applied to several dairy management practices that were considered and subsequently included in the total maximum daily load implementation plan for the North Bosque River (Keplinger, 2003). The CEEOT can be applied to evaluate a wide array of CPs or policies in any watershed in the United States where data are available.

Farmer-Friendly Tool

As CPs are installed on farmland, it is desirable that tools be developed that can be used by the farmers themselves to estimate the impacts of the CPs and the potential credits that can be derived from their implementation. This section discusses a readily available tool that was primarily designed for that purpose, the Nutrient Tracking Tool.

Nutrient Tracking Tool

The NTT (Saleh et al., 2011) is an enhanced version of an earlier model that was developed by the USDA, the Nitrogen Trading Tool (Delgado et al., 2008, 2010). Both the Nitrogen Trading Tool and NTT are primarily designed as tools that can assist farmers or technical service providers to estimate eligible water quality credits when implementing CPs as part of a water quality credit-trading program. In addition to N loss estimates, which are provided by the Nitrogen Trading Tool, NTT also provides estimates of P and sediment losses, flow, crop yields, and other environmental and agronomic indicators relevant to farm management by using APEX as its core simulation model.

In addition, the NTT facilitates the simulation of numerous management scenarios including structural and nonstructural CPs. The NTT is a web-based computer program that allows farmers or their crop advisors, technical service providers, USDA-NRCS staff, Soil and Water Conservation District personnel, or others to estimate flow and nutrient and sediment loss impacts of alternative practices or management scenarios implemented on a specific field or land management area. The NTT also provides estimates of crop yields and other indicators of interest to farmers.

The NTT provides users with the opportunity to compare the effects of two CPs, practice combinations, or other alternative conditions on indicators of interest. The alternative conditions being compared are broadly referred to as scenarios. A scenario can consist of any feasible combination of practices, policies, biophysical parameters, or other conditions. In the NTT, scenarios generally refer to CP combinations.

The Role and Implications of Precision Conservation

Given the wide range of variability in CP effectiveness, significant drawbacks exist in the use of these practices in legally binding trades in water quality markets. By reducing the variability of CP performance and improving their overall effectiveness, precision conservation methods can play a significant role in enhancing water quality markets for agriculture. The following sections discuss the role and implications of precision conservation as applied in the context of water quality trading.

Improvement in Effectiveness

Precision farming techniques will generally enhance the effectiveness of CPs. By targeting customized resource conservation measures to subareas of fields rather than implementing a blanket treatment across the entire field or farm, precision conservation will generally improve the effectiveness of both structural and managerial practices.

Reductions in Variability

One of the most significant contributions of precision conservation to water quality markets is that it generally reduces the variability of CP effectiveness. Since practices are customized based on site-specific soil, weather, and landscape attributes, a more consistent reduction in sediment and nutrient losses is expected. Consequently, precision conservation reduces the degree of uncertainty associated with credits generated from practice implementation sites. As a result, a smaller margin of safety is needed to address concerns related to uncertainty of practice effects.

Improvement in Cost-Effectiveness

Precision conservation has the potential to optimize conservation impact on water quality by identifying and targeting the level of contribution from each land use within a watershed, a farmland, or area of interest. For instance, a small portion of land within the area of interest could contribute the majority of pollutants. Therefore, targeting the optimal placement of selected CPs within the area of interest can resolve the issue. Consequently, precision conservation enables landowners and watershed planners to achieve their goals at a lower cost than traditional implementation using the same amount of total CP expenditures. Economic models, such as the FEM (Osei et al., 2000b), are used in conjunction with nutrient tracking tools, such as NTT or APEX, to determine the cost-effectiveness of alternative implementations of CPs.

Case Study: Application of the Nutrient Tracking Tool to a Sample Farm in Carroll County, Maryland

To illustrate the scope of NTT applications, selected CPs were run on a sample of soil series representative of soils in Maryland, United States. A total of nearly 2000 soils were used in these simulations representing soils where field crops were grown recently in Maryland based on USDA's 2013 Cropland Data Layer. In this NTT application, 11 alternative scenarios were evaluated (Table 4) for each of the soil types. All the scenarios included here have an impact on erosion losses but also highlight NTT's capabilities for estimating nutrient and sediment losses and various indicators of relevance for farmers. Table 4 describes the key features of each of these alternative scenarios as well as the baseline scenario.

Each alternative scenario was evaluated assuming a conventional-till field with corn (*Zea mays* L.)–soybean [*Glycine max* (L.) Merr.] rotation, which is one of the main cropping systems in Carroll County, MD. Table 5 shows the management practices for the corn–soybean cropping system assumed for the simulations.

To determine the effectiveness of precision conservation as compared with traditional CP, (i) the physical and chemical characteristics of each of the 2000 representative soils in Maryland and local weather data were obtained using the data inquiry functions in NTT, and (ii) The APEX nested in NTT was used to simulate each of the roughly 2000 soil types from the period of 1987 through 2006 under two scenarios: (i) no CPs (baseline) and (ii) traditional implementation of an individual practice (without precision conservation)

Effectiveness of the selected precision CPs was determined by comparing scenarios the two scenarios for each practice. Results of simulations conducted with NTT for various CPs are shown in Table 6. The results clearly indicate that practice impacts vary considerably across soils, implying that precision conservation methods will likely improve the effectiveness of each of the practices.

Table 4. List of conservation practices for sample farm Carroll County, MD.

Scenario	Full name of conservation practice	Conservation practice description
1	Baseline	Typical no-till corn–soybean rotation with field operations as indicated in Table 5
2	Cover crop	Wheat cover crop after corn
3	Irrigation (sprinkler)	Automatic irrigation efficiency set at 0.8
4	Terraces	Gradient terraces
5	Nutrient management	Incorporation of nutrient applications within 48 h
6	Riparian forest buffer	FFPQ† = 0.8; width = 9 m; slope relative to upland = 0.25; grass portion = 25%; trees and grass = 75%
7	Grassed waterway (4.6 m each side)	FFPQ = 0.8; width = 9 m (4.5 m on each side of field channels)
8	Stream-bank stabilization	Changing the stream surface cover and erodibility from medium to very low level
9	Filter strip	FFPQ = 0.8; width = 15 m width of switchgrass
10	Contour buffer strip	6 m of dense grass, 24 m of crop
11	Wetland	0.8 ha wetland

† Fraction of floodplain flow.

Table 5. Operations for corn–soybean rotation.

Year	Date	Operation	Application type, rate
			kg ha⁻¹
1	15 April	Fertilizer N	Surface, 180
1	15 April	Fertilizer P	Surface, 67
1	5 May	Planting corn	–
1	10 October	Harvest corn	–
1	11 October	Kill corn	–
2	14 May	Fertilizer P	Surface, 45
2	15 May	Plant soybeans	–
2	15 October	Harvest soybeans	–
2	16 October	Kill soybeans	–

Table 6. Summary Nutrient Tracking Tool estimates of the impacts of scenarios (percentage changes from baseline values).

	Flow	Sediment	Total N	Total P	Organic N	Organic P	Soluble N	Soluble P	Corn yield	Soybean yield
	mm	Tg ha⁻¹				— kg ha⁻¹ —			— Tg ha⁻¹ —	
Baseline values	340.61	1.82	9.05	1.27	6.896	1.04	2.16	0.24	10.7.6	3.7
				% change from corresponding baseline value						
Cover crop	-4.4	-14.3	-9.5	-6.5	-9.1	-7.5	-9.8	-4.8	-2.2	0.7
Irrigation	3.9	4.6	-0.2	1.7	4.9	2.0	-6.2	1.1	3.4	1.4
Terraces	-0.1	-48.8	-18.5	-21.4	-33.6	-33.5	-1.1	-0.1	0.2	0.4
Nutrient management	-0.1	-5.5	-28.2	-35.5	2.1	-17.3	-63.1	-67.4	2.6	1.6
Stream-bank stabilization	0.0	-17.2	-3.5	-4.6	-6.6	-7.2	0.0	-0.2	0.0	0.1
Riparian forest buffer	-25.5	-74.4	-29.7	-30.2	-47.5	-46.6	-9.2	-1.5	0.4	0.4
Grassed waterways	-13.2	-74.5	-25.8	-27.2	-45.5	-43.5	-3.2	1.4	0.4	0.0
Filter strip	-14.9	-67.0	-24.8	-25.8	-42.6	-40.7	-4.3	0.5	0.4	0.0
Contour buffer strip	-58.4	-84.7	-65.4	-61.9	-72.3	-72.7	-57.4	-43.0	0.2	0.4
Wetland	-3.0	-23.1	-28.9	-22.5	-35.7	-35.7	-21.1	0.7	0.4	-0.5

Results and Discussion

To determine the effectiveness of precision conservation as compared with traditional CP, 11 scenarios were simulated: the baseline and 10 alternative scenarios. Each scenario was simulated over a 47-yr period (from 1960 to 2006) based on available weather data associated with the location of the Maryland soil series being used in the simulations. The results presented here are annual averages and other statistical summaries across all soil types based on the last 45 yr of simulation. The first crop rotation period (2-yr) of the 47 yr of simulation output was not included in the scenario results because it was assumed that a significant period of time is required for the relevant indicators to reflect conditions attributable to the scenario being simulated. The average impacts of the 10 scenarios on flow, sediment, and nutrient losses, relative to the baseline, are shown in Table 6. The results shown in the table indicate that the NTT simulation output captures the essential features of each practice based on the general consensus in academic publications on CP effectiveness and prevailing expert opinions. In addition to the annual averages reported in Table 6, the range of impacts of each CP on flow and sediment, total N (NO_3–N and organic N), and total P (PO_4–P and particulate P) losses are also represented in box and whisker plots in Fig. 16 through 19. In these plots, each box represents the range in values of the percentage changes from the first quartile to the third quartile with the median in the middle of the range. Thus, the vertical placement of each box on the figures represents 50% of the range of impacts of the scenario on the indicator. The lines (whiskers) below and above the boxes represent the extent to the minimum and maximum percentage impacts, respectively.

Among all scenarios, only irrigation management and nutrient management had the most impacts on crop yields on the portion of fields that were actually cropped (Table 6). Some of the other scenarios, such as filter strip, riparian forest buffer, and grassed waterways, have significant impact on crop yields only in the sense that land had to be taken out of production to implement them. However, these yield reductions are not reflected in the table, since the yields indicated in the table only refer to yield impacts on land that was actually cropped with corn and soybeans.

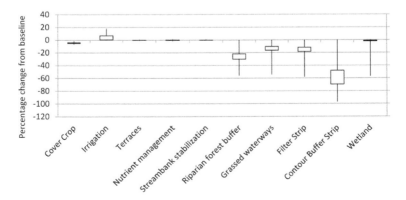

Fig. 16. Range of simulated impacts of conservation practices on flow.

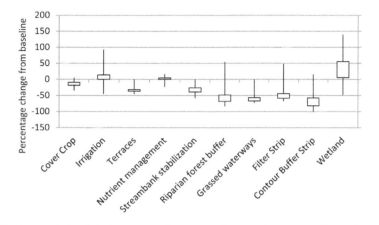

Fig. 17. Range of simulated impacts of conservation practices on sediment losses.

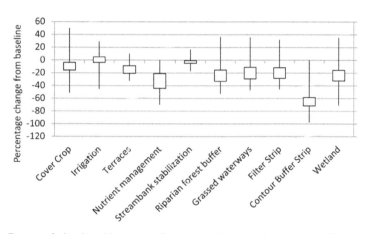

Fig. 18. Range of simulated impacts of conservation practices on total N losses.

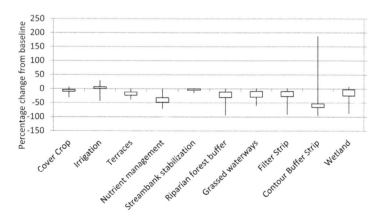

Fig. 19. Range of simulated impacts of conservation practices on total P losses.

The following describes the effect of each CP on water quality and crop production:

Cover Crops: Simulation results across all Maryland soils likely to have field crops (cultivated soils) indicate that cover crops would have significant impact on reducing runoff (flow), sediment losses, and all nutrient losses on average. Box plots in Fig. 16 through 19 also indicate that flow and sediment losses are very likely to be reduced under cover crops regardless of soil type. For the most part, N and P losses are also projected to decline under cover crop implementation except for a few soils, particularly in the case of total N losses. Reductions in runoff and nutrients are due in part to cover crop water and nutrient uptake.

Irrigation Management: Irrigation is projected to increase crop yields as expected, but increases in runoff, sediment, and P losses are also indicated according to the average results presented in Table 6. These increases are also reflected in the range of impacts across all soils, with virtually all soils showing increases in flow and sediment and nutrient losses under irrigation. However, lower losses of sediment and nutrients are expected in situations with higher irrigation efficiency where less runoff occurs. The irrigation effect on crop yield is naturally higher in dryer climate and shallower soils.

Terraces: Gradient terraces are projected to have very little impact on flow, but significant reductions in sediment and nutrient losses are indicated when terraces are installed. The annual average across all soils, as well as the range of impacts, suggests that terraces will generally reduce sediment and nutrient losses.

Nutrient Management: Nutrient management consists of a wide variety of management practices. In this chapter, we simply simulated nutrient incorporation by using a knifing operation for the nutrient management scenario. Fertilizer was incorporated during application instead of being surface applied. This scenario resulted in consistent reductions in nutrients as a result of placement of fertilizer below the soil surface. A small increase in crop yield is associated with higher availability of nutrients to crops as a result of incorporation. The results show that this practice could reduce nutrient losses considerably, but sediment losses may increase in many situations.

Stream and Riparian Management: The stream and riparian management systems (stream-bank stabilization, riparian forest buffer, grassed water ways, and filter strips) are projected to result in significant reductions in losses of sediment and nutrients (Table 6). In addition to a reduction in sediment losses, both total N and total P losses are projected to decline when either a 15 m (49 feet) filter strip or a 9 m (30 feet) riparian forest buffer is put in place at the downslope edge of the field or 4.5-m (15 feet) grassed waterways are placed on each side of field channels. Stabilization of stream banks is also projected to result in significant sediment loss reduction and moderate nutrient loss reduction. These four scenarios, with the exception of the stream-bank stabilization scenario, also result in reductions in edge-of-field runoff (flow). Among all four scenarios, grassed waterways had the greatest reduction in all four indicators. The lower runoff level mainly was due to water uptake by the grass and higher infiltration in the riparian zones. The range of impacts shown in Fig. 16 through 19 also indicates that all four scenarios in this group had consistent reductions, on average, in all indicators across all simulated soils except for few highly sloped soils (Fig. 17, 18).

Contour Buffer Strips: Of all the scenarios simulated, contour buffer strips had the most significant impact on all indicators. Total sediment and nutrient reductions are projected to vary from 60 to 70%, on average, while flow reductions are also found to be close to 50% on average runoff from uptake of grass and higher infiltration in the riparian zones. Furthermore, the box and whisker plots indicate that these reductions are generally consistent except in the case of sediment and total P reductions on a few soils (Fig. 17, 19). Contour buffer strip implementation is anticipated to result in mostly consistent and significant reductions in runoff, sediment, and nutrient losses when implemented on a corn–soybean field.

Wetlands: The final scenario simulated across all Maryland cultivated soils used for field crops in 2013 entailed a 0.8-ha (2 acre) wetland at the downslope edge of the 100-ha (247 acre) field. The wetland is projected to result in reduction in flow volumes while also leading to significant decreases in sediment and nutrient losses on average. However, even though the average area-weighted impact of wetlands entails a reduction in sediment losses, results indicate (Fig. 17) that for some fields (typically those with low slope), wetlands may result in an increase in sediment losses. This implies that very large reductions would be achieved on a few fields (with high slope) and very small increases on many soils (with low slope) with a net effect of a reduction in sediment losses on average across all soils.

Design and Implementation of Precision Conservation for Water Quality Markets

Optimal implementation of precision CPs for generating water quality credits entails careful consideration of economic impacts and water quality effects of site-specific practice combinations. A detailed economic simulation model, such as FEM, is used in conjunction with biophysical tools such as APEX or NTT to determine the optimal practice implementation profile. The following steps will enable farmers, watershed planners, and other stakeholders to maximize the benefits of resource conservation dollars for water quality markets.

- Specify the target goal
- Determine resource limitations
- Assemble data
- Use available tools to determine optimal implementation profile
- Implement the practices
- Monitor, evaluate, and adjust the implemented practices as necessary

In the following sections, each of the steps outlined above will be discussed briefly.

Goal Specification

The first step in applying precision conservation to generate credits for water quality markets is to specify the goal that needs to be achieved. In a water quality market setting, the goal is typically to generate credits for trading. It is important to specify whether the credits will be for sediment load reduction or N or P reduction, since various practices will have different efficacies for various indicators of interest. For instance, nutrient management practices are generally very effective at reducing N or P losses but have little impact on sediment losses in runoff. On the other hand, reduced tillage methods may lead to significant reductions in sediment losses but might actually result in increases in N losses. The goal should also specify the amount of credits that need to be generated, since the optimal practice combination will depend on how much of each pollutant indicator needs to be reduced.

Determination of Resource Limitations

This is a very important step in precision conservation. Before the optimal precision conservation portfolio is determined, the amount of funds available for implementing the practices must be determined. Otherwise, once the optimal practice implementation has been determined, the farmer or stakeholder may be presented with a situation where the optimal combination of practices cannot be implemented because of dollar constraints. That would result in a more time-consuming iterative process of determining another optimal combination that is within the resource constraints.

Data Assembly

Data needed to determine the credit generation potential of alternative suites of CPs are largely available in the farmer-friendly NTT tool. However, for the spatial resolution necessary for precision conservation, farmers may need to assemble more spatially refined data pertaining to their fields, specifically, the spatial distribution of soil types, slopes, and nutrient contents of subareas within their fields. Armed with this data, farmers can delineate the subareas within each field in NTT for all fields on their farms.

Specific economic data can also be assembled. Farmers with contract pricing may need to assemble data on the relevant prices for their cash crops or livestock, as these may be different from the defaults used in the FEM model or other economic tools they might be using.

Simulation of Practices Using the Tools

Once all data have been assembled, farmers or other stakeholders can use the appropriate tools to determine the cost-effectiveness of alternative practice implementations. By comparing various practice implementations, the optimal precision conservation portfolio can be determined as the most cost-effective among all viable options. The current version of NTT allows such iterative comparisons of cost-effectiveness. A subsequent version of NTT will actually perform optimizations, allowing the user to more easily determine the optimal precision conservation implementation for their farm.

Implementation of Optimal Practice

Once the optimal practice implementation has been determined, it is simply implemented at this step. Assuming enough resources are available to implement the chosen practice portfolio, it must be implemented as determined through the optimization process. However, if funds are insufficient, it would be necessary to repeat the previous step to determine the optimal practice portfolio that is pertinent to the available funds and would still allow for upgrades when the remaining funds needed for the full implementation become available.

Monitoring and Adjustment of Practice Implementation

After the practices have been installed or implemented, it is necessary to monitor their performance to determine whether they are in fact as effective as predicted by the tools. Generally, it is hoped that the actual effectiveness in the field is close enough to the projections of the tool. If not, some adjustments may be necessary. Nonstructural CPs are more amenable to adjustment than structural practices that cost more to implement and are generally fixed. Monitoring the practice implementation may take a few years before a final version is arrived at for calculating the actual credits being generated.

Conclusions

Conservation practices are projected to play a key role in water quality markets into the future. Given the wide variation in the effectiveness of these practices in actual implementations, precision conservation provides an opportunity for enhancing the role of these practices in water quality markets. Precision conservation will reduce the variability of CP effectiveness and will also enhance the gross effectiveness of these practices as well as their cost-effectiveness. By using tools that are readily available, such as the NTT, any user can determine the optimal practice implementation for their farms. The NTT provides farmers, government officials, researchers, and others an efficient, web-based, and user-friendly method of evaluating the impacts of proposed and existing CPs on water quality and quantity. A series of simulations using farm practices and soil and weather data from Maryland was used to assess the effectiveness of NTT in simulating major USDA-NRCS CPs in terms of runoff, nutrients, sediment, and crop yield. The NTT was used to evaluate the impacts of 10 CPs (of the many possible individual CPs or CP combination scenarios that users can specify) when implemented for applicable soils in Maryland with typical corn–soybean management. The results obtained from NTT, which were based on simulation of soils used for field crops in 2013, indicate that the tool is capable of simulating numerous CPs reasonably well given the documented effectiveness of the CPs. The NTT is a valuable tool by which users could make the most effective decisions to increase their farm productivity while reducing nutrient and sediment losses to water bodies.

Optimal practice implementations entail precision conservation approaches that maximize the benefit from each dollar of CP investment. Using precision conservation will reduce the margin of safety that is necessary for water quality trades thereby improving water quality trading options for farmers.

References

Agriculture and Agri-Food Canada. 2013. Controlled tile drainage: Increasing yields and helping the environment. Government of Canada. http://publications.gc.ca/collections/collection_2013/aac-aafc/A22-516-2010-eng.pdf (accessed 19 Apr. 2014).

American Society of Agricultural Engineers. 1995a. Agricultural machinery management. ASAE Engineering Practice: EP496.1, ASAE Standards. ASAE, St. Joseph, MI.

American Society of Agricultural Engineers. 1995b. Agricultural machinery management data. ASAE Data: ASAE D497.1, ASAE Standards. ASAE, St. Joseph, MI.

Arnold, J.G., R. Srinivasan, R.S. Muttiah, and J.R. Williams. 1998. Large area hydrologic modeling and assessment; Part I: Model development. J. Am. Water Resour. Assoc. 34:73–89. doi:10.1111/j.1752-1688.1998.tb05961.x

Barfield, B.J., R.L. Blevins, A.W. Flofle, C.E. Madison, S. Inamder, D.I. Carey, and V.P. Evangelou. 1992. Water quality impacts of natural riparian grasses: Empirical studies. ASAE Paper No. 922100. American Society of Agricultural Engineers, St. Joseph, MI.

Brooke, A., D. Kendrick, and A. Meeraus. 1992. GAMS: A user's guide, Release 2.25. World Bank, Washington, DC.

Cole, J.T., J.H. Baird, and B.T. Basta. 1997. Influence of buffers on pesticide and nutrient runoff from Bermudagrass turf. J. Environ. Qual. 26:1589–1598. doi:10.2134/jeq1997.00472425002600060019x

Delgado, J.A., C.M. Gross, H. Lal, H. Cover, P. Gagliardi, S.P. McKinney, E. Hesketh, and M.J. Shaffer. 2010. A new GIS nitrogen trading tool concept for conservation and reduction of reactive nitrogen losses to the environment. Adv. Agron. 105:117–171. doi:10.1016/S0065-2113(10)05004-2

Delgado, J.A., M.J. Shaffer, H. Lal, S. McKinney, C.M. Gross, and H. Cover. 2008. Assessment of nitrogen losses to the environment with a nitrogen trading tool. Comput. Electron. Agric. 63:193–206. doi:10.1016/j.compag.2008.02.009

Dillaha, T.A., R.B. Reneau, S. Mostaghimi, and D. Lee. 1989. Vegetative filter strips for agricultural nonpoint source pollution control. Trans. ASAE 32:0513–0519. doi:10.13031/2013.31033

Dosskey, M.G., S. Neelakantan, T. Mueller, and Z. Qiu. 2016. Vegetative filters. In: J.A. Delgado, G.F. Sassenrath, and T. Mueller, editors, Precision conservation: Geospatial techniques for agricultural and natural resources conservation. Agron. Monogr. 59. ASA, CSSA, and SSSA, Madison, WI. doi:10.2134/agronmonogr59.2013.0019 (in press.)

Doyle, R.C., G.C. Stanton, and D.C. Wolf. 1977. Effectiveness of forest and grass buffer filters in improving the water quality of manure polluted runoff. ASAE Paper no. 77-2501. ASAE, St. Joseph, MI.

Feiner, P., and K. Auerswald. 2016. Grassed waterways. In: J.A. Delgado, G.F. Sassenrath, T. Mueller, editors, Precision conservation: Geospatial techniques for agricultural and natural resources conservation. Agron. Monogr. 59. ASA, CSSA, and SSSA, Madison, WI. doi:10.2134/agronmonogr59.2013.0021 (in press.)

Fletcher, T.D., H.P. Duncan, P. Poelsma, and S.D. Lloyd. 2003. Stormwater flow, quality and treatment: Literature review, gap analysis and recommendations report. NSW Environment Protection Authority and Institute for Sustainable Water Resources, Dep. of Civil Engineering, Monash Univ., Sydney.

Grismer, M.E., A.T. O'Geen, and D. Lewis. 2006. Vegetative filter strips for nonpoint source pollution control in agriculture. Publ. 8195. Univ. of California. Div. of Agric. and Natural Resources. http://anrcatalog.ucdavis.edu/pdf/8195.pdf (accessed 19 Apr. 2014).

Keith, G., S. Norvell, R. Jones, C. Maguire, E. Osei, A. Saleh, P. Gassman, and J. Rodecap. 2000. Livestock and the environment: A national pilot project: CEEOT-LP modeling for the Upper Maquoketa River Watershed, Iowa: Final report. Report No. PR0003. Texas Institute for Applied Environmental Research, Tarleton State Univ., Stephenville, TX. http://tiaer.tarleton.edu/pdf/PR0003.pdf (accessed 19 Apr. 2014).

Kelly, D., K. Hoffsis, N. Lucore, and S. Wallace. 2014. Tile drains: Exploration of best management practices and remediation techniques. The Univ. of Vermont, Burlington, VT. http://www.uvm.edu/~wbowden/Teaching/Risk_Assessment/Resources/Public/Projects/Project_docs2014/Tiles/Reports/Tile_Drains_Remediation_Report.pdf (accessed 19 Apr. 2014).

Keplinger, K. 2003. The economics of total maximum daily loads. Texas Institute for Applied Environmental Research. Project Report PR0301. Nat. Res. J. 43: 1057–1091.

Keplinger, K., and J. Abraham. 2002. Project summary: CEEOT-LP modeling for Duck Creek Watershed, Texas. Report No. PR0206. Texas Institute for Applied Environmental Research, Tarleton State University, Stephenville, TX.

McNitt, J., R. Jones, E. Osei, L. Hauck, and H. Jones. 1999. Livestock and the environment: Precedents for runoff policy: Policy option-CEEOT-LP. Report PR9909. Texas Institute for Applied Environmental Research, Tarleton State University, Stephenville, TX.

NRCS. 2010. Conservation practice standard: Constructed wetlands. http://www.nrcs.usda.gov/Internet/FSE_DOCUMENTS/nrcs143_025770.pdf (accessed 19 Apr. 2014).

NRCS. 2016. National conservation practice standards. http://www.nrcs.usda.gov/wps/portal/nrcs/detail/national/technical/?cid=NRCSDEV11_001020 (accessed 19 Apr. 2014).

Osei, E., P.W. Gassman, L.M. Hauck, R.D. Jones, L.J. Beran, P.T. Dyke, D.W. Goss, J.D. Flowers, and A.M.S. McFarland. 2003a. Economic costs and environmental benefits of manure incorporation on dairy waste application fields. J. Environ. Manage. 68:1–11. doi:10.1016/S0301-4797(02)00226-8

Osei, E., P.W. Gassman, L.M. Hauck, S. Neitsch, R.D. Jones, J. McNitt, and H. Jones. 2003b. Economic and environmental impacts of pasture nutrient management. J. Range Manage. 56:218–226. doi:10.2307/4003810

Osei, E., P.W. Gassman, R.D. Jones, S.J. Pratt, L.M. Hauck, L.J. Beran, W.D. Rosenthal, and J.R. Williams. 2000a. Economic and environmental impacts of alternative practices on dairy farms in an agricultural watershed. J. Soil Water Conserv. 55:466–472.

Osei, E., P. Gassman, and A. Saleh. 2000b. Livestock and the environment: A national pilot project: Economic and environmental modeling using CEEOT. Report No. PR0002. Texas Institute for Applied Environmental Research, Tarleton State University, Stephenville, TX.

Parsons, J.E., R.D. Daniels, J.W. Gilliam, and T.A. Dillaha. 1990. Water quality impacts of vegetative filter strips and riparian areas. Presented at the 1990 International Winter Meeting of the Am. Soc. Agric. Eng., Paper no. 90-2501. ASAE, St. Joseph, MI.

Pratt, S., R. Jones, and C.A. Jones. 1997. Livestock and the environment: A national pilot project; Expanding the focus: CEEOT-LP. Report No. PR9603. Texas Institute for Applied Environmental Research, Tarleton State University, Stephenville, TX.

Saleh, A., J.G. Arnold, P.W. Gassman, L.M. Hauck, W.D. Rosenthal, J.R. Williams, and A.M.S. McFarland. 2000. Application of SWAT for the North Bosque River Watershed. Trans. ASAE 43:1077–1087. doi:10.13031/2013.3000

Saleh, A., and O. Gallego. 2007. Application of SWAT and APEX using the SWAPP (SWAT-APEX) program for the Upper North Bosque River Watershed in Texas. Trans. ASAE 50:1177–1187. doi:10.13031/2013.23632

Saleh, A., O. Gallego, E. Osei, H. Lal, C. Gross, S. McKinney, and H. Cover. 2011. Nutrient tracking tool: A user-friendly tool for calculating nutrient reductions for water quality trading. J. Soil Water Conserv. 66:400–41. doi:10.2489/jswc.66.6.400

Spaulding, J.T. 1996. Best management practices for urban stormwater runoff. State of New Hampshire Dep. of Environ. Serv, Water Supply and Pollution Control Div. NHDES-WSPCD-95-3. Chapter 7. http://des.nh.gov/organization/divisions/water/wmb/tmdl/bmps_manual_usr.htm (accessed 19 Apr. 2014).

Thompson, A. 2016. Terraces. In: J.A. Delgado, G.F. Sassenrath, and T. Mueller, editors, Precision conservation: Geospatial techniques for agricultural and natural resources conservation. Agron. Monogr. 59. ASA, CSSA, and SSSA, Madison, WI. doi:10.2134/agronmonogr59.2016.0010 (in press.)

USDA–USEPA. 2013. USDA–EPA partnership supports water quality trading to benefit environment, economy. http://yosemite.epa.gov/opa/admpress.nsf/ec5b6cb1c087a2308525735900404445/41888687a8e96fdb85257c36005830d4!opendocument (accessed 19 Apr. 2014).

USEPA. 1993. Natural wetlands and urban stormwater: Potential impacts and management. USEPA, Office of Wetlands, Oceans and Watersheds. Wetlands Division. Washington, D.C. http://www.epa.gov/owow/wetlands/pdf/stormwat.pdf (accessed 19 Apr. 2014).

Williams, J.R. 1990. The erosion productivity impact calculator (EPIC) model: A case history. Philos. Trans. R. Soc. Lond. 329:421–428. doi:10.1098/rstb.1990.0184

Williams, J.R., J.G. Arnold, and R. Srinivasan. 2000. The APEX model. BRC Report No. 00-06, Oct. 2000.

Young, R.A., R. Huntrods, and W. Anderson. 1980. Effectiveness of vegetative buffer strips in con-trolling pollution from feedlot runoff. J. Environ. Qual. 9:483–487. doi:10.2134/jeq1980.00472425000900030032x

Field- and farm-scale assessment of soil greenhouse gas mitigation using COMET-Farm

Keith Paustian,* Mark Easter, Kevin Brown, Adam Chambers, Marlen Eve, Adriane Huber, Ernie Marx, Mark Layer, Matt Stermer, Ben Sutton, Amy Swan, Crystal Toureene, Sobha Verlayudhan and Steve Williams

Abstract

Achieving more sustainable production of food, fiber and energy and reducing environmental burdens from agricultural systems is a global challenge. Meeting this challenge will create new opportunities for producers to provide a broader range of ecosystem services, including reducing greenhouse gas emissions (and sequestering more carbon) in their production systems. To meet these new objectives, land managers will need new decision tools and performance metrics.

The COMET-Farm system was designed to fill this need by incorporating state-of-the-art greenhouse gas quantification methods into a web-based tool that can be used by farmers, ranchers, land managers and others. The system is capable of doing a full greenhouse gas assessment for CO_2, CH_4, and N_2O, from all major on-farm emission sources (and CO_2 removal into biomass and soil sinks), including land management of annual and perennial crops, pasture, range and agroforestry systems, as well as emissions from livestock and on-farm energy use. The system uses a fully spatial mapping and menu-driven graphical user interface (GUI) to facilitate data entry and evaluation of user-defined conservation practices.

In this paper we provide an overview of the system and a description of the user interface and integrated databases in the system. We follow this with a brief description of the models and data requirements for the major emission source categories in the system. We illustrate the application of the system using examples of emission reductions from adoption of different conservation management practices and discuss how the tool can help meet needs for different policy- and market-driven greenhouse gas reduction efforts.

Abbreviations: GHG, greenhouse gas; GUI, graphical user interface.
K. Paustian, K. Brown, A. Huber, M. Layer, B. Sutton, and S. Verlayudhan, Department of Soil and Crop Sciences, Colorado State University, Fort Collins, CO, 80523; K. Paustian, M. Easter, E. Marx, M. Stermer, A. Swan, C. Toureene, and S. Williams, Natural Resource Ecology Laboratory, Colorado State University, Fort Collins, CO, 80523; A. Chambers, USDA-NRCS West National Technology Support Center, 1201 NE Lloyd Blvd., Portland OR 97232; M. Eve, USDA-ARS Natural Resources and Sustainable Agricultural Systems, 5601 Sunnyside Ave, Beltsville, MD 20705.*Corresponding author (keithp@nrel.colostate.edu)

doi:10.2134/agronmonogr59.2013.0033

Precision Conservation: Geospatial Techniques for Agricultural and Natural Resources Conservation
J.A. Delgado, G.F. Sassenrath, and T. Mueller, editors. Agronomy Monograph 59.

Improving the sustainability of agricultural production by reducing environmental burdens on air, water, and soil resources while still meeting the growing global demand for food and fiber production are now firmly established as guiding principles for agricultural management. These principles have been formalized within the framework of 'ecosystem services', in which agriculture is tasked with providing *provisioning* (e.g., food, fiber and energy products), *supporting* (e.g., water and nutrient cycling), *regulating* (e.g., greenhouse gas mitigation, carbon storage, water filtration) and *cultural* (e.g., historical landscape, flora and fauna) services (Robinson et al., 2014). Although maximizing economic returns from food, fiber, and energy production remains the dominant goal of most agricultural producers today, the value of other ecosystem services is increasingly being appreciated and monetized, which introduces much greater complexity in the decision-making and management choices of farmers and ranchers.

Effective management requires performance metrics to evaluate whether management objectives have been achieved and to compare alternative management systems and help choose the best options, for both short and long time horizons. When provisioning services were the singular objective of agricultural producers, the needed performance metrics (i.e., yields and economic returns on harvested commodities) were self-evident and readily quantifiable. In contrast, quantifying other ecosystem services such as greenhouse gas (GHG) mitigation is not straightforward and requires tools and technologies that have not previously been available to agricultural producers. If farmers and ranchers are to expand their management decision-making to incorporate ecosystem services as part of their bottom-line (or 'triple bottom-line') then they need rigorous yet accessible and easy-to-use metrics to inform their decision making.

The COMET-Farm system has been designed to meet this need by putting the capability of state-of-the-art estimation of greenhouse gas emissions and emission reductions, including carbon (C) sequestration in biomass and soils, into the hands of producers as well land management agencies, C offset project developers and other interested users. The concept of greenhouse gas or 'carbon footprint' calculators to assess climate impacts of various management activities is not new, and a variety of different calculators have been put forth for different GHG emission sectors, including agriculture (Denef et al., 2012). What sets COMET-Farm apart from other systems is that it implements the official USDA GHG inventory guidelines for entity-scale reporting (Eve et al., 2014) and it does so within a fully spatial user interface operating at field and subfield scales. The USDA guidelines were the result of a 3-yr effort that included multiple working groups comprised of leading experts on all land use-related GHG emissions, coming from academia, government and industry. It included multiple rounds of peer-review, internal USDA review and public comments, which were incorporated into the final document. Methods within the USDA guidelines include a number of empirical emission-factor models, as well as the DayCent model (Parton et al., 2001, Del Grosso et al., 2010), which is a widely used and well-vetted process-based simulation model of C and N dynamics, including trace gas [CO_2 (carbon dioxide), N_2O (nitrous oxide), and CH_4 (methane)] emissions. The COMET-Farm system executes these models in real time and then compiles results at field and full-farm level, computing a full-net GHG balance, including future projections, for the entire farm operation.

While COMET-Farm includes modules (Fig. 1) for GHG emissions from live-stock (i.e., enteric methane emissions, methane and nitrous oxide emissions from manure management), biomass C stock changes for woody biomass in agroforestry, perennial crops (i.e., orchards, vineyards) and woodlots, and fossil C emissions from on-farm energy use, here we present only methods and example case studies for carbon dioxide (CO_2), methane (CH_4) and nitrous oxide (N_2O) emissions and removal in soils, focusing on annual cropping systems. A brief overview of the system components is given, followed by a description of the user interface and details about the integrated databases, followed by an outline of the main soil emission source categories [which are described in more detail in Eve et al. (2014)]. We then illustrate the application of the system using examples of adoption of conservation management practices and the projected impacts on GHG emissions. We then close with a brief discussion of ongoing and future developments of COMET-Farm.

COMET-Farm Description

System overview

The basic principle of any model-based GHG inventory system, including COMET-Farm, is the integration of *activity data*, that is, land use and management practices that impact GHG emission/removal rates, with *emission rate models* in which the rates are a function of the activity data as well as climatic and edaphic (e.g., soil physical properties such as texture, depth, porosity) conditions at a par-ticular location. The activity data is primarily supplied by the user, who has the specific knowledge of the management practices applied to a particular field, pasture, wood lot, or livestock herd for their farm or ranch, while the emission models and spatial databases of climatic and edaphic conditions are research-derived information that is embedded in the COMET-Farm system (Fig. 1).

To assess GHG mitigation options, the user needs to supply activity data for their current practices and one or more mitigation project scenarios. Their cur-rent management, continued into the future (i.e., with no change in practices),

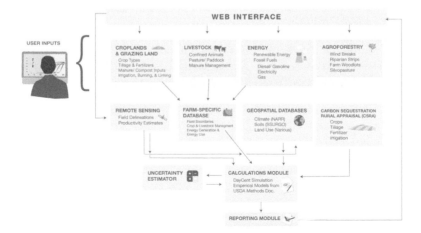

Fig. 1. Overview of components of the COMET-Farm system.

serves as a reference condition or 'baseline scenario'. The mitigation project scenario represents a change in one or more management practices to reduce emissions and/or sequester CO_2. The GHG mitigation estimate is then the difference between the net emissions in project versus baseline scenarios.

For soil and biomass GHG source categories, an individual field is defined as the basic management entity and users can enter detailed management information for individual fields or may choose from regionally-representative default management practices using a new drag-and-drop feature (see below). At present, COMET-Farm uses 2000 to 2016 as the current period for which detailed field-scale management can be specified. The mitigation project scenario is a 10-yr projection period following the adoption of one or more mitigation practices. Future enhancements of COMET-Farm will include user-selectable baseline and projection periods (see discussion below).

Software Architecture

The COMET-Farm system runs on current versions of standard web browsers (Firefox, Chrome, Safari, Internet Explorer) and is freely available to any user (http://cometfarm.nrel.colostate.edu). Users who register using a secure username and password can store several projects, and can run up to 10 scenarios on up to 50 field parcels per project. User information is protected according to USDA-NRCS policies regarding personally identifiable information (PII). Project information generated by the user is available only to that user. Users may also choose to use the tool without registering; however, at the end of an unregistered user's session (when they close their browser or after 30 min of inactivity) their projects are automatically deleted.

The web environment is an ASP.NET application running HTML5 utilizing C# components on an IIS 7 web server and ExtJS-enabled javascript components on the user's web browser. Data are stored in a Linux-hosted PostgreSQL 8.4 database server running PostGIS 1.4 for spatial analysis. DayCent model runs are completed through a custom Python process queue management system running on a Linux parallel processor with 96 cores.

Graphical User Interface

The system is driven by a fully spatial GUI that allows the user to locate and delineate individual field boundaries (which denote specific management units), using drawing tools or dropped pins on Google Map images and then an overlay with Web Soil Survey (WSS) map units (Soil Survey Staff, 2015) produces a set of distinct land parcels (soil type × management combinations) for which calculations are performed individually (Fig. 2). Soil attributes needed in various emission source models are taken from the WSS database and include soil texture, pH, coarse fragment content and rooting depth. From these soil attributes, other soil properties such as field capacity, wilting point, bulk density and saturated hydraulic conductivity are calculated from pedotransfer functions and used in the models.

Information on historic land use, predating the start of the current period (beginning in 2000), is requested from the user to represent the longer-term trajectory of soil organic carbon (SOC) stocks. This is mainly to capture major changes in land use that have taken place in the last few decades, such as plow-out of pasture to annual cropping or converting annual cropland to permanent grassland or woody vegetation. Such major shifts in vegetation and land use typically result in large changes in SOC stocks which attenuate over several decades. Users need only enter a very limited amount of information for past management history before the year 2000, using drop-down menus, from which they select the most representative choice for a given field (Fig. 3). From 2000 to the present, defines the 'current practices' time period for which the user can specify, for each year, details of management practices used on a specific field or management unit and the same annual resolution for management practices are used for future projections. A summary of the management inputs for each of the defined management time series is as follows:

- *Historic period (1950–1980):* User selects (Fig. 3) the main land use type (e.g., annual cropland, grazing land, orchards, forest)
- *Modern period (1980–2000):* User selects (Fig. 3) from very broadly defined management systems (e.g., continuous row crops, continuous hay, irrigated vs. non-irrigated) that are representative for the region. The specific management practices used for these generic systems are derived from a USDA-NRCS survey (the Carbon Sequestration Rural Appraisal, Brenner et al., 2002) and from information derived from the USDA National Resources Inventory (NRI) and the

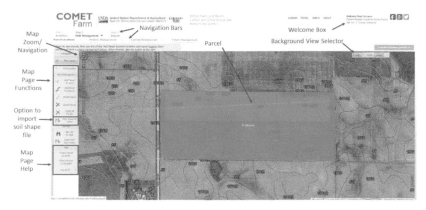

Fig. 2. COMET-Farm map and/or area of interest page, showing the overlay of land cover and soil maps, showing navigation and field delineation and mapping tools.

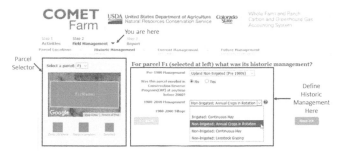

Fig. 3. COMET-Farm historic management page.

Cropping Practices Survey (USDA, 2016). They are specific to the Major Land Resource Region (MLRA) within which the parcel is found. In addition to selecting a representative cropping system, users specify the dominant tillage intensity (intensive, reduced, no till) used during this period and indicate whether the parcel was enrolled in the Conservation Reserve Program (CRP) after 1986.

- *Current period (2000-current year):* More detailed information is necessary for this time period. Users enter the crops grown in each year, dates for planting and harvests, type and timing of major tillage events, and if applicable: amounts and timing of nitrogen fertilizer, manure and/or compost additions, irrigation, lime and residue burning events, for each field. Alternatively, the user can use the drag-and-drop feature (Fig. 4) to quickly construct a management system which includes default practice data, based on the USDA data sources described above for the 'modern period'. The user then has the option of editing any of the management input data (Fig. 5).

- *Future ('project scenario') period (current year to 10 yr in the future):* The same types of information entered for the current period is required for the future period.

Data entry for the current (Fig. 5) and future periods has been further streamlined by allowing users to describe management practices for a single year, and then quickly copy the information from 1 yr to other years in the cropping sequence (with the option of editing any year in the sequence). For example, in a corn–soybean rotation, the user only needs to describe their management practices for 1 yr of corn and 1 yr of soybeans, and copy the data to the appropriate years where those crops are grown. Even when using the copy function to build the management time series, users can still edit individual years and management events as they choose.

The COMET-Farm system is designed to allow users to rapidly assess multiple conservation scenarios. Users may build up to eight future mitigation scenarios in their project. As an example, users may wish to evaluate a change in tillage, a change in nitrogen management, a switch from synthetic fertilizers to manure or compost, or combinations of these. The tool is designed to allow users to copy the cropping practices from the user's current period into a future scenario, give the new scenario a descriptive name (e.g., "conversion to no-till") and

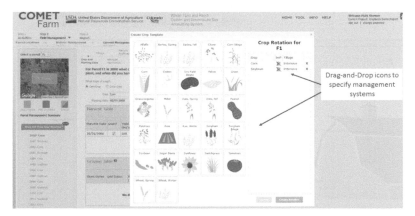

Fig. 4. COMET-Farm Current Management page, showing the drag-and-drop feature that allows for rapid definition of management systems.

then simply change the applicable management practices in the new scenario to represent the change.

Backend Databases

Historic Practices

A database of crop and pasture rotations and management practices prior to 2000 was built for each MLRA and are stored in the PostgreSQL database. These representative sets of the most common management practices in each MLRA were developed to allow users to easily choose representative historical management conditions on their field, to capture **major** changes in past management (e.g., conversion of pasture or rangeland to annual crops or from annual crops to pasture or rangeland, change from dryland to irrigated crops, etc.) that can have longer term impacts on soil C trajectories. This information is used in the 'spinup' (i.e., initialization) of the DayCent ecosystem model, prior to the current period, and thus the exact details of historical management practices are not necessary.

Climate Data

The DayCent simulation model, which is used to estimate soil C stock change in COMET-Farm, requires daily input data for minimum temperature, maximum temperature and precipitation. Currently, these data are derived from the National Centers for Environmental Prediction (NCEP) North American Regional Reanalysis (NARR) climate database (Mesinger et al., 2006). NARR is a gridded weather data product with a 32-km resolution with data available from 1979 to present. Site- and date-specific weather data are used for the current period in COMET-Farm. For future scenarios, the NARR weather data are recycled (i.e., current climate variables are also used in the projections). Thus, differences in estimated GHG emissions are due to management changes and are not confounded by differences in the weather data used in the simulations. Starting in 2017, climate data will be from the PRISM data product (Daly et al., 2008), which provides a higher resolution (4 km) gridded product.

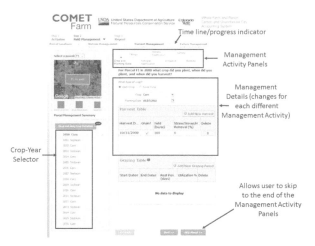

Fig. 5. COMET-Farm Current Management page, showing the pop-up menus for inputting management scheduling for each year and crop type.

Soils Data

COMET-Farm relies on the SSURGO database of soils information compiled by the National Cooperative Soil Survey (Soil Survey Staff, 2015). SSURGO provides a nearly complete coverage of soils data for the conterminous United States at approximately 1:15,000 to 1:20,000 scale. SSURGO data for soil texture, pH, percent rock fragments and rooting depth are preprocessed for all map units in the WSS. During preprocessing, the soil profile is reconfigured into prescribed layers for DayCent input. Pedotransfer functions (Saxton et al., 1986) are used to estimate bulk density, field capacity, wilting point and saturated hydraulic conductivity for each soil layer. The resulting soils data are archived and accessed during the COMET-Farm session.

Soil Emission Source Models

All of the GHG emission models are described in full in the USDA entity-scale inventory guidelines (Eve et al., 2014) and thus only a brief overview is provided here.

Soil Carbon

Mineral Soils

Initial SOC values in mineral soils are based on DayCent simulations of historical management that provide estimates of stocks and the distribution of organic carbon among the pools represented in the model (active, slow, and passive soil organic matter pools) for each of the map units within a field or subfield parcel. First, the model is run to a steady-state condition (e.g., equilibrium) under native vegetation, historical climate data, and the soil physical attributes for the land parcel. The next step is to simulate a period of time from the 1800s to 1980 for the broad set of land use histories of the region (i.e., continuous grassland, grassland conversion to cropland, etc.). Historic trends in management (such as residue removal, fertilizer addition, dominant crops, drainage improvements) are simulated for each region. The simulated carbon stocks at the end of these spinups provide the initial values for COMET-Farm user simulations. The model is then run in real time, during the user session, using the management inputs provided for the modern and current time periods. To simulate the impact of mitigation project options, the model is run for a 10 yr future projection, using the management inputs provided for the future project scenarios. Additionally the current management practices are simulated for the same 10-yr projection period to provide a baseline (i.e., no change in management) estimate. Net emissions or removals as CO_2 from mineral soils are then computed as the soil C stock difference between the mitigation and the baseline scenarios at the end of the 10-yr projection period, with the molar conversion (3.667) for CO_2 and reported as an average annual value.

Organic Soils

Organic soils (i.e., histosols), such as peat and muck soils, are a special case in which conversion to agricultural use, with drainage and often liming, leads to very high rates of carbon mineralization and carbon dioxide emissions. Currently the DayCent model is not capable of simulating C stock changes for organic soils and thus the method used (Eve et al., 2014) is an empirical model consisting of an emission factor multiplied by the area of drained organic soils. The method assumes that soils mapped as organic soils have a significant organic horizon remaining in the soil over the baseline and projection time periods.

Soil Nitrous Oxide

Mineral Soils

COMET-Farm estimates of N_2O emissions from mineral soils include both direct and indirect sources. Direct N_2O emissions are a result of nitrification and denitrification processes acting on N applied (i.e., as mineral or organic fertilizer and manures) at the field location. Indirect N_2O emissions, represent off-site emissions, derived from N lost from the field via leaching, runoff and volatilization that are then deposited elsewhere, giving rise to N_2O from the off-site N enrichment.

To estimate direct soil N_2O emissions, the USDA GHG methods use a hybrid approach that includes results from process-based models in combination with empirical scaling factors based on U.S. specific field measurement data on a seasonal timescale (Eve et al., 2014). Process-based modeling (an ensemble approach using the DayCent and the DNDC (Li, 2000) models) was used to derive base emission rates for dominant crops and soil texture classes in each USDA Land Resource Region (LRR). Base emission rates were modeled using average nitrogen fertilizer rates for each crop and region, based on agricultural statistics. An additional set of model runs were made for each crop and region with zero N additions. Since the methodology targets emissions from added N (as mineral fertilizer, manures and other organic N additions), emission factors are calculated after subtracting emissions estimated for the zero N addition case.

In COMET-Farm, the base emission rate for N_2O base is then computed for each field according to the user's specified nitrogen inputs. If the user enters an N input value greater than the typical fertilizer rate from the initial DayCent-DNDC simulations, then the user's base emission rate is adjusted upward using an empirically-derived adjustment factor. The base emission rate remains constant with N input values below the typical fertilizer rate.

After the base emission rate has been adjusted for the actual rate of N addition, direct N_2O estimates are scaled to account for management practices including tillage (or no-till), crop residue removal and the use of slow release nitrogen fertilizers or nitrification inhibitors. The scaling factors were empirically derived from experiments throughout the United States. Tillage and nitrification inhibitor scaling factors vary with climate region. Tillage scaling factors also vary depending on the length of time no-till has been implemented. Full details on the methodology are provided in Eve et al. (2014).

Because only dominant crops were included in the initial DayCent-DNDC simulations, not all scenarios entered by the COMET-Farm user can be estimated using the method described above. If a user selects a minor crop not included in the full method development, then the USDA methodology defaults to a modified IPCC Tier 1 approach (de Klein et al., 2006) in which the base emission factor is set at one percent of N inputs and then adjusted to account for management (i.e., tillage, residue management, slow release fertilizers or nitrification inhibitors) using the same scaling factors as those used for the dominant crops.

Indirect N_2O emissions from N lost via leaching and runoff are estimated using a modified IPCC Tier 1 method (Eve et al., 2014). Where precipitation plus irrigation water inputs are greater than 80% of potential evapotranspiration, leaching and runoff losses are estimated as 30% of nitrogen inputs (IPCC Tier 1 default). Where cover crops are used, the estimated fraction of N lost to leaching and runoff is reduced to 18% (leguminous cover crops) or 9% (non-leguminous cover crops). The amount of N lost via leaching and runoff is then multiplied by the IPCC Tier 1 indirect emission factor of 0.0075 to estimate the amount of indirect N_2O emissions from this source.

For land parcels where precipitation plus irrigation water input is less than 80% of potential evapotranspiration, N leaching and runoff are considered negligible and indirect N_2O emissions from leaching and runoff are assumed to be zero.

Indirect N_2O emissions from volatilization are estimated by applying the IPCC Tier 1 method to N added as synthetic or organic fertilizers (Eve et al., 2014). This method uses default values of 10% of added synthetic N fertilizers and 20% of added organic N inputs as the fraction of applied nitrogen lost to volatilization. The IPCC Tier 1 emission factor of 1% of volatilized N is multiplied by the amount of N lost to volatilization to arrive at an estimate of indirect N_2O emissions.

Indirect N_2O emissions from leaching, runoff and volatilization are added to direct N_2O emission estimates to arrive at a total soil N_2O emission estimate, which is then converted to CO_2 equivalents and entered into the COMET-Farm report.

Organic Soils

As described above, organic soils (i.e., histosols), when drained and managed for crop production, are subject to high rates of organic matter decomposition and mineralization of the organically-bound nitrogen, which is then subject to N_2O emissions via nitrification and denitrification. The method used (Eve et al., 2014) is an empirical model consisting of an emission factor (0.008 Mg N_2O-N ha^{-1} yr^{-1}) multiplied by the area of drained organic soils.

Rice Nitrous Oxide and Methane Emissions

Methods used for nitrous oxide and methane from wetland rice are those described by the Intergovernmental Panel on Climate Change (IPCC), as originally developed by Lasco et al. (2006) for CH_4 and de Klein et al. (2006) for N_2O. The baseline emission of CH_4 represents fields that are continuously flooded during the cultivation period, not flooded at all during the 180 d prior to cultivation, and receive no organic amendments. Differences between the baseline continuously flooded fields without organic amendments are accounted for by scaling factors, including adjustments for water regime (i.e., duration and frequency of flooding both before and during the cultivation period) and the type and amount of organic amendments (Lasco et al., 2006). Baseline emission rates for N_2O and a scaling factor to account for drainage effects are from Akiyama et al. (2005). The management choices entered by the COMET-Farm user for each parcel determine the scaling factors applied to the base emission rates for both methane and nitrous oxide.

Liming Carbon Dioxide Emissions

Carbon dioxide emissions due to liming of soils are based on a modified IPCC approach using U.S. specific emissions factors (Eve et al., 2014), in which the COMET-Farm user indicates the type of liming material and rate of application.

System Verification and Validation

The COMET-Farm team utilizes a three-level verification process to ensure that the system is functioning properly and producing the correct results based on the underlying emission source models. The level one verification process verifies that all equations provided in the USDA methods document are implemented correctly in the software code while the level two verification ensures that results for each equation implemented in the tool are being calculated correctly. This entails a separate 'off-line' set of calculations performed periodically for each equation provided in the USDA inventory guidelines and comparing to the output values reported in COMET-Farm. Finally, for the level three verification, a set of virtual farms have been created to test the functionality and accuracy of the tool. These virtual sites simulate common farming and livestock practices throughout the United States and include all the emission sources and multiple management alternatives in the methodology (Table 1). This detailed process of verification ensures that results remain consistent following software updates, improvements to the GUI, and back-end database revisions. Data export tools allow the team to export modeled data and internal calculations from COMET-Farm™ to verify results against offline calculated empirical and dynamic simulation model results.

Case Study Examples

To illustrate the application of COMET-Farm, we show some examples for adoption of conservation scenarios for common land use practices. For convenience, the first two examples are among an extensive set of tutorials that are set up within COMET-Farm, accessible through the help pages, to guide new users of the tool.

Cropland

We show the potential carbon sequestration benefits of converting from intensive tillage to no tillage, for a typical corn–soybean rotation in the Des Moines Lobe soil region of Iowa. The site is a 60 acre field on the Allee Demonstration Farm operated by Iowa State University. Corn was intensively tilled in the current period and fertilized with 160 lbs of nitrogen per acre using anhydrous ammonia, planted the first week of May and harvested the end of October. Soybeans were reduce-tilled in the spring, receiving no fertilizer. Neither of these crops were irrigated. They received no manure, compost or lime, and residues were not burned. COMET-Farm predicted that no-till adoption increased soil C stocks by to 33 Mg CO_2 eq yr^{-1} (0.6 Mg CO_2 eq ac^{-1} yr^{-1}) over the 10 yr after the tillage practice change (Fig. 6). This rate of soil C gain is somewhat lower than reported for two nearby long-term experiments with adoption of no-till in corn-soybean rotations [e.g., 1– 2 Mg CO_2 eq acre^{-1} yr^{-1} (2.5– 5 Mg CO_2 eq ha^{-1} yr^{-1}) over 10 yr (Al-Kaisi et al. 2014) and approximately 0.8 Mg CO_2 eq acre^{-1} yr^{-1} (2 Mg CO_2 eq ha^{-1} yr^{-1}) over 7 yr (Al-Kaisi et al. 2005)], but within the range from larger regional meta-analyses of 0.1 to 1 Mg CO_2 eq acre^{-1} yr^{-1} (0.15–2.5 Mg CO_2 eq ha^{-1} yr^{-1}), depending on climate and soil conditions (Denef et al. 2011).

For the purpose of this example, only one management change was evaluated; a shift from more intensive tillage to no tillage. In most instances, practices that reduce soil disturbance and/or increase carbon inputs to soil will enhance soil carbon sequestration (Ogle et al. 2005). Other practices that users may evaluate in COMET-Farm with potential to increase soil carbon sequestration include:

Table 1. Cropland module virtual farm sites, scenarios tested and methods verified.

Location	Cropping system	Scenarios	Methods verified
Jackson County, CO	Grass Hay	Varied: · Tillage practices · Fertilizer type, application rates and methods · Organic matter addition types and rates	· Biomass carbon stock · Change in SOC stocks · Soil N2O emissions · CO2 emissions from urea fertilization
Wibaux County, MT	Winter Wheat–Fallow	Varied: · Tillage practices · Irrigation · Fertilizer type, application rates and methods · Organic matter addition types and rates	· Change in SOC stocks · Soil N2O emissions · CO2 emissions from urea fertilization
Madison County, OH	Corn–Soy	Varied: · Tillage practices · Irrigation · Fertilizer type, application rates and methods · Organic matter addition types and rates · Liming addition material and rates	· Biomass carbon stock · Change in SOC stocks · Soil N2O emissions · CO2 emissions from urea fertilization · Change in soil carbon stocks from lime application
Arkansas County, AR	Continuous Rice	Varied: · Fertilizer type, application rates and methods · Organic matter addition types and rates · Water regime before and during cultivation	· Change in SOC stocks · Flooded rice methane emissions · Soil N2O emissions from flooded rice
Shiawassee County, MI	Alfalfa–Corn	Varied: · Tillage practices · Irrigation · Fertilizer type, application rates and methods · Organic matter addition types and rates	· Biomass carbon stock · Change in SOC stocks · Soil N2O emissions
Chase County, NE	Continuous Corn	Varied: · Tillage practices · Irrigation · Fertilizer type, application rates and methods · Organic matter addition types and rates	· Change in SOC stocks · Soil N2O emissions

increased cropping intensity (e.g., reduced fallow frequency), perennial crops in rotation, use of cover crops, or addition of manure or compost. In the case study presented, soil N_2O emissions were relatively unchanged as the rate of nitrogen fertilizer was unchanged between scenarios and tillage has a minimal impact on N_2O in this region (Eve et al. 2014). However, the results from COMET-Farm (Fig. 6) demonstrate the large impact N_2O can have on overall GHG emissions, with N_2O emissions exceeding soil carbon sequestration in both scenarios, when compared on a CO_2 equivalent basis. There are several approaches for reducing soil N_2O emissions from croplands that farmers may evaluate in COMET-Farm, including reducing nitrogen fertilizer rate and using enhanced efficiency products, such as nitrification inhibitors and slow release fertilizers. Other emissions sources from croplands include non-CO_2 emissions from burning crop residues,

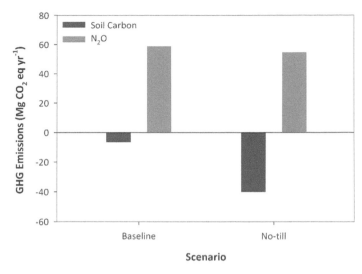

Fig. 6. Results from the cropland case study, comparing the effect of no-till adoption on soil GHG emissions in a corn–soybean rotation on a 60-acre field in Iowa.

CO_2 from liming, CH_4 (as well as N_2O) from rice cultivation, and CO_2 from urea fertilizer.

Agroforestry

Carbon sequestration in biomass is often overlooked as a potential way to mitigate atmospheric greenhouse gas emissions in agricultural systems while providing co-benefits such as protecting soils by preventing wind erosion and protecting waterways by intercepting nutrients prior to entering water bodies. We show an example of an agroforestry practice consisting of a three-row windbreak installation in Polk County, IA that contained young trees in 2015. The 300-foot section of the windbreak represented in the demo project is predicted to sequester approximately 0.7 Mg CO_2 eq yr^{-1} in aboveground and belowground biomass after 10 yr of additional tree growth (Fig. 7). COMET-Farm does not currently estimate soil carbon changes for conversion of annual cropland to permanent woody vegetative cover (but this process will be included in the next release of COMET-Farm). However, setting aside annual cropland to permanent vegetation is predicted to increase soil carbon stocks by a factor of 1.1 to 1.3 times the cultivated soil carbon stock (Eve et al., 2014). This case study example examined a windbreak; however users may evaluate other agroforestry systems, such as riparian buffers, alley cropping or farm woodlots.

Riparian Buffer

In addition to the examples included in the COMET-Farm tutorials, we illustrate how COMET-Farm may be used to assess conservation practices such as the installation of riparian buffers. Organizations such as USDA's Natural Resources Conservation Service promote riparian herbaceous cover for improving water quality, reducing soil erosion, enhancing wildlife habitat and a number of other benefits (NRCS, 2010). For this example, we assessed a hypothetical 26-acre herbaceous riparian buffer in Otoe County, NE (Fig. 8). Under the baseline scenario,

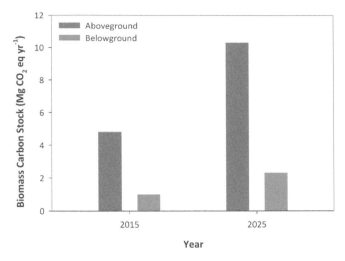

Fig. 7. Results from the agroforestry case study in Iowa, showing gains in aboveground and belowground biomass carbon stocks over 10 yr for a 300-foot three-row windbreak section.

this area was previously under long-term cropping, with a typical corn–soybean rotation for this region. For future management, we simulated conversion to permanent herbaceous cover, by planting perennial grasses and excluding tillage, fertilizer applications and other agronomic practices used in the baseline annual crop scenario. Converting long-term cropping to permanent grass cover was projected to sequester approximately 13 Mg CO_2 eq yr^{-1} (0.5 Mg CO_2 eq ac^{-1} yr^{-1}) as soil C over 10 yr (Fig. 9). Since this land was retired from annual cropping and no longer receiving nitrogen fertilizer applications, N_2O emissions were projected to decrease by approximately 19 Mg CO_2 eq yr^{-1} (0.7 Mg CO_2 eq ac^{-1} yr^{-1}).

Analogous field measurements for comparison are limited; however a study in Iowa found that native grass strips planted in footslope positions had significant gains in carbon over time, relative to similar cropland soils (Perez-Suarez et al., 2014). A 15-yr study near Lincoln, NE estimated soil carbon accumulation rates in upland switchgrass barriers of 1.3 Mg CO_2 eq ac^{-1} yr^{-1} (3.1 Mg CO_2 eq ha^{-1} yr^{-1}) (Blanco-Canqui et al. 2014). In the Prairie Pothole Region of North Dakota, multiple on-farm comparisons showed approximately 13% higher surface (0–20 cm) SOC stocks in Conservation Reserve Program lands relative to paired croplands (Phillips et al. 2015). These field experiments and on-farm measurements are consistent with meta-analyses conducted at the national scale, which predict carbon sequestration rates from land retirement to range from 0.6 to 1.4 Mg CO_2 eq ac^{-1} yr^{-1} (1.6–3.4 Mg CO_2 eq ha^{-1} yr^{-1}) (Denef et al., 2011).

For this example, we examined the impacts of planting herbaceous riparian cover on previously cropped land. However, other NRCS conservation practices that convert cultivated land to perennial herbaceous cover may be evaluated in the same way, such as conservation cover, vegetative barriers, field borders, filter strips or grassed waterways. Alternatively, a land owner could also examine the GHG impacts of removing existing conservation practices, when considering reverting that land back to cropping.

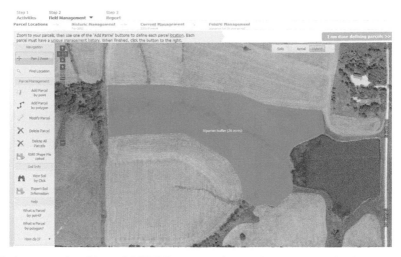

Fig. 8. An example of how COMET-Farm may be used to evaluate the impacts of adopting conservation practices, such as riparian buffers.

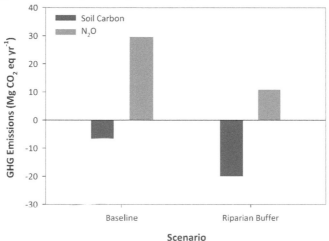

Fig. 9. Results from the riparian buffer case study, showing gains in soil carbon and reductions in N₂O emissions compared to the baseline scenario.

A major strength of the COMET-Farm tool is that it provides a high level of flexibility to the user to design and combine suites of conservation practices that can be uniquely tailored to their land, soil characteristics and management goals, in a highly visual environment. Further, the fact that the system considers **all** major greenhouse gases and emission sources (as well as C sinks) helps to avoid outcomes in which adoption of management practices to achieve mitigation for one gas or source category (e.g., increasing soil C) might inadvertently lead to increased emissions from other sources (e.g., soil N₂O emissions), and thus provide little or no net mitigation benefit.

Future Work and Applications

Parts of the COMET-Farm system are still under active development, including compiling data and refining methods for estimating emissions associated with management and production of specialty crops, including orchard species and high value vegetable crops, focusing on California, with the goal to add these systems to COMET-Farm in 2017. Implementing the uncertainty methods in the USDA guidelines (Eve et al., 2014) is also in progress.

An applications area that could be exploited in the future by COMET-Farm is the inclusion of precision agriculture management systems (Berry et al., 2003) as a mitigation option, particularly for N_2O. Currently, COMET-Farm treats a user-delineated field as having uniform management and uses those management inputs, as specified by the user, to drive the calculation of emissions for the various GHG source categories. However, to account for heterogeneity in soil physical properties, COMET- Farm subdivides single fields into discrete areas as given by the overlay with SSURGO soil map units. GHG emission calculations are then done separately for each map-unit segment within the field and then area-weight averaged in the emission report section. COMET- Farm already provides the user the capability to upload a custom soil map that can specify sub-field map units. That capability could also allow users to define spatially-explicit management zones which are a core concept in precision agriculture. If modifications were made to the GUI to allow management practices to vary by these zones (for example, the amount of N applied in a variable rate precision agriculture scheme) GHG emissions could be calculated separately for each of the defined management zones. Within-field variable rate fertilization practices have been shown to be an effective means to increase N use efficiency by enabling spatially-variable rate reductions while maintaining or increasing yields (Berry et al., 2003, Delgado and Bausch 2005, Khosla et al., 2002). Since N_2O emissions can vary nonlinearly as a function of N addition rates (Millar et al., 2010), accounting for spatially varying N rates would provide a more accurate estimate than using field-averaged fertilizer rates.

Applications of the COMET-Farm system can continue to expand as the awareness of agriculture's potential to contribute to GHG mitigation continues to increase. To take advantage of this potential, an expansion of policies and business strategies that incentivize agricultural producers to implement 'low-carbon' management practices will be needed. These incentives can include expanded opportunities to participate in GHG offset and other environmental markets (both voluntary and compliance markets), with payments for the ecosystem services provided (Bernoux and Paustian, 2015). Another area of increasing interest is the marketing of 'low carbon' agricultural products to consumers, who may choose to purchase products with reduced environmental footprints (Lavallée and Plouffe, 2004). Finally, producers of agricultural biofuels have an interest in minimizing the GHG footprint of their product (for which the within farm-gate feedstock production makes up a large portion of the emissions footprint) to meet renewable fuel standards (Field et al. 2016). All of these organizations, that is, offset project developers and GHG market registries, retailers marketing low C products and bioenergy producers, require rigorous and consistent science-based GHG metrics that can be tied directly to the activities and performance of the on-farm producers. Tools such as COMET-Farm provide a means to directly engage the farmers and ranchers themselves, by harnessing their unique knowledge of how they manage the land and assisting them in developing and adopting GHG-friendly management practices.

Acknowledgments

We would like to thank the system testers in the 'Cave' for their diligence and hard work and the many COMET-Farm users who have supplied valuable feedback and suggestions on how to improve the system. Support for development and implementation from USDA Natural Resource Conservation Service (under USDA CESU AGREEMENT 68-7482-15-507), USDA's Climate Change Program Office (under USDA Contract No. GS23F8182H) and California Department of Food and Agriculture (Agreement No. SC14061A) is gratefully acknowledged.

References

Akiyama, H., K. Yagi, and X. Yan. 2005. Direct N_2O emissions from rice paddy fields: Summary of available data. Global Biogeochem. Cycles 19:GB1005. doi:10.1029/2004GB002378

Al-Kaisi, M.M., A. Douelle, and D. Kwaw-Mensah. 2014. Soil microaggregate and macroaggregate decay over time and soil carbon change as influenced by different tillage systems. J. Soil Water Conserv. 69:574–580. doi:10.2489/jswc.69.6.574

Al-Kaisi, M.M., X. Yin, and M.A. Licht. 2005. Soil carbon and nitrogen changes as influenced by tillage and cropping systems in some Iowa soils. Agric. Ecosyst. Environ. 105:635–647. doi:10.1016/j.agee.2004.08.002

Bernoux, M., and K. Paustian. 2015. Climate change mitigation. In: S.A. Banwart, E. Noellemeyer, and E. Milne, editors, Soil carbon: Science, management and policy for multiple benefits SCOPE series. Vol. 71. CAB International, Wallingford, UK. p. 119–131.

Berry, J.K., J.A. Delgado, R. Khosla, and F.J. Pierce. 2003. Precision conservation for environmental sustainability. J. Soil Water Conserv. 58:332–339.

Blanco-Canqui, H., J.E. Gilley, D.E. Eisenhauer, P.J. Jasa, and A. Boldt. 2014. Soil carbon accumulation under switchgrass barriers. Agron. J. 106:2185–2192. doi:10.2134/agronj14.0227

Brenner, J., K. Paustian, G. Bluhm, K. Killian, J. Cipra, B. Dudek, S. Williams, and T. Kautza. 2002. Analysis and reporting of carbon sequestration and greenhouse gases for conservation districts in Iowa. In: J.M. Kimble, R. Lal, and R.F. Follett, editors, Agriculture practices and policies for carbon sequestration in soil. Lewis Publishers, CRC Press, Boca Raton, Fl. p. 127–140. doi:10.1201/9781420032291.pt3

Daly, C., M. Halbleib, J.I. Smith, W.P. Gibson, M.K. Doggett, G.H. Taylor, J. Curtis, and P.P. Pasteris. 2008. Physiographically sensitive mapping of climatological temperature and precipitation across the conterminous United States. Int. J. Climatol. 28:2031–2064. doi:10.1002/joc.1688

de Klein, C., R.S.A. Novoa, S. Ogle, K.A. Smith, P. Rochette, T.C. Wirth, B.G. McConkey, A. Mosier, and K. Rypdal. 2006. Chapter 11: N2O emissions from managed soil, and CO2 emissions from lime and urea application. In: 2006 IPCC guidelines for national greenhouse gas inventories, Vol. 4: Agriculture, forestry and other land use. In: S. Eggleston, L. Buendia, K. Miwa, T. Ngara and K. Tanabe (eds.). Japan: IGES, IPCC National Greenhouse Gas Inventories Program. Hayama, Japan.

Delgado, J.A., and W. Bausch. 2005. Potential use of precision conservation techniques to reduce nitrate leaching in irrigated crops. J. Soil Water Conserv. 60:379–387.

Del Grosso, S.J., S.M. Ogle, W.J. Parton, and F.J. Breidt. 2010. Estimating uncertainty in N2O emissions from US cropland soils. Global Biogeochem. Cycles 47:GB1009.

Denef, K., K. Paustian, S. Archibeque, S. Biggar, and D. Pape. 2012. Greenhouse gas accounting tools for agriculture and forestry sectors. http://www.usda.gov/oce/climate_change/techguide/Denef_et_al_2012_GHG_Accounting_Tools_v1.pdf. (accessed 1 Aug. 2015) (verified 5 June 2017).

Denef, K., S. Archibeque, and K. Paustian. 2011. Greenhouse gas emissions from U.S. agriculture and forestry: A review of emission sources, controlling factors, and mitigation potential. http://www.usda.gov/oce/climate_change/techguide/Denef_et_al_2011_Review_of_reviews_v1.0.pdf. (accessed 1 Aug. 2015) (verified 5 June 2017).

<unknown>ERS. 2004. Cropping Practices Surveys (now ARMS Farm Financial and Crop Production Practices Surveys). Economic Research Service. USDA.</unknown>

Eve, M., D. Pape, M. Flugge, R. Steele, D. Man, M. Riley-Gilbert, and S. Biggar. 2014. Quantifying greenhouse gas fluxes in agriculture and forestry: Methods for entity-scale inventory. USDA Office of the Chief Economist, Climate Change Program Office. USDA Technical Bulletin 1939. Viewed at http://www.usda.gov/oce/climate_change/estimation.htm. (accessed 1 Aug. 2015)(verified 5 June 2017).

Field, J.L., E. Marx, M. Easter, P. Adler, and K. Paustian. 2016. Ecosystem model parameterization and adaptation for sustainable cellulosic biofuel landscape design. Glob. Change Biol. Bioenergy 8:1106–1123. doi:10.1111/gcbb.12316

Khosla, R., K. Fleming, J.A. Delgado, T. Shaver, and D. Westfall. 2002. Use of site specific management zones to improve nitrogen management for precision agriculture. J. Soil Water Conserv. 57:513–518.

Lasco, R.D., S.M. Ogle, J. Raison, L. Verchot, R. Wassmann, K. Yagi, S. Bhattacharya, J.S. Brenner, J.P. Daka, S.P. Gonzalez, T. Krug, L. Yue, D.L. Martino, B.G. McConkey, P. Smith, S.C. Tyler, and W. Zhakata. 2006. Chapter 5: Cropland. In: Eggleston, L. Buendia, K. Miwa, T. Ngara and K. Tanabe, editors. 2006 IPCC Guidelines for National Greenhouse Gas Inventories, Vol. 4: Agriculture, forestry and other land use. Japan: IGES, IPCC National Greenhouse Gas Inventories Program. Hayama, Japan.

Lavallée, S., and S. Plouffe. 2004. The ecolabel and sustainable development. Int. J. Life Cycle Assess. 9(6):349–354. doi:10.1007/BF02979076

Li, C. 2000. Modeling trace gas emissions from agricultural ecosystems. Nutr. Cycling Agroecosyst. 58:259–276. doi:10.1023/A:1009859006242

Mesinger, F., G. DiMego, E. Kalnay, K. Mitchell, P.C. Shafran, W. Ebisuzaki, D. Jovic, J. Woollen, E. Rogers, E.H. Berbery, M.B. Ek, Y. Fan, R. Grumbine, W. Higgins, H. Li, Y. Lin, G. Manikin, D. Parrish, and W. Shi. 2006. North American regional reanalysis. Bull. Am. Meteorol. Soc. 87:343–360. doi:10.1175/BAMS-87-3-343

Millar, N., G.P. Robertson, P.R. Grace, R.J. Gehl, and J.P. Hoben. 2010. Nitrogen fertilizer management for nitrous oxide (N2O) mitigation in intensive corn (maize) production: An emissions reduction protocol for US Midwest agriculture. Mitig. Adapt. Strategies Glob. Change 15:185–204. doi:10.1007/s11027-010-9212-7

NRCS. 2010. Conservation practice standard: Riparian herbaceous cover. Code 390. USDA, National Resources Conservation Service. Washington, DC. http://www.nrcs.usda.gov/Internet/FSE_DOCUMENTS/nrcs143_026183.pdf (accessed 1 Aug. 2015) (verified 5 June 2017).

Ogle, S.M., F.J. Breidt, and K. Paustian. 2005. Agricultural management impacts on soil organic carbon storage under moist and dry climatic conditions of temperate and tropical regions. Biogeochemistry 72:87–121. doi:10.1007/s10533-004-0360-2

Parton, W.J., E.A. Holland, S.J. Del Grosso, M.D. Hartman, R.E. Martin, A.R. Mosier, D.S. Ojima, and D.S. Schimel. 2001. Generalized model for NOx and N2O emissions from soils. J. Geophys. Res. Atmos. 106:17403–17419. doi:10.1029/2001JD900101

Perez-Suarez, M., M.J. Castellano, R. Kolka, H. Asbjornsen, and M. Helmers. 2014. Nitrogen and carbon dynamics in prairie vegetation strips across topographical gradients in mixed Central Iowa agroecosystems. Agric. Ecosyst. Environ. 188:1–11. doi:10.1016/j.agee.2014.01.023

Phillips, R.L., M.R. Eken, and M.S. West. 2015. Soil organic carbon beneath croplands and re-established grasslands in the North Dakota Prairie Pothole Region. Environ. Manage. 55:1191–1199. doi:10.1007/s00267-015-0459-3

Robinson, D.A., I. Fraser, E.J. Dominati, B. Davíðsdóttir, J.O.G. Jónsson, L. Jones, S.B. Jones, M. Tuller, I. Lebron, K.L. Bristow, D.M. Souza, S. Banwart, and B.E. Clothier. 2014. On the value of soil resources in the context of natural capital and ecosystem service delivery. Soil Sci. Soc. Am. J. 78:685–700. doi:10.2136/sssaj2014.01.0017

Saxton, K.E., W.J. Rawls, J.S. Romberger, and R.I. Papendick. 1986. Estimating generalized soil-water characteristics from texture. Soil Sci. Soc. Am. J. 50:1031–1036. doi:10.2136/sssaj1986.03615995005000040039x

Soil Survey Staff. 2015. Web soil survey. USDA, Natural Resources Conservation Service. http://websoilsurvey.nrcs.usda.gov/. (accessed 17 July 2015).

USDA. 2016. Agricultural resource management survey (ARMS). USDA, Economic Research Service. https://www.ers.usda.gov/data-products/arms-farm-financial-and-crop-production-practices.aspx (accessed 1 Dec. 2016).

USEPA. 2011. Inventory of U.S. greenhouse gas emissions and sinks: 1990-2009. Environmental Protection Agency, Washington, DC.

Precision Conservation and Precision Regulation

G. F. Sassenrath and J.A. Delgado

"Good soils make up for poor management"– Georgia farmer

Touring research plots in the southeastern United States, I (G.F. Sassenrath) remarked how the soil looked like beach sand – and how challenging it must be to grow a crop in such rapidly-draining soil. A farmer from Georgia, overhearing my comments, asked where I was from. At the time, I was working in the deep alluvial silt-loam soils of the Mississippi Delta. In response, the farmer wisely commented, "Ah, yes, good soils make up for poor management."

Indeed, many farmers in North America have been blessed with good soils, and in response have neglected adopting best management practices for soil and water conservation (Rodriguez et al., 2008; Ribaudo et al., 2011; Carlisle, 2016). This inaction can continue no longer, as even good soils have suffered a loss of productive capacity. Soil degradation and erosion reduce crop production (Bakker et al., 2004, 2007) and contaminate waterways with sediment and nutrient runoff. Without implementing policies for conservation practices to confront the challenges that humanity faces in its future, including a changing climate, we will not be able to achieve sustainable levels of food production needed to feed the growing world population (Delgado et al., 2011b).

Traditional production methods in North America were unsophisticated (Fig. 1a). The tenant farmer would not be familiar with the computer used for crop and yield monitoring that his great-grandson takes for granted. Advances in genetics, engineering, and mechanics have led to significant improvements in the methods used for agricultural production and have allowed farmers to adopt precision conservation (Fig. 1b; Sassenrath et al., 2008). Farmers being trained today are accustomed to having computers in the cab of the tractor or combine. It is interesting to note that the current average age of U.S. farmers is 64, born around 1954 (NASS, 2014). These farmers came of age during a period of incredible

Abbreviations: ACPF, Agricultural Conservation Planning Framework; DEM, digital elevation map; GHG, greenhouse gases; GIS, geographic information systems; GPS, global positioning systems; LiDAR, Light Detection and Ranging; NTT, nitrogen trading tool; PCM, precision conservation management; RS, remote sensing; RTK, real-time kinetic; sUAS, small unmanned aircraft systems;

G.F. Sassenrath, Agronomy Dept., Kansas State University, Manhattan, KS; J.A. Delgado, Soil Management and Sugarbeet Research, USDA-ARS, Fort Collins, CO. *Corresponding author (gsassenrath@ksu.edu)

doi:10.2134/agronmonogr59.2018.0002

Precision Conservation: Geospatial Techniques for Agricultural and Natural Resources Conservation
J.A. Delgado, G.F. Sassenrath, and T. Mueller, editors. Agronomy Monograph 59.

Fig. 1. Tenant farmer working Lambert Field, circa 1930s. Now the St. Louis Lambert International Airport (a). Computers have become common in combines and tractors as methods to spatially monitor crop growth and yield, and spatially manage inputs such as fertilizer, circa 2000s (b).

technological advances that resulted in persistent and substantial improvements in crop yields, with virtually no changes in crop acreage (Fig. 2).

With the ever-increasing world population, a changing climate, and higher demand for water and soil resources, humanity faces great challenges in the 21st century. Agricultural systems must significantly increase agricultural production to meet humanity's growing production needs and dietary shifts while confronting new challenges created from a changing and uncertain climate (Delgado et al., 2011b; Cassidy et al., 2013). Although the previous yield gains have been impressive, it is doubtful whether they can be maintained and advanced sufficiently in the long term to address future needs. Moreover, the acreage available for farming is now decreasing, facing rapid conversion to urbanization (Sorensen et al., 2018). Conservation of soil and water resources will be critical to ensure humanity's food security in the 21st century and address the ever-growing challenges of an increasing global population, a changing climate, extreme weather events, greater demands for water resources, depletion of aquifers, desertification, and other environmental and societal challenges (Ascough et al., 2018). It has been estimated that it takes 500 to thousands of years to form as little as one inch of soil, yet this fragile layer could be lost via erosion in a single extreme event.

In the intervening seventy years between the two images in Fig. 1, crop yields have increased more than fourfold (Fig. 2). Nevertheless, associated with this increase was a loss of 10 to 20 tons of soil per acre per year [20 to 40 Mg (ha year)$^{-1}$]. In many areas in the United States more than 1400 tons (2800 Mg) of topsoil have been lost. This would be approximately between 5 and 10 inches (13 to 25 cm) of soil lost during this 70-yr period. Rates of erosion have decreased with implementation of conservation practices. However, much more aggressive conservation measures must be implemented to move beyond simply slowing rates of soil degradation. We must restore the natural resource base. Conservation of soil resources across all continents is a serious global challenge that will need to be addressed to achieve food security for humanity in the 21st century (Delgado et al., 2011b; Fig. 3).

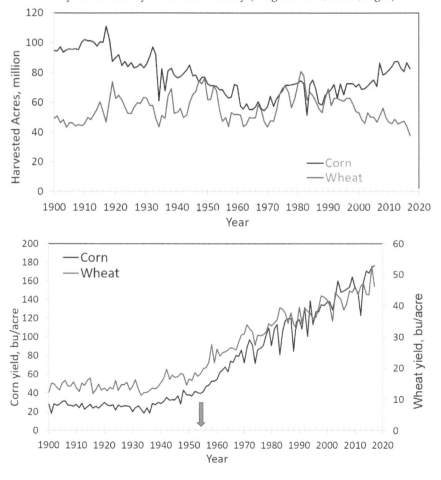

Fig. 2. Historical changes in harvested acres and yield for two major grain crops in the U.S., corn and wheat, 1900 to present. Blue arrow marks the birth year of the average U.S. farmer (NASS, 2014).

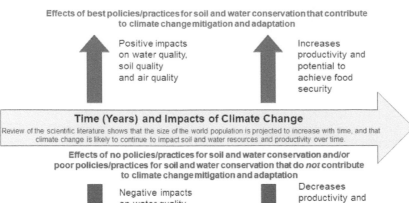

Fig. 3. There is a close relationship between climate change, limited global water and soil resources, population growth and food security. As climate change impacts the world's soil and water resources, it threatens to negatively impact food production (i.e., decrease food production and/or food production potential). As the climate changes, conservation practices have the potential to help us achieve maximum sustainable levels of food production, which will be essential to efforts to feed the world's growing population. Good policies and/or practices for soil and water conservation will contribute to positive impacts on soil and water quality, soil productivity, and efforts toward achieving and/or maintaining food security. These good policies/practices will contribute to climate change mitigation and adaptation. Poor policies/practices for soil and water conservation (or a lack of policies/practices) will contribute to negative impacts on soil and water quality, soil productivity, and efforts toward achieving and/ or maintaining food security (From Delgado et al., 2011b).

The information presented in this book details precision conservation technologies as key approaches to increase the effectiveness of conservation practices to conserve and/or improve air, soil, and water quality and increase the potential for adaptation to a changing climate. Taking the business-as-usual approach will be insufficient to reduce the nutrient losses or erosion from agricultural systems. For example, over four decades after the United States Clean Water Act was established, a national water quality assessment by the U.S. Environmental Protection Agency (EPA) found that greater than fifty percent of the assessed waters still fail to meet water quality standards (USGAO, 2013). Human activities continue to contribute to greenhouse gas emissions and soil degradation. The precision conservation approaches presented in this book demonstrate how to increase conservation effectiveness by identifying sensitive areas within a field and across a watershed and then designing conservation practices to target those key areas.

First defined by Berry et al. (2003), precision conservation is a method of coordinating spatial technologies and procedures to accommodate spatial and temporal variability in natural and agricultural systems and implement conservation management practices. We now have the tools to address and correct soil degradation and erosion, improving both the productive capacity of the land and preserving the soil and water natural resource base. Improvements in information technologies, sensor systems, and modeling of ecosystem processes have further advanced our abilities to detect problems and design solutions (Delgado et al., 2011a). Integration across disciplines has brought social awareness and goals into solution-building and delivery for better participatory planning and longer-term support of conservation programs (Sassenrath et al., 2010). New economic structures have developed natural resource markets and "impact investing" that offer direct monetary rewards for implementing conservation management (Jackson, 2013; Barman, 2015). Precision conservation has the potential to significantly increase the effectiveness of conservation efforts and the sustainability of the agroecosystem by improving soil and water conservation for agricultural and nonagricultural areas. There are additional opportunities to use these approaches to increase voluntary application of precision conservation and to trade reductions in agrochemicals and erosion losses in air and water quality trading markets, increasing economic benefits for farmers. Precision conservation can increase the impact of each dollar applied in conservation practices by improving best management practices outlined in the USDA-Natural Resources Conservation Service (NRCS) national conservation practice standards (Delgado and Berry, 2008). Moreover, precision conservation can help to increase the efficiency of the applied inputs, maintain and increase yields, increase economic returns, and contribute to increased wildlife biodiversity, all of which enhance the potential for voluntary use of this approach.

The Right Approach– Adopting the 7 R's

The 4Rs are a key management approach of precision agriculture to reduce nutrient losses from farming systems (the *right* product, at the *right* rate, at the *right* time, and at the *right* place; Roberts, 2007). However, Delgado (2016) reported on examples in which the 4 Rs of precision farming were applied, but significant losses of soil (via erosion) or nutrients still occurred (Fig. 4). Berry et al. (2003, 2005) described the need to manage and connect the fluxes of nutrients from fields to fluxes outside of the field using spatially- and temporally-variable information (Fig. 5). They demonstrated how a precision conservation approach can be used to identify where field buffers, sediment traps, denitrification traps, and riparian buffers can be placed to minimize the losses of nutrients from the watershed. This concept of precision conservation was simplified by Cox (2005) as applying the *right* conservation practice, at the *right* place, at the *right* time, and at the *right* scale (the 4 Rs of conservation). Delgado (2016) combined the 4 Rs of precision farming with the 4 Rs of precision conservation to reduce the losses of nutrients from the field and across the watershed, increase nutrient use efficiency, and increase conservation effectiveness with a precision conservation approach that establishes 7 Rs for nutrient management and conservation. By applying the *right* product (fertilizer), at the *right* fertilizer rate, with the *right* method of fertilizer application, with the *right* conservation practice, and with conservation practices at the *right* place, and at the *right* scale of conservation practice, and applying both the product and the conservation practice at the *right* time, (applying the 7 Rs of nutrient management and conservation: *right* product, *right* rate, *right* method, *right* practice, *right* place, *right* scale, and *right* time), we will address both productivity and conservation concerns by minimizing nutrient losses from the field and across the watershed while ensuring adequate productivity (Delgado, 2016). To maximize the effectiveness of conservation practices and nutrient management, the 4 Rs of nutrient management and the 4 Rs of

Fig. 4. Application of the 4 Rs of nutrient management can result in significant losses of soil and nutrients from fields. Integrating the 4 Rs of nutrient management with the 4 Rs of conservation to create the 7 Rs of nutrient management and conservation stewardship is critical to reduce transport of sediments and nutrients from fields and avoid development of ephemeral gullies. Using a precision conservation approach, we can fine-tune the placement of contour furrows, grass waterways, buffers, and other conservation practices. A. photo source: NRCS Photo, development of ephemeral gullies. B. Development of ephemeral gullies in tilled (B, C) and no-tilled (D) crop production fields and field edges (E, F). Photo source: G.F. Sassenrath, K-State Research and Extension.

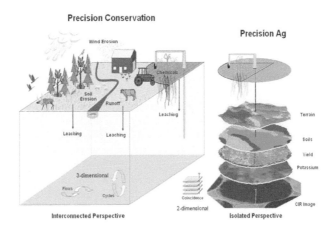

Fig. 5. Berry et al. (2003) reported that precision conservation should consider a three-dimensional approach where we could conduct assessments of the inflows and outflows from fields to the watershed and regional scales to determine how to implement conservation practices to increase the effectiveness of conservation efforts (From Berry et al., 2003).

precision conservation, or the 7 Rs for nutrient management and conservation, need to be applied in the field and off the field.

This book details precision conservation approaches for increasing yields, economic returns, and nutrient use efficiencies at a field level, and reducing the losses and transport of nutrients across watersheds. It shows that a voluntary precision regulation approach can be used to encourage farmers, nutrient managers, conservationists, wildlife biologists, consultants, non-governmental organizations, and other users of these new technologies to apply these conservation practices across the watershed to achieve reductions in nutrient losses, greenhouse gas (GHG) emissions, and erosion, while increasing carbon sequestration and other ecosystem benefits that could potentially be traded in voluntary air and water quality markets.

The challenges of voluntary conservation programs have been described previously by King et al. (2009), Cox et al. (2012), and Cox and Hug (2012). These programs can utilize an approach to increase conservation effectiveness by using model analyses, identifying critically sensitive areas, and assigning conservation programs that are focused on these critical areas. Delgado (2016) reported that precision conservation using the 7 Rs approach could be used for voluntary precision regulation. Delgado (2016) reported that these new ecotechnologies with geospatial capabilities are allowing nutrient managers to change their management approach across a watershed from a concept of managing nonpoint sources to a concept that identifies hot spots that disproportionately contribute to nutrient and sediment contamination within the watershed, and then implementing voluntary approaches to preserve the air, soil and water quality.

Technological Advances in Remote Sensing, GPS, GIS, Drones, and Artificial Intelligence

A critical need for implementing sound conservation practices is the identification of sensitive areas. Precision conservation integrates positioning technologies such as GPS or GNSS and sensing systems to detect and record spatial patterns, application equipment for variable rate and positioning of inputs, and knowledge systems such as models to interpret and develop decisions (Berry et al., 2003). Advances in GPS now allow centimeter accuracy for precise sensing of conditions and placement of management inputs (Fulton and Darr, 2018). GPS–guided systems are integrated with automated navigation for precise control of management inputs, improving accuracy and productivity. The tools and technologies that have contributed to advances in precision methodologies in agriculture and conservation ultimately improve the profitability and decision-making ability of the user. Fulton and Darr (2018) demonstrated how these technologies can be used to increase yields, and also how to manage tillage, fertilizers, pesticides, and even water more effectively, reducing inputs per unit area, enabling farmers and operators to avoid sensitive areas across the landscape, minimizing overlapping application of these agrochemicals, increasing input application efficiency, and reducing the potential for agrochemical losses from the field site. Variable rate detection technologies integrate spatially sensed crop data, such as nutrient status or yield, with system models to develop spatially precise maps for targeted management of inputs, allowing the application of a 7R nutrient management and conservation stewardship approach, better utilizing inputs such as fertilizers, fungicides, or tillage and reducing environmental contamination. There is also potential to use new tools such as small unmanned aircraft systems (sUAS), robots, smartphone apps, and other new technologies to help with scouting of fields and decision-making in application of agrochemicals.

Soil Erosion and Nutrient Management Modeling

Models are powerful scientific tools allowing identification of critical areas, exploration of alternative management practices and environmental impacts, and quantitation of the impacts of alternative practices (Ascough et al., 2018). Accurate monitoring of soil and nutrient losses and their impacts on water bodies are time consuming, expensive, and under many conditions, unrealistic to adequately capture the problems. Simulation models can be used in concert with targeted monitoring to better assess problems and impacts of conservation programs after implementation while reducing monitoring costs. Many types of models are available for erosion modeling and conservation planning (Ascough et al., 2018). Ascough et al. (2018) noted that although there are several erosion models that are powerful tools for assessing erosion for a given watershed or management scenario, the estimation of nitrogen loss at the watershed scale is complex due to the biogeochemical transformations of nitrogen. Additionally, moving from individual fields to watersheds introduces considerable additional complexity. Models have been developed to simulate nitrogen (N) flows within fields and watersheds, which is particularly challenging given the complexities and difficulties of adequately simulating nutrient flows (Yuan et al., 2018). Yuan et al. (2018) described several of the models that can conduct evaluations of nitrogen losses at a watershed level and noted that these models can make reasonable predictions of total N loading following model calibration. They reported that since the nitrogen cycle is so complex and there are so many pathways for nitrogen loss, including gaseous, surface and leaching losses, some of the models exhibit stronger performance in predicting one pathway than another. Following calibration for a given watershed there is potential to use the models in conservation planning. These models can be used at a watershed level to evaluate the implementation of conservation practices to reduce nitrogen losses from the hot spots across the watershed or to apply targeted practices at given positions in the watershed to reduce losses of nitrogen. The spatial maps developed from the models should be useful in identifying areas of high N loss and implementing precision conservation strategies to increase N use efficiency while reducing N losses.

Because ravines are a primary source of sediment contamination to waterways, spatial localization and morphometric characterization allow development of strategic control methods to ameliorate sediment impairments (Mulla and Belmont, 2018). A GIS analytical technique using high-resolution Light Detection and Ranging (LiDAR) digital elevation map (DEM) data can detect ravines with 90% accuracy (Mulla and Belmont, 2018). These analytical procedures allow determination of ravine volume, area, and relief; all important factors contributing to the amount of sediment loading from ravines. This technique has been used to identify regions within the Minnesota River Basin that have the potential to contribute disproportionately large amounts of sediment to the Minnesota River, allowing implementation of strategic control methods for optimal improvement of water quality (Mulla and Belmont, 2018).

These precision tools improve the accuracy of management decisions, the efficiency of production by targeting inputs to areas of need, and the environmental stewardship by reducing over-application or applications in sensitive areas.

Physical Structures

A prime consideration in reducing the impact of agricultural production on the environment is the reduction of sediments, nutrients, and agricultural chemicals that move from fields to waterways with runoff. Physical modification of the crop field and surrounding areas is a common approach that reduces sediment and nutrient transport. Precision conservation tools provides a method to reduce these losses.

Terraces and contours are structures that have been used for millennia to manage the flow of water over the field surface. The spatial precision of terrace design and placement is enhanced with GIS and GPS technologies such as high-resolution DEM from LiDAR or real-time kinematic (RTK) GPS equipment (Thompson and Sudduth, 2018). Precision technologies have improved the design and placement of terraces and contours to better control movement of water over crop fields. Thompson and Sudduth (2018) used a web-based conservation planning tool (WebTERLOC) to provide better visualization and evaluation, integrating field conditions, farming practices, soil characteristics, erosion potential, and expense to develop more accurate conservation plans. They showed that precision design systems can reduce the terrace lengths and construction costs by 15%, while reducing erosion and controlling gully formation. This precision conservation approach considers the spatial and temporal variability, which contributes to increased effectiveness of conservation practices, contributing to more viable farm operations and increased economic sustainability (Thompson and Sudduth 2018).

In contrast to grassed or vegetative filter strips that are placed at field borders, grassed waterways are areas within a field designed to control the movement of water off the field. Precision conservation technologies can be used to precisely implement grassed waterways (Mueller et al., 2005; Fig. 6). Fiener and Auerswald (2003, 2018) reported that well-established grassed waterways effectively prevent gully erosion, reduce sediment and agrochemical delivery, and dampen peak

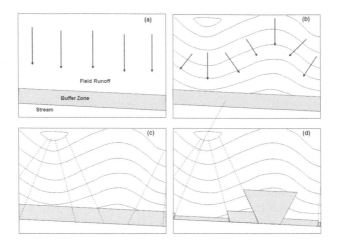

Fig. 6. Dosskey et al. (2005) illustrated how crop-field runoff can exhibit different patterns: (a) uniform runoff flow and buffer; (b) non-uniform flow to a uniform buffer; (c) non-uniform runoff areas and uniform-width buffer; (d) non-uniform runoff areas and the variable-width buffer areas that they flow to (From Dosskey et al., 2005).

discharge rates. The loss of cropland set aside for formation of grassed waterways is more than offset by the benefits gained (Fiener and Auerswald 2018). They described the need to improve modeling assessments of the benefits of using grass waterways, but reported that there are beneficial on-site and off-site effects of using this practice. Grassed waterways provide additional benefits through increased biodiversity, soil carbon stocks, and expanded habitat for wildlife.

Dosskey et al. (2018) explored how precision conservation can be used to better design and implement vegetative filters, such as filter strips and riparian buffers, to reduce the off-site transport of sediment and chemical runoff from fields to waterways. Dosskey et al. (2002, 2005, 2007, 2018) reported on the nonuniformity of flows from agricultural fields and the need to better design practices to handle these flows (Fig. 7). Dosskey et al. (2018) recommended the use of a precision conservation approach using digital elevation models, geographic information systems (GIS), and modeling with algorithms to more precisely develop mitigation strategies for agricultural runoff. The models are designed as field planning tools to reduce agricultural pollution by improving the size and configuration design of filter areas. Vegetative filter strips designed with AgBufferBuilder capture sediment, nutrients, and agricultural chemicals from crop fields, reducing their contamination of waterways and improving water quality. The team demonstrated how precision conservation technologies can be used to find alternative configurations that substantially improve the water quality performance of vegetative filters (Dosskey et al., 2018). These off-field practices will be needed for climate change adaptation and to increase conservation effectiveness, protecting streams and reducing off-site transport of sediments and agrochemicals from fields during extreme events, which will increase with a changing climate (Nearing, 2001; Nearing et al., 2004; Delgado et al., 2013).

Subsurface management of nutrient runoff from agricultural fields through implementation of riparian buffers is possible with a new siting tool (Tomer et al.,

2018). These riparian buffers slow water efflux from tile-drained fields, reducing nutrient levels though biological uptake and denitrification prior to releasing the water into waterways. Tomer et al. (2018) developed the Agricultural Conservation Planning Framework (ACPF) tool to help identify locations of the watershed where conservation practices could be applied to increase conservation effectiveness across the watershed (Fig. 7 and 8). Agricultural Conservation Planning Framework can be used to increase precision conservation, reduce the losses of nutrients and erosion across the watershed, and identify hot spots that are more susceptible to nutrient loss. Implementing these practices can result in a more resilient and sustainable system. They also reported that ACPF helps identify the best areas to place saturated riparian buffers. Previous reports have documented the targeted use of saturated riparian buffers, denitrification traps, and even denitrification barriers underground to reduce nutrient transport across the watershed (Tomer et al., 2003, 2007; Berry et al., 2003, 2005; Delgado et al., 2010b; and Hunter, 2001, 2013). Similarly, phosphorus removal traps can also be placed at targeted locations using precision conservation (Delgado and Berry 2008).

Targeted Manure and Irrigation Management Systems

Precision conservation uses spatially registered agronomic information, but also integrates the dynamics and flows of inputs across the field and connects them with the surrounding areas to help managers make decisions that can be applied to reduce the offsite transport of nutrients and chemicals. Both precision irrigation and precision manure management are key approaches used to increase nutrient cycling efficiency. Precision manure management involves careful manipulation and interaction of practices that span a livestock production system and should consider additional implications beyond the farm gate (Kleinman et al., 2018). Improvements in three key areas can contribute to advances in manure management: manure generation and storage (feed inputs, barnyard management, manure storage and handling), land application (crop utilization: source, rate, timing, and method), and export and import of manures. Opportunities exist that develop holistic strategies for precision manure management that balance agronomic productivity and conservation priorities.

Management of water onto and off of fields is critically important to ensure sufficient water for crop growth without removing sediments and nutrients that can contaminate waterways. Technology is commercially available to use precision irrigation to apply the right amount of water at the right time and location to make the best use of irrigation water (Bjorneberg et al., 2018). However, user-friendly decision tools are needed to quantify specific irrigation needs and control water application within fields. Precision irrigation not only manages the amount of water needed by the crop, but can also provide information on irrigation system capacity and plant water requirements. Precision irrigation systems integrate monitoring of crops and soils through remote imagery from satellite or aerial vehicles with decision support programs to apply the right amount of water at the right time and place. Precision manure and precision irrigation management could be implemented to reduce surface and leaching losses of nutrients, but there is a need to also consider weather and hydrology at each site as well as the spatial and temporal variability at each site.

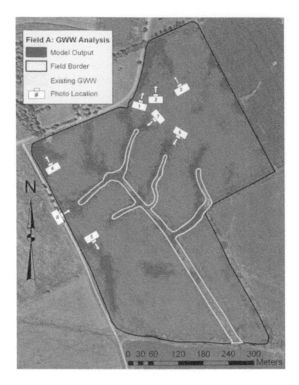

Fig. 7. Aerial photograph of field identifying areas where the model identified a probability of erosion with values greater than 0.5 in red, existing grassed waterways (GWWs), and locations of photographed eroded areas (From Luck et al., 2010).

Precision Management for Drainage and Soil Health

Additional software is available to use GIS–GPS data for the planning and design of drainage systems and for guidelines for managing drainage systems (Shedekar and Brown, 2018). Shedekar and Brown (2018) describe a precision conservation approach using GPS, remote sensing, DEM, and new software and/or technologies to account for spatial and temporal variability of soil type and hydrology to manage drainage at a given site. Agricultural drainage systems include surface and subsurface drains (or tile drains), as well as the drainage ditches that connect to the water system, and can reduce crop losses due to water logging, as well as provide reductions in nonpoint-source pollution. Recent technological advances integrate spatial and temporal information on soil type, topography, crop management, location and elevation of natural outlets, and land use to develop and implement precision drainage systems, making this an important tool for higher productivity and improved conservation during the 21st century. Higher yields will be obtained when we are improving drainage management (Skaggs et al., 2012).

Managing for soil health is a relatively new concept. However, evidence has demonstrated the importance of soil health, and more specifically soil organic carbon (SOC), for the productive capacity of the soil (Doran, 2002; Lal, 2016). Improving SOC improves the soil health, increases the resilience of the system, and reduces soil erosion (Clay and Mishra, 2017). Precision conservation can be used to improve

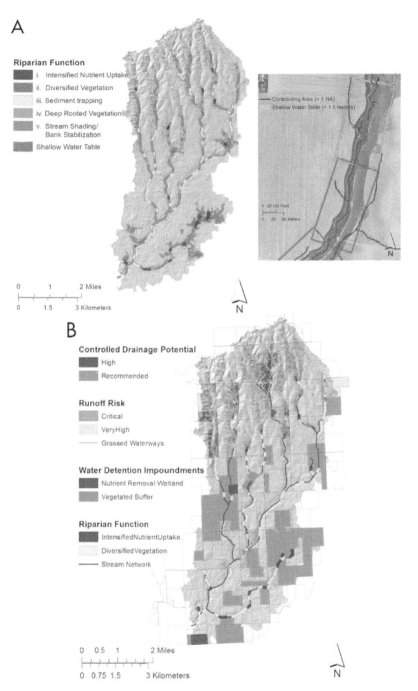

Fig. 8. A. Distribution of potential riparian buffer functions in Lime Creek. B. A potential conservation planning scenario for Lime Creek. This is an example of how we can implement a precision conservation approach (a.k.a. a targeted approach) for a watershed (From Tomer et al. 2013).

soil health and increase carbon sequestration across the field by integrating spatial and temporal variability using GIS, GPS, modeling, and other layers of information (Clay et al., 2018). Clay et al. (2018) documented the impact of crop residue harvest across landscapes, its impact on soil carbon sequestration and soil health, and how to go from point carbon budget measurements to precision conservation assessments across the landscape. The authors also provided examples of how to convert point carbon budget measurements into a precision conservation assessment.

Markets and Emerging Approaches to Improve Conservation Metrics and/or Ecosystem Services

Ecosystem service markets are a new agroecological industry that can enhance the adoption of conservation practices by providing a source of income to farmers. Trading reductions in greenhouse gas emissions and/or increases in carbon sequestration could contribute to climate change mitigation. More importantly, ecosystem service markets could also increase the potential for climate change adaptation, since the best conservation practices that contribute to climate change adaptation also contribute to ecosystem services via reduction of the transfer of agrochemicals to water bodies while reducing greenhouse gas emissions and increasing carbon sequestration.

The tools, technologies, and processes associated with precision agriculture can be adapted through strategic, objective-driven conservation to inform conservation practice adoption when wildlife objectives are explicitly incorporated into farm- and landscape-level decision frameworks for precision conservation (McConnell and Burger, 2018). McConnell and Burger (2018) argued that identifying economic opportunities for conservation enrollment will increase adoption by landowners. They identified the need for greater synergy between professionals working in the areas of soil conservation, wildlife biology, and agricultural economics to confront the challenges that come with agricultural intensification. They reported that scientists, landowners, consultants, and others must work collectively to "...continue to develop new, innovative ways to protect and enhance the ecological integrity of agricultural landscapes" and that a precision conservation approach can benefit wildlife populations. They noted that, for example, conservation practices could increase the breeding abundance of a given species, and that a precision conservation approach can be applied to any wildlife species or ecosystem service to improve the benefits of conservation practices for wildlife. A precision conservation approach could use new and advanced technologies to produce economic benefits for farmers and ranchers, conserve wildlife, and improve conservation across the watershed.

Saleh and Osei (2018) describe the use of calibrated models to evaluate the effectiveness of conservation practices at a given field considering spatial and temporal variability. Analyses with robust simulation models can contribute to evaluations of what conservation practice or combination of conservation practices can be applied to a given field to quantify the reduction of nitrogen or other nutrient losses, reductions that can potentially be traded in water quality markets. The concept of a nitrogen trading tool (NTT) was proposed by Delgado et al. (2008) to quantitatively assess reductions in nitrogen losses via emissions, leaching and surface runoff; these savings (reductions in losses) can potentially be traded by the farmer or nutrient trader in air and water quality markets (Delgado, 2012; Delgado et al., 2008; 2010a; Lal et al., 2009, Ribaudo et al., 2005, Saleh et al., 2011, Saleh and Osei 2018). The original version of the NTT was published by

Delgado et al. (2008, 2010a). Saleh et al. (2011) and Saleh and Osei (2018) improved the NTT by incorporating erosion modeling capabilities, as well as economic analysis and assessment of losses of other nutrients.

Simulation tools such as the Comprehensive Economic Environmental Optimization Tool (CEEOT) at the watershed-scale level, and the more user-friendly Nutrient Tracking Tool (NTT) at the farm and small-watershed level, can be used to determine the optimal practice implementation for farms that embodies precision conservation approaches to maximize the benefit from each dollar of conservation practice investment (Saleh and Osei 2018). Nutrient tools like the Nutrient Trading Tool, and its successor, the Nutrient Tracking Tool, could be calibrated for trading reductions in nutrient losses from using precision conservation practices in water quality markets and for the development and implementation of voluntary precision regulation (Delgado et al., 2008; 2011a; Saleh and Osei 2018). This tool can help policymakers evaluate what policies to apply or support, and/or encourage voluntary precision regulation using precision conservation for ecosystem trading in markets (Lal et al., 2009, Delgado et al., 2008; Ribaudo et al., 2011). Saleh and Osei (2018) show that there is potential to apply specific conservation practices at a hot spot and quantify the reduction in nutrient losses. These savings (reduced nutrient losses), quantified by the Nutrient Tracking Tool, could potentially be traded in water quality trading markets (Saleh and Osei 2018). Saleh and Osei (2018) reported that the Nutrient Tracking Tool is a valuable tool by which users could make the most effective decisions to increase their farm productivity while reducing nutrient and sediment losses to water bodies using precision conservation.

Nitrous oxide (N_2O) contributes to about 78% of the GHG emissions from the agricultural sector in the United States when direct and indirect N_2O emissions are expressed in CO_2 equivalent emissions (Walthall et al., 2014, USEPA, 2010), and there is potential to apply precision conservation to reduce the direct and indirect emissions of N_2O (Delgado and Berry 2008; Berry et al., 2005). Paustian et al. (2018) discuss the potential to apply calibrated simulation models such as the COMET-Farm system to assess management at a farm level considering spatial and temporal variability to reduce GHG emissions. The COMET-Farm system is a state-of-the art GHG modeling quantification method in a web-based tool that can be used to assess the effects of precision conservation practices on CO_2, methane (CH_4), and N_2O emissions from all major sources on the farm (Paustian et al., 2018). Paustian et al. (2018) described how this tool could assess the effects of conservation practices such as land management of annual and perennial crops, pasture, range and agroforestry systems, as well as emissions from livestock and on-farm energy use. There is potential to use this tool to implement voluntary application of conservation practices to reduce GHG emissions and trade these reductions in air quality markets. Paustian et al. (2018) also reported that consumers are interested in products that could be grown with low GHG emissions and this web tool can be used to assess management practices and show when farmers and ranchers are being good stewards of air, soil, and water quality, and are implementing these conservation practices to produce products with lower emissions and a smaller carbon footprint. This tool could also be used to assess the carbon footprint of biofuel cropping systems (Paustian et al., 2018). This system can help meet different policy- and market-driven greenhouse gas reduction efforts.

There is also the potential to develop precision conservation plans that meet desired sustainability and biofuel production goals (Chaubey et al., 2018). Chaubey et al. (2018) described how precision conservation can be used across the landscape to produce food, feed and bioenergy crops in a manner that takes into account social, economic and environmental factors. Chaubey et al. (2018) reported how precision conservation can be used to spatially locate various biofuel crops such as perennial grasses in vegetated filter strips, highly erodible lands, or at specific landscape positions across the watershed to conserve water quality, soil health and ecosystems. They described precision conservation methods to assess what conservation practice can be used to target specific locations of a watershed to increase the effectiveness of conservation efforts. This is in agreement with the Berry et al. (2003), Tomer (2010), and Tomer et al. (2013) recommendation to use modeling, data mining, and map analysis to assess the hot spots across the watershed.

Precision Regulation

Precision regulation is a methodology that goes beyond only reducing soil and nutrient losses to the environment by targeting identified critical areas. *Voluntary* precision regulation is the use of voluntary programs to encourage the use of best precision conservation practices to reduce environmental impacts. It also addresses other environmental impacts such as effects on wildlife, GHGs, salinity, efficiency in use of water resources, efficiency in use of nutrients, etc. It has been proposed that a voluntary precision regulation approach could be implemented to increase the effectiveness of conservation practices across the watershed and that precision conservation could be part of precision regulation to manage the hot spots across a watershed (Cox 2009; Cox, 2010; King et al., 2009). Delgado (2016) reported the use of precision conservation for precision regulation is a good approach, which moves beyond the concept of nonpoint sources to increase our capacity to assess the transport of nutrients from the field to the watershed and manage the spatial variability in soils, slopes, variable hydrology, vegetation and other components of the agricultural system. This book presents examples of how precision conservation is being used to implement a voluntary precision regulation approach. It has been shown that voluntary conservation measures could help wildlife and landowners use USDA NRCS–prescribed conservation practices to improve their lands in ways that benefit wildlife (USDA, 2015). Voluntary precision conservation measures could help wildlife and improve conservation across the watershed (McConnell and Burger, 2018). Additionally, several chapters in this book explored the economic benefits of using a precision conservation approach, which can include higher yields, increased efficiency in application of inputs, reduced nutrient losses, and/or the potential to trade environmental benefits in air and water quality trading markets.

The USDA has voluntary conservation programs such as the Environmental Quality Incentives Program (EQIP), which helps producers invest in improved practices for soil, air, and water conservation for their agricultural operations (USDA-NRCS, 2018). There are private companies that are also supporting voluntary regulation and are encouraging the use of a precision conservation approach to revolutionize soil and water management (Buman 2016a; Buman 2016b; Hammes, 2016). Using this precision conservation approach, the private industry could help implement variable buffers, denitrification traps, and other new technologies that could be used to manage the flows (Berry et al., 2003, Buman, 2016a; Buman, 2016b; Hammes, 2016; Delgado and Berry, 2008; Heartland Science and Technology Group, 2017; Illinois Sustainable Ag Partnership, 2018). This book covered the advances in precision conservation best management practices that could be used by USDA personnel, consultants, farmers, wildlife biologists, and other users interested in minimizing losses of nutrients, sediment, and agrochemicals, while also benefitting from other advantages. These tools and approaches described by Berry et al. (2003) and the available literature from the last 15 yr, support the use of these approaches as part of the ongoing voluntary regulation approach, and precision regulation has been shown to be a viable approach (Delgado and Berry, 2008). Delgado and Berry (2008) reported on how precision conservation could be used to improve best management practices across NRCS national conservation practice standards.

Voluntary precision regulation is already taking place; for example, producers in Illinois could apply for the USDA–NRCS special funding for precision conservation management (PCM) techniques (USDA NRCS, 2017). This program aims to help producers in Illinois to "combine financial farm business planning with precision conservation technology" to address conservation challenges such as soil health (USDA NRCS, 2017). Another example of voluntary precision regulation using precision conservation can be found in the Chesapeake Bay, where the Chesapeake Conservancy is working to develop tools and datasets for precision conservation (Chesapeake Conservancy, 2018). These are just a couple of examples among several other current efforts across the United States of the implementation of precision conservation and precision regulation. See also Buman (2016a) and Buman (2016b).

Conclusions

We are undergoing a paradigm shift in agriculture. While traditional agriculture involves the conversion of natural resources (soil, water, air, and sunlight) into consumable products to support society, new "farms" include energy farms that convert incoming solar radiation or wind energy into sustainable energy. New marketing and investing tools such as "impact investing" now allow investors to directly invest in conservation and environmental programs (Bugg-Levine and Emerson, 2011; Jackson, 2013). These markets and emerging approaches are targeted on improving conservation metrics and ecosystem services by generating "environmental value" to the investors (Barman, 2015). While rural populations and the number of farmers have decreased steadily in the United States (NASS, 2014), this investment strategy integrates social segments disconnected from traditional agricultural activities. While many have called for integrated, multi- and cross-disciplinary solutions, these emerging markets and investment tools are directly integrating social values with conservation and agricultural production.

Rapid development of new information technologies and knowledge systems has given us the ability to better coordinate and strategically design conservation systems. Traditional conservation practices are not enough to preserve our soil, water, and air resources and increase the productive capacity of our soil. We need more aggressive, efficient, and targeted conservation efforts. Integrating precision agriculture with precision conservation through the 7 R approach can contribute to better decision-making processes that increase the efficiency of inputs, and increase yields and economic returns while minimizing the losses of soil, water, and agrochemicals; conserving air, soil, and water quality in the field and across the watershed; and enhancing wildlife habitat. There is potential to use GPS, GIS, remote sensing, modeling, robotics, drones, artificial intelligence, and software applications for a new, *smart approach to soil and water conservation (smart soil and water conservation).* We can use these advanced technologies to increase the efficiency of conservation practices and increase the conservation return per dollar applied. Additionally, the reductions in nutrient, soil, sediment, and other losses could potentially be traded in air and water quality markets. The information presented in these book chapters provides information on the latest advances in technology that can assist agricultural managers and integrate multiple layers of spatial information to make management decisions that increase nutrient use efficiency, economic returns, production efficiency and resiliency, yields, and environmental conservation.

Precision conservation, defined as "the integration of spatial technologies such as global positioning systems (GPS), remote sensing (RS) and geographic information systems (GIS) and the ability to analyze spatial relationships modeling, data mining and map analysis" (Berry et al., 2003), is being used as one of the tools and approaches of voluntary precision regulation to target hot spots across the watershed that are contributing to higher nutrient, soil, and/or sediment losses. However, low tech approaches have also been used to apply precision conservation agriculture across the Sub-Saharan region of Africa (FAO, 2009; Silici, 2010; Thiombiano and Meshack, 2009; Jenrich, 2011). This book covers the advantages of applying a precision conservation approach for increasing yields, economic returns, and nutrient use efficiencies at a field level, and reducing the losses and transport of nutrients at the field edge, from the farm gate, and across the watershed. Examples are presented that show that a voluntary precision regulation approach can be used to encourage farmers, nutrient managers, conservationists, wildlife biologists, consultants, non-governmental organizations, and other users of these new technologies to apply conservation practices across the watershed to achieve reductions in nutrient losses, GHG emissions, and erosion, while increasing carbon sequestration and other ecosystem benefits that could potentially be traded in voluntary air and water quality markets.

This book shows that there is potential to use these precision conservation tools to assess the benefits of conservation practices that account for spatial and temporary variability to increase conservation effectiveness and increase the potential to trade reductions in nitrogen losses (by converting the reductions in nitrogen fluxes that contribute to lower direct and indirect N_2O emissions to reductions in CO_2 equivalent emissions), as well as reductions in GHG emissions and agrochemicals and increases in soil organic matter sequestration, in environmental quality markets. Better designed practices could contribute to increased conservation effectiveness, protecting streams and reducing off-site transport of sediments and agrochemicals from fields.

Precision conservation can be used to assess the flows across landscape to better identify where grass waterways, buffers, sediment ponds, nutrient traps, cover crops, and other conservation practices could be placed and even improve their design at the field to reduce the losses of soils, nutrients and agrochemicals. It could contribute to increased economic returns to farmers in the form of reduced usage of nutrient inputs, lower costs to establish a given practice, and/or increased yields. For example, precision conservation can be used to improve the efficiency of terrace design and layout time, providing savings in time and money. Precision conservation could be used for water management (irrigation and/or drainage), manure management, nutrient management and other farm operations, and could be used to produce economic benefits for farmers and ranchers, conserve wildlife, and improve conservation across the watershed.

Voluntary precision regulation is already taking place, and precision conservation may contribute to the implementation of these voluntary regulations. With the challenges that humanity is confronting during the 21st century such as a growing global population, a changing climate, extreme weather events, and other environmental challenges, conservation of soil resources will be critical for humanity's future. Social stability and national security depend on food security, and soils, which are a key natural resource for humanity. Precision conservation could be a key tool in our toolbox for achieving sustainable systems and food security in the 21st century.

Advances in conservation technologies and knowledge systems enhance our ability to preserve our natural resources, increase the productive capacity of the agronomic system, and integrate social values into agronomic production. The challenge is to increase the adoption of these innovations.

Acknowledgments
This research was supported in part by the USDA National Institute of Food and Agriculture, Hatch project 1003478. Mention of a trade name or proprietary product does not constitute an endorsement by the USDA. Details of specific products are provided for information only and do not imply approval of a product to the exclusion of others that may be available.

References

Ascough, J.C., II, D.C. Flanagan, J. Tatarko, M.A. Nearing, and H. Kipka. 2018. Soil erosion modeling and conservation planning. In: J.A. Delgado, G. Sassenrath, and T. Mueller, editors, Precision conservation: Geospatial techniques for agricultural and natural resources conservation. Agronomy Monograph 59. ASA, CSSA, SSSA, Madison, WI.

Bakker, M.M., G. Govers, and M.D.A. Rounsevell. 2004. The crop productivity-erosion relationship: An analysis based on experimental work. Catena 57:55–76. doi:10.1016/j.catena.2003.07.002

Bakker, M.M., G. Govers, R.A. Jones, and M.D.A. Rounsevell. 2007. The effect of soil erosion on Europe's crop yields. Ecosystems 10:1209–1219. doi:10.1007/s10021-007-9090-3

Barman, E. 2015. Of principle and principal: Value plurality in the market of impact investing. LiU Electronic Press. doi:10.3384/VS.2001-5592.15319 http://valuationstudies.liu.se (accessed 30 Aug. 2018).

Berry, J.K., J.A. Delgado, R. Khosla, and F.J. Pierce. 2003. Precision conservation for environmental sustainability. J. Soil Water Conserv. 58(6):332–339.

Berry, J.K., J.A. Delgado, F.J. Pierce, and R. Khosla. 2005. Applying spatial analysis for precision conservation across the landscape. J. Soil Water Conserv. 60:363–370.

Bjorneberg, D.L., R.G. Evans, and E.J. Sadler. 2018. Irrigation management. In: J.A. Delgado, G. Sassenrath, and T. Mueller, editors, Precision conservation: Geospatial techniques for Agricultural and Natural Resources Conservation. Agronomy Monograph 59. ASA, CSSA, SSSA, Madison, WI.

Bugg-Levine, A., and J. Emerson. 2011. Impact investing: Transforming how we make money while making a difference. Innovations 6(3):9–18 https://www.mitpressjournals.org/doi/pdf/10.1162/INOV_a_00077. doi:10.1162/INOV_a_00077

Buman, T. 2016a. Testimony of Tom Buman, CEO of Agren. For the House Agriculture Sub-Committee on Conservation Forestry. HearingTopic: Solutions through Voluntary and/or Locally Led Conservation Efforts. House Committee on Agriculture, Washington, D.C. https://agriculture.house.gov/uploadedfiles/buman_testimony.pdf (accessed 31 Aug. 2018).

Buman, T. 2016b. Voluntary conservation solutions. Agren Tools, Carroll, IA. https://www.agrentools.com/voluntary-conservation-solutions/ (accessed 30 Aug. 2018).

Carlisle, L. 2016. Factors influencing farmer adoption of soil health practices in the United States: A narrative review. Agroecology and Sustainable Food Systems 40:583–613.

Cassidy, E.S., P.C. West, J.S. Gerber, and J.A. Foley. 2013. Redefining agricultural yields: From tonnes to people nourished per hectare. Environ. Res. Lett. 8:034015. doi:10.1088/1748-9326/8/3/034015

Chaubey, I., R. Cibin, and Q. Feng. 2018. Precision conservation for biofuel production. In: J. A. Delgado, G. Sassenrath, and T. Mueller, editors.Precision Conservation: Geospatial techniques for agricultural and natural resources conservation. Agronomy Monograph 59. ASA, CSSA, SSSA, Madison, WI.

Chesapeake Conservancy. 2018. Precision conservation. Chesapeake Conservancy, Annapolis, MD. http://chesapeakeconservancy.org/conservation-innovation-center/precision-conservation/ (accessed 30 Aug. 2018).

Clay, D.E., and U. Mishra. 2017. The importance of crop residues in maintaining soil organic carbon in agroecosystems. In: Z. Qin, U. Mishra, and A. Hastings, editors, Geophysical Monograph Series: Bioenergy and land use change. https://doi.org/10.1002/9781119297376.ch8 doi:10.1002/9781119297376.ch8

Clay, D.E., J. Chang, G. Reicks, S.A. Clay, and C. Reese. 2018. Calculating soil organic turnover at different landscape position in precision conservation. In: J.A. Delgado, G. Sassenrath, and T. Mueller, editors, Precision conservation: Geospatial techniques for agricultural and natural resources conservation. Agronomy Monograph 59. ASA, CSSA, SSSA, Madison, WI.

Cox, C. 2005. Precision conservation professional. J. Soil Water Conserv. 60:134A.

Cox, C. 2009. Agriculture and nutrient pollution. U.S. Environmental Protection Agency. http://water.epa.gov/learn/training/wacademy/upload/2009_12_01_slides.pdf (accessed May 22 2014).

Cox, C. 2010. Agriculture and nutrient pollution. Illinois Nutrient Summit, 13-14 Sept. 2010. Environmental Working Group, Washington, D.C. http://www.epa.state.il.us/water/nutrient/presentations/craig_cox.pdf (accessed 22 May 2014).

Cox, C., A. Hug, and N. Bruzelius. 2012. Losing ground. Environmental Working Group, Washington, D.C. http://static.ewg.org/reports/2010/losingground/pdf/losingground_report.pdf (accessed 30 Aug. 2018).

Cox, C., and A. Hug. 2012. Murky waters: Farm pollution stalls cleanup of Iowa streams. Environmental Working Group, Washington, D.C. http://static.ewg.org/reports/2012/murky_waters/Murky_Waters.pdf (accessed 30 Aug. 2018).

Delgado, J.A., and J.K. Berry. 2008. Advances in precision conservation. Adv. Agron. 98:1–44. doi:10.1016/S0065-2113(08)00201-0

Delgado, J.A., M.J. Shaffer, H. Lal, S. McKinney, C.M. Gross, and H. Cover. 2008. Assessment of nitrogen losses to the environment with a Nitrogen Trading Tool (NTT). Comput. Electron. Agric. 63:193–206. doi:10.1016/j.compag.2008.02.009

Delgado, J.A., C.M. Gross, H. Lal, H. Cover, P. Gagliardi, S.P. McKinney, E. Hesketh, and M.J. Shaffer. 201a0. A new GIS nitrogen trading tool concept for conservation and reduction of reactive nitrogen losses to the environment. Adv. Agron. 105:117–171. doi:10.1016/S0065-2113(10)05004-2

Delgado, J.A., S.J. Del Grosso, and S.M. Ogle. 2010b. 15N Isotopic crop residue cycling studies suggest that IPCC methodologies to assess N2O-N emissions should be reevaluated. Nutr. Cycling Agroecosyst. 86:383–390. doi:10.1007/s10705-009-9300-9

Delgado, J.A., R. Khosla, and T. Mueller. 2011a. Recent advances in precision (target) conservation. J. Soil Water Conserv. 66(6):167A–170A. doi:10.2489/jswc.66.6.167A

Delgado, J.A., P.M. Groffman, M.A. Nearing, T. Goddard, D. Reicosky, R. Lal, N.R. Kitchen, C.W. Rice, D. Towery, and P. Salon. 2011. Conservation practices to mitigate and adapt to climate changeb. J. Soil Water Conserv. 66:118A–129A. doi:10.2489/jswc.66.4.118A

Delgado, J.A. 2012. Nitrogen trading tool. In: S.E. Jorgensen, editor, Encyclopedia of environmental management. Taylor & Francis, New York. p. 1772–1784.

Delgado, J.A., M.A. Nearing, and C.W. Rice. 2013. Conservation practices for climate change adaptation. Adv. Agron. 121:47–115. doi:10.1016/B978-0-12-407685-3.00002-5

Delgado, J.A. 2016. 4 Rs are not enough. We need 7 Rs for nutrient management and conservation to increase nutrient use efficiency and reduce off-site transport of nutrients. In: R. Lal and B.A. Stewart, editors, Soil specific farming: Precision agriculture. Advances in Soil Science series. CRC Press, Boca Raton, Fl. p. 89–126.

Doran, J.W. 2002. Soil health and global sustainability: Translating science into practice. Agric. Ecosyst. Environ. 88:119–127. doi:10.1016/S0167-8809(01)00246-8

Dosskey, M.G., M.J. Helmers, D.E. Eisenhauer, T.G. Franti, and K.D. Hoagland. 2002. Assessment of concentrated flow through riparian buffers. J. Soil Water Conserv. 57:336–343.

Dosskey, M.G., D.E. Eisenhauer, and M.J. Helmers. 2005. Establishing conservation buffers using precision information. J. Soil Water Conserv. 62:349–354.

Dosskey, M.G., M.J. Helmers, and D.E. Eisenhauer. 2007. An approach for using soil surveys to guide the placement of water quality buffers. J. Soil Water Conserv. 61:344–354.

Dosskey, M.G., S. Neelakantan, T. Mueller, and Z. Qiu. 2018. Vegetative filters. In: J.A. Delgado, G. Sassenrath, and T. Mueller, editors, Precision conservation: Geospatial techniques for agricultural and natural resources conservation, Agronomy Monograph 59. ASA, CSSA, SSSA, Madison, WI.

FAO (Food and Agriculture Organization of the United Nations). 2009. Agriculture and consumer protection department. Conservation Agriculture. Food and Agriculture Organization of the United Nations, Rome, Italy. http://www.fao.org/ag/ca/.

Fiener, P., and K. Auerswald. 2003. Effectiveness of grassed waterways in reducing runoff and sediment delivery from agricultural watersheds. J. Environ. Qual. 32:927–936. doi:10.2134/jeq2003.9270

Fiener, P., and K. Auerswald. 2018. In: J. A. Delgado, G. Sassenrath, and T. Mueller, editors, Grassed waterways. Precision conservation: Geospatial techniques for agricultural and natural resources conservation. Agronomy Monograph 59. ASA, CSSA, SSSA, Madison, WI.

Fulton, J., and M. Darr. 2018. GPS, GIS, guidance, and variable-rate technologies for conservation management. In: J.A. Delgado, G. Sassenrath, and T. Mueller, editors, Precision conservation: Geospatial techniques for agricultural and natural resources conservation. Agronomy Monograph 59. ASA, CSSA, SSSA< Madison, WI.

Hammes, A. 2016. Precision conservation management program receives national grant. AgSolver, Ames, IA. https://blog.agsolver.com/precision-conservation-management-program-receives-national-grant-97404c374ca2

Heartland Science and Technology Group. 2017. Precision conservation management. Heartland Science and Technology Group, Champaign, IL. https://farmerportal.precisionconservation.org/authentication/signup

Hunter, W.J. 2001. Remediation of drinking water for rural populations. In: R.F. Follett and J.T. Hatfield, editors, Nitrogen in the environment: Sources, problems, and management. Elsevier Science B.V., Amsterdam, the Netherlands. p. 433–453. doi:10.1016/B978-044450486-9/50019-4

Hunter, W.J. 2013. Pilot-scale vadose zone biobarriers removed nitrate leaching from cattle corral. J. Soil Water Conserv. 68:52–59. doi:10.2489/jswc.68.1.52

Illinois Sustainable Ag Partnership. 2018. Precision conservation management. East Peoria, IL. http://ilsustainableag.org/programs/precision-conservation-management/ (accessed 30 Aug. 2018).

Jackson, E.T. 2013. Evaluating social impact bonds: Questions, challenges, innovations, and possibilities in measuring outcomes in impact investing. Community Dev. (Columb.) 44(5):608–616. doi:10.1080/15575330.2013.854258

Jenrich, M. 2011. Potential of precision conservation agriculture as a means of increasing productivity and incomes for smallholder farmers. J. Soil Water Conserv. 66:171A–174A. doi:10.2489/jswc.66.6.171A

King, E.S., C. Cox, and W.L. Baker. 2009. An urgent call to action: Nutrient Innovations Task Group report. U.S. Environmental Protection Agency, Washington, D.C. http://water.epa.gov/learn/training/wacademy/upload/2009_12_01_slides.pdf (accessed May 22 2014).

Kleinman, P.J.A., A.R. Buda, A.N. Sharpley, and R. Khosla. 2018. Elements of precision manure management. J.A. Delgado, G. Sassenrath, and T. Mueller, editors, Precision conservation: Geospatial techniques for agricultural and natural resources conservation. Agronomy Monograph 59. ASA, CSSA, SSSA, Madison, WI.

Lal, H., J.A. Delgado, C.M. Gross, E. Hesketh, S.P. McKinney, H. Cover, and M. Shaffer. 2009. Market-based approaches and tools for improving water and air quality. Environ. Sci. Policy 12:1028–1039. doi:10.1016/j.envsci.2009.05.003

Lal, R. 2016. Soil health and carbon management. Food and Energy Security 5(4):212–222. doi:10.1002/fes3.96

Luck, J.D., T.G. Mueller, S.A. Shearer, and A.C. Pike. 2010. Grassed waterway planning model evaluated for agricultural fields in the western coal field physiographic region of Kentucky. J. Soil Water Conserv. 65:280–288. doi:10.2489/jswc.65.5.280

McConnell, M.D., and L.W. Burger, Jr. 2018. Precision conservation to enhance wildlife benefits in agricultural landscapes. In: J.A. Delgado, G. Sassenrath, and T. Mueller, editors, Precision conservation: Geospatial techniques for agricultural and natural resources conservation. Agronomy Monograph 59. ASA, CSSA, SSSA, Madison, WI.

Mueller, T.G., H. Cetin, R.A. Fleming, C.R. Dillon, A.D. Karathanasis, and S.A. Shearer. 2005. Erosion probability maps: Calibrating precision agriculture data with soil surveys using logistic regression. J. Soil Water Conserv. 62:462–468.

Mulla, D.J., and S. Belmont. 2018. Identifying and characterizing ravines with GIS terrain attributes for precision conservation. In: J.A. Delgado, G. Sassenrath, and T. Mueller, editors, Precision Conservation: Geospatial Techniques for Agricultural and Natural Resources Conservation. Agronomy Monograph 59. ASA, CSSA, SSSA, Madison, WI.

National Agriculture Statistics Service. 2014. Farm demographics: U.S. farmers by gender, age, race, ethnicity, and more. ACH12-3, May 2014. NASS, Washington, D.C. https://www.agcensus.usda.gov/Publications/2012/Online_Resources/Highlights/Farm_Demographics/Highlights_Farm_Demographics.pdf (accessed 30 Aug. 2018).

Nearing, M.A. 2001. Potential changes in rainfall erosivity in the U.S. with climate change during the 21st century. J. Soil Water Conserv. 56:229–232.

Nearing, M.A., F.F. Pruski, and M.R. O'Neal. 2004. Expected climate change impacts on soil erosion rates: A review. J. Soil Water Conserv. 59:43–50.

Paustian, K., M. Easter, K. Brown, A. Chambers, M. Eve, A. Huber, E. Marx, M. Layer, M. Stermer, B. Sutton, A. Swan, C. Toureene, S. Verlayudhan, and S. Williams. 2018. Field- and farm-scale assessment of soil greenhouse gas mitigation using COMET-Farm. In: J.A. Delgado, G. Sassenrath, and T. Mueller, editors, Precision conservation: Geospatial techniques for agricultural and natural resources conservation. Agronomy Monograph 59. ASA, CSSA, SSSA, Madison, WI.

Ribaudo, M.O., R. Heimlich, and M. Peters. 2005. Nitrogen sources and Gulf hypoxia: Potential for environmental credit trading. Ecol. Econ. 52:159–168. doi:10.1016/j.ecolecon.2004.07.021

Ribaudo, M., J. Delgado, L. Hansen, M. Livingston, R. Mosheim, and J. Williamson. 2011. Nitrogen in agricultural systems: Implications for conservation policy. Economic Research Report # 127. USDA-ERS, Washington, D.C. doi:10.2139/ssrn.2115532

Roberts, T.L. 2007. Right product, right rate, right time and right place…the foundation of best management practices for fertilizer. In: Fertilizer best management practices: General principles, Strategies for their adoption and voluntary initiatives vs regulations. IFA International Workshop on Fertilizer Best Management Practices, Brussels, Belgium, 7-9 March 2007. International Fertilizer Industry Association, Paris, France. p. 29-32.

Rodriguez, J.M., J.J. Molnar, R.A. Fazio, E. Sydnor, and M.J. Lowe. 2008. Barriers to adoption of sustainable agriculture practices: Change agent perspectives. Renew. Agric. Food Syst. 24:60–71. doi:10.1017/S1742170508002421

Saleh, A., and E. Osei. 2018. Precision conservation and water quality markets. J.A. Delgado, G. Sassenrath, and T. Mueller, editors, Precision Conservation: Geospatial Techniques for Agricultural and Natural Resources Conservation. Agronomy Monograph 59. ASA, CSSA, SSSA, Madison, WI.

Saleh, A., O. Gallego, E. Osei, H. Lal, C. Gross, S. McKinney, and H. Cover. 2011. Nutrient Tracking Tool—a user-friendly tool for calculating nutrient reductions for water quality trading. J. Soil Water Conserv. 66:400–410. doi:10.2489/jswc.66.6.400

Sassenrath, G.F., P. Heilman, E. Luschei, G.L. Bennett, G. Fitzgerald, P. Klesius, W. Tracy, J.R. Williford, and P.V. Zimba. 2008. Technology, complexity and change in agricultural production systems. Renew. Agric. Food Syst. 23(4):285–295. doi:10.1017/S174217050700213X

Sassenrath, G.F., J.D. Wiener, J. Hendrickson, J. Schneider, and D. Archer. 2010. Achieving effective landscape conservation: Evolving demands, adaptive metrics. In: P. Nowak and M. Schnepf, editors, Managing Agricultural landscapes for environmental quality: Achieving more effective conservation. Soil and Water Conservation Society, Ankeny, IA. p. 107–120.

Shedekar, V.S., and L.C. Brown. 2018. GIS and GPS applications for planning, design and management of drainage systems. In: J. A. Delgado, G. Sassenrath, and T. Mueller, editors, Precision conservation: Geospatial techniques for agricultural and natural resources conservation. Agronomy Monograph 59. ASA, CSSA, SSSA, Madison, WI.

Silici, L. 2010. Conservation agriculture and sustainable crop intensification in Lesotho. Integrated Crop Management Vol. 10-2010. Food and Agriculture Organization of the United Nations, Rome, Italy. http://www.fao.org/docrep/012/i1650e/i1650e00.pdf (accessed 30 Aug. 2018).

Skaggs, R.W., N.R. Fausey, and R.O. Evans. 2012. Drainage water management. J. Soil Water Conserv. 67:167A–172A. doi:10.2489/jswc.67.6.167A

Sorensen, A.A., J. Freedgood, J. Dempsey, and D.M. Theobald. 2018. Farms Under Threat: The State of America's Farmland. American Farmland Trust, Washington, D.C. https://www.farmland.org/initiatives/farms-under-threat (accessed 30 Aug. 2018).

Thiombiano, L., and M. Meshack. 2009. Scaling-up conservation agriculture in Africa: Strategy and approaches. Addis Ababa: FAO Subregional Office for Eastern Africa. Food and Agriculture Organization of the United Nations (FAO) Sub Regional Office for Eastern Africa, Addis Ababa, Ethiopia. http://www.fao.org/ag/ca/doc/Kenya%20Workshop%20Proceedings.pdf (accessed 31 Aug. 2018).

Thompson, A., and K. Sudduth. 2018. Terracing and contour farming. J.A. Delgado, G. Sassenrath, and T. Mueller, editors, Precision conservation: Geospatial techniques for agricultural and natural resources conservation. Agronomy Monograph 59. ASA, CSSA, SSSA, Madison, WI.

Tomer, M.D., D.E. James, and T.M. Isenhart. 2003. Optimizing the placement of riparian practices in a watershed using terrain analysis. J. Soil Water Conserv. 58:198–206.

Tomer, M.D., T.B. Moorman, J.L. Kovar, D.E. James, and M.R. Burkart. 2007. Spatial patterns of sediment and phosphorous in a riparian buffer in western Iowa. J. Soil Water Conserv. 62:329–338.

Tomer, M.D. 2010. How do we identify opportunities to apply new knowledge and improve conservation effectiveness? J. Soil Water Conserv. 65:261–265. doi:10.2489/jswc.65.4.261

Tomer, M.D., S.A. Porter, D.E. James, K.M.B. Boomer, J.A. Kostel, and E. McLellan. 2013. Combining precision conservation technologies into a flexible framework to facilitate agricultural watershed planning. J. Soil Water Conserv. 68:113A–120A. doi:10.2489/jswc.68.5.113A

Tomer, M.D., D.B. Jaynes, S.A. Porter, D.E. James, and T.M. Isenhart. 2018. Identifying riparian zones best suited to installation of saturated buffers: A preliminary multi-watershed assessment. In: J.A. Delgado, G. Sassenrath, and T. Mueller, editors, Precision conservation: Geospatial techniques for agricultural and natural resources conservation. Agronomy Monograph 59. ASA, CSSA, SSSA, Madison, WI.

USDA. 2015. An ag outlook audience learns how voluntary conservation can help at-risk wildlife and reduce the need for regulation. USDA, Washington, D.C. https://www.usda.gov/media/blog/2015/02/26/ag-outlook-audience-learns-how-voluntary-conservation-can-help-risk-wildlife (accessed 30 Aug. 2018).

USDA-NRCS. 2017. NRCS & Precision Conservation Management. https://www.nrcs.usda.gov/wps/portal/nrcs/detail/?navtype=SUBNAVIGATION&ss=161017&cid=NRCSEPRD1372142&navid=105100000000000&pnavid=105000000000000&position=News&ttype=detail (accessed 31 Aug. 2018).

USDA-NRCS. 2018. Environmental Quality Incentives Program. National Resources Conservation Service, Washington, D.C. https://www.nrcs.usda.gov/wps/portal/nrcs/main/national/programs/financial/eqip/ (accessed 30 Aug. 2018).

USEPA. 2010. Inventory of U.S. Greenhouse Gas Emissions and Sinks: 1990-2008. Washington, D.C., USEPA. http://www.epa.gov/climatechange/emissions/downloads10/508_Complete_ GHG_1990_2008.pdf (accessed 30 Aug. 2018).

USGAO (United States Government Accountability Office). 2013. CLEAN WATER ACT: Changes needed if key EPA program is to help fulfill the nation's water quality goals. U.S. Government Accountability Office, Washington, DC.

Walthall, C., J.A. Delgado and S. Del Grosso. 2014. ARS NP212 Climate Change, Soils, & Air Emissions Program Update. Agricultural Air Quality Task Force meeting, November, 2014. https://www.nrcs.usda.gov/wps/PA_NRCSConsumption/download/?cid=stelpr db1265389&ext=pdf

Yuan, Y., R. Bingner, and H. Momm. 2018. Nitrogen component in nonpoint-source pollution models. In: J. A. Delgado, G. Sassenrath, and T. Mueller, editors, Precision conservation: Geospatial techniques for agricultural and natural resources conservation. Agronomy Monograph 59. ASA, CSSA, SSSA, Madison, WI.